Elaine Toms

Advanced Information Processing

Springer

Berlin
Heidelberg
New York
Hong Kong
London
Milan
Paris
Tokyo

Y. Ohsawa • P. McBurney (Eds.)

Chance Discovery

With 121 Figures and 32 Tables

 Springer

Prof. Dr. Yukio Ohsawa
Graduate School of Business Sciences, University of Tsukuba
GSSM, University of Tsukuba, 3-29-1 Otsuka, Bunkyo-ku
Tokyo 112-0012 Japan
E-mail: osawa@gssm.otsuka.tsukuba.ac.jp

Dr. Peter McBurney
Agent Applications, Research and Technology (Agent ART) Group
Department of Computer Science
University of Liverpool
Liverpool L69 7ZF U.K.
Email: P.J.McBurney@csc.liv.ac.uk

Cataloging-in-Publication Data applied for
A catalog record for this book is available from the Library of Congress.

Bibliographic information published by Die Deutsche Bibliothek.
Die Deutsche Bibliothek lists this publication in the Deutsche
Nationalbibliographie; detailed bibliographic Data is available in the
Internet at http://dnb.ddb.de.

ACM Subject Classification G.3, I.2, J.2

ISBN 3-540-00549-8 Springer-Verlag Berlin Heidelberg New York

Springer-Verlag Berlin Heidelberg New York
a member of BertelsmannSpringer Science+Business Media GmbH

© Springer-Verlag Berlin Heidelberg 2003
Printed in Germany

Cover Design: KünkelLopka, Heidelberg
Typesetting: Computer to film by author's data
Printed on acid-free paper   45/3142XO   5 4 3 2 1 0

Foreword

Some events or items can be very difficult to predict or find. When information about these events or items is also valuable for some reason, as in the cases of earthquake prediction or drug discovery, finding information about them becomes important, and effort is devoted to developing specialized methods to this end.

The new field of chance discovery is devoted to thinking about the problem of difficult-to-discover information. The hope is that this problem will have general attributes and solutions that will reach across many specialized fields and problems. The field is now in a very exciting early stage, with experts in different specialized applications coming together to exchange ideas for the first time.

My own first interaction with scholars interested in chance discovery is illustrative in this regard. I study the sources of innovation, with innovation by product and service users being a particular interest. When a user, a firm or individual, develops an innovation that might be generally useful, that user does not necessarily have any incentive to call this innovation to the attention of others.

I therefore face a difficult puzzle in my research: how to efficiently identify generally valuable user-developed innovations that may be thinly distributed in user populations numbering in the millions. The problem is compounded by the fact that information about the general value of an innovation is not necessarily known by or even of interest to the user developing it. I attended a November 2002 seminar on chance discovery that had been organized by the editors of this book. I presented my problem to scholars from a range of fields, and explained the methods I currently used. My reward was many new and interesting initial ideas, and potentially interesting research relationships as well.

This book offers access to early thoughts and ideas in the field of chance discovery from a number of scholars. I hope that it triggers interest in many. Chance Discovery is certainly a very exciting field in the making!

Boston, December, 2002
Massachusetts Institute of Technology

Eric von Hippel

Preface

The Scope of Chance Discovery

The manager of a store is waiting for profitable customers, i.e. chances to sell high-profit products. From all sources of information s/he hears about new opportunities to sell products, but loses way in the flood of information. The manager wants to know which product makes highest profit for the store, and how the opportunity should be managed to gain income. To do this, s/he has to understand who desires each product and to whom and how the sales clerks should talk to customers in the store. However, it is impossible to investigate in detail all the products coming out and their market before selling them – because time flies. Therefore, it becomes essential to select good products to investigate, before selecting the best products. How can the manager start, continue, and stop this recursive selection process? Where is the essential information for his/her novel decision – on the Web, or already in his/her own mind?

What we call *chance discovery* is the discovery *of* a chance, not *by* chance. Here, a "chance" is defined as a piece of information about an event or a situation which significantly impacts on the decision-making of humans, agents, and robots (following the second meaning of "chance"– *a suitable time or occasion to do something* – in the Oxford Advanced Learner's Dictionary, assuming that the situation or event occurring at a certain time is more significant than the time itself, and that a decision to do something precedes the action of doing it). In this book and in previous literature, the first editor defined chance discovery as the awareness of a *chance*, defined above, and gave an explanation of its significance. This triggered researchers in various domains to explore challenging methods for chance discovery even if the chance is rare and its significance has been unnoticed. These researchers are going much beyond previous data mining and statistics by introducing various ideas.

An essential aspect of a chance is that it can be a new seed of significant future changes and actions. The discovery of a rare opportunity may lead to a benefit not experienced before, because people have hardly noticed the benefit and were accustomed to frequent past opportunities. The discovery of a new hazard is indispensable for minimizing risk, because existing solutions that worked for frequent past hazards may not work this time. Besides data-mining methods for finding rare but important events from time series, it is also important to draw humans' attention to such events and to knowledge for dealing with them, i.e. to make humans ready to catch and manage chances. In this sense, the complex interactions of human–environment, human–human, human–agent (either hardware or software), agent–agent (either hardware or software), and agent–environment are all essential to chance discovery, as well as interactions among various components of the environment. Colleagues in the chance discovery field have been examining discoveries of such chances as:

- signs of great earthquakes in the future, as a complex system where a small cause results in a great effect
- key genomes triggering diseases, etc.
- triggers of human, robot, and agent behaviors in dynamic environments
- new products and consumers worth promoting in sales
- signs of critical events in the future of nature or society
- signs of social changes from the data of questionnaires
- words on the WWW which attract people
- communications for creating new value criteria for evaluating chances
- shared context and leading opinions in online communities
- anomalies with significant impact on the economy
- visual information aiding human discovery of chances

and so on. This book presents theories and methods of chance discovery, based on topics about innovation, creative communications, data mining with a visual interface, and other central points, for enabling the reader to realize his/her own decisions in dynamic environments. Each chapter by leading researchers will navigate researchers, graduate/undergraduate students, and business people to the techniques for winning creative life both personally and socially.

Let us express many thanks to the people who contributed to publishing this book. First of all, we thank the authors of all chapters for presenting ideas, logics, and results, reflecting both the established contexts and the up-to-date interests relevant to chance discovery. We also thank Ralf Gerstner and the copy editors of Springer-Verlag for their understanding and helpful support in editing. And we thank Junichiro Mori and Naomi Okazaki in the University of Tokyo for fixing the text and the figures in the entire book. The final checking by Masashi Taguchi, Yo Nakamura, and Hiroki Yoshimura in the laboratory of chance discovery in Tokyo helped us in completing the details. We hope these efforts will be of fruitful influence on the decision-making of readers.

Tokyo, June, 2003   *Yukio Ohsawa* and *Peter McBurney*

Table of Contents

List of Contributors

Yukio Ohsawa
Graduate School of Business Sciences,
University of Tsukuba, 3-29-1
Otsuka, Bunkyo-ku, Tokyo
112-0012, Japan
email: osawa@gssm.otsuka.tsukuba.ac.jp

Peter McBurney
Department of Computer Science
University of Liverpool
Chadwick Building, Peach Street
Liverpool, L69 7ZF, UK
email: p.j.mcburney@csc.liv.ac.uk

Akinori Abe
NTT MSC R&D,
No. 43 000, Jalan APEC, 63 000 Cyberjaya Selangor Darul Ehsan, Malaysia
email: ave@cslab.kecl.ntt.co.jp

Peter Bruza
Distributed Systems Technology Centre, University of Queensland, Brisbane,
Australia 4072
email: bruza@dstc.edu.au

Eric Dietrich
Philosophy Department,
Binghamton University,
Binghamton, NY 13 902-6000, USA
email: dietrich@binghamton

Hisashi Fukuda
Applied Meteorology Department,
CRC Solutions Corp.,
Kotoh, Tokyo 136-8581, Japan
email: h-fukuda@crc.co.jp

David E. Goldberg
Illinois Genetic Algorithms Laboratory
(IlliGAL), Department of General Engi-
neering, University of Illinois at Urbana-
Champaign, Urbana, IL 61 801, USA
email: deg@uiuc.edu

Mitsuru Ishizuka
Department of Information and Communication Engineering, Graduate School
of Information Science and Technology, University of Tokyo, 7-3-1, Hongo,
Bunkyo-ku, Tokyo 113-8656, Japan
email: ishizuka@miv.t.u-tokyo.ac.jp

Paul Jefferies
Physics Department, Oxford University,
Oxford OX1 3PU, UK
email. linc0227@hcrald.ox.ac.uk

Henrik J. Jensen
Department of Mathematics,
Imperial College, 180 Queen's Gate,
London SW2BZ, UK
email: h.jensen@ic.ac.uk

Neil F. Johnson
Oxford University,
Oxford OX1 3LB, UK
email: n.johnson@physics.ox.ac.uk

David Lamper
Oxford Centre for Industrial
and Applied Mathematics,
24-29 St. Giles' Oxford OX1 3LB, UK
email: lamper@maths.ox.ac.uk

Arthur B. Markman
Psychology Department,
University of Texas,
Austin, TX 78 712, USA
email: markman@psy.utexas.edu

Kenji Mase
ATR Media Information Science Laboratories, 2-2-2 Hikaridai, Seika-cho,
Kyoto 619-0288, Japan
email: mase@atr.co.jp

Naohiro Matsumura
Graduate School of Engineering,
University of Tokyo, 7-3-1 Hongo,
Bunkyo-ku, Tokyo 113-8656, Japan
email: matumura@miv.t.u-tokyo.ac.jp

Yutaka Matsuo
Cyber Assist Research Center,
National Institute of Advanced Indus-
trial Science and Technology,
Aomi 2-41-6, Tokyo 135-0064, Japan
email: y.matsuo@aist.go.jp

Robert McArthur
Distributed Systems Technology Centre, University of Queensland, Brisbane,
Australia 4072
email: mcarthur@dstc.edu.au

Makoto Mizuno
R&D Division, Hakuhodo Inc.,
Tokyo 108-8088, Japan
email: mmizuno@hakuhodo.co.jp

Yumiko Nara
Osaka Kyoiku University,
Osaka 582-8582, Japan
email: nara@cc.osaka-kyoiku.ac.jp

Simon Parsons
Department of Computer and Informa-
tion Science, Brooklyn College, City
University of New York, 2900 Bedford
Avenue, Brooklyn, NY 11210, USA
email: parsons@sci.brooklyn.cuny.edu

Helmut Prendinger
Department of Information and Communication Engineering, Graduate School
of Information Science and Technology, University of Tokyo, 7-3-1, Hongo,
Bunkyo-ku, Tokyo 113-8656, Japan
email: helmut@miv.t.u-tokyo.ac.jp

Kumara Sastry
Illinois Genetic Algorithms Laboratory
(IlliGAL),
Department of General Engineering,
University of Illinois at Urbana-
Champaign, Urbana, IL 61 801, USA
email: ksastry@uiuc.edu

Hiroko Shoji
Department. of Information and Communication Sciences, Faculty of Education, Kawamura Gakuen Women's University, Abiko, Chiba 275-1138, Japan
email: hiroko@da2.so-net.ne.jp

C. Hunt Stilwell
Psychology Department,
University of Texas,
Austin, TX 78 712, USA
email: stilwell@psy.utexas.edu

Yasuyuki Sumi
ATR Media Information Science Laboratories, 2-2-2 Hikaridai, Seika-cho,
Kyoto 619-0288, Japan
email: sumi@atr.co.jp

Wataru Sunayama
Osaka University, 1-3 Machikaneyama,
Toyonaka, Osaka 560-8531, Japan
email: sunayama@sys.es.osaka-u.ac.jp

Yasufumi Takama
Tokyo Metropolitan Institute of Technology, Hino,
Tokyo 191-0065, Japan
email: ytakama@cc.tmit.ac.jp

Michael Winkley
Philosophy Department,
Binghamton University,
Binghamton, NY 13 902-6000, USA
email: mwinkley@binghamton.edu

Fumiko Yoshikawa
Hiroshima Shudo University,
1-1-1 Ozukahigashi, Asaminami-ku,
Hiroshima 731-3195, Japan
email: fumiko-y@shudo-u.ac.jp

Part I
Chance Discovery in the Complex Real World

This part introduces the methodological frameworks of chance discovery. Chance discovery is a challenge to discover events which may enable us to foresee the future of a complex and dynamic system, such as the social and natural environment in which we make various decisions. In aspects from artificial intelligence, physics, sociology, and linguistics, we require interactions with the environment for discovering chances. In this part, let us show how these interactions can be realized using tools available now and in the near future.

1. Modeling the Process of Chance Discovery

Yukio Ohsawa[1,2]

[1] PRESTO, Japan Science and Technology Corporation, 2-2-11 Tsutsujigaoka, Miyagino-ku, Sendai, Miyagi 983-0852, Japan
[2] Graduate School of Business Sciences, University of Tsukuba, 3-29-1 Otsuka, Bunkyo-ku, Tokyo 112-0012, Japan
email: osawa@gssm.otsuka.tsukuba.ac.jp

Summary.

   The fundamental philosophy of chance discovery is introduced. By comparison with the cyclic model of knowledge discovery, this chapter describes the essentials for realizing chance discovery. From these discussions, three keys for chance discovery are proposed, i.e. communication, context shifting, and data mining. As a result, the double helix and the subsumption architecture are presented as methods for realizing chance discovery.

1.1 Which Do You Want, a Rule or an Opportunity?

Do you like a rule 'if it is Sunday morning, one must go out jogging'? When I asked this question to colleagues, some disliked jogging because of its heavy feeling. Some preferred a rest or a holiday, and others wanted to decide what to do by flexibly reacting to the dynamic changes in the environment, e.g. the weather. They look out the window to see the weather conditions, or watch weather-forecast programs on TV.

   In the recent business literature, the word 'knowledge' has been appearing with increasing frequency. For example, 'knowledge management' and 'knowledge discovery' are the most popular terms bundled with keywords on business information technology. However, the definition of 'knowledge' has been hardly consistent among researchers. In [1.9], knowledge has been distinguished into implicit (not symbolized, although existing in the mind of humans) knowledge and explicit (symbolized) knowledge. In the case above of jogging on Sunday, what kind of knowledge should be activated for deciding to stay home, when we look outdoors? A person who decided so may answer:

   'I used the explicit knowledge that one should stay home when it is raining.' However, she could have used other knowledge in this situation, e.g.

   'I go out even if it is raining, as long as it is not thundering.'

   Each piece of knowledge used by humans in real situations is a part of complex knowledge with a large number of conditions (i.e. the terms in the 'IF ...' part), if the knowledge can be symbolized as an IF–THEN rule. Rather than believing in her comment, it would be more realistic to say she has had experiences of weather-based actions, and decides her next action based on these experiences and the on-site information relating to the current weather. In other words, she prefers to act reactively to changes in the environment rather than be bound by a rule much simpler than the dynamics of the real world, even if it has thousands of conditions. In this chapter, let us assume that there are situations where humans seek a *chance*, i.e. information about an event or a situation significant for making decisions in a dynamic environment.

A short note on dynamic environments

$$Y = f(X) \qquad\qquad (1.1)$$

In (1.1), let function f define a deterministic system, which generates a unique output Y for each input X. Even if a part of the process from the input to the output is non-deterministic, i.e. has multiple decision candidates, the overall process is deterministic if the final output is uniquely determined. For example, the process from the input of external information by a sensor of a robot to its action, is deterministic.

Here, let us represent input X by an n-dimensional vector $(x_1, x_2, \ldots, x_n)$, where variables x_1, x_2, $\ldots$, and x_n represent the features of the input to a deterministic system, and consider obtaining Y for a certain situation corresponding to a set of values of variables in X. Without loss of generality, we can denote the controllable or observable part of X by $X0$: $(x_1, x_2, \ldots, x_m)$ $(m \le n)$. In case f is identified and $m = n$, we call f a programmed decision-making system. Otherwise, i.e. if f is not identified or if $m < n$, we call f a non-programmed decision-making system. This is a simple formalization of the programmed and the non-programmed decision-making systems presented in [1.17]. In previous system-identification methods, the targets were in the case where f is not identified and $m = n$. This condition applies for artificial systems whose behaviors are determined by limited input factors.

The process of humans starting from information acquisition and ending in a determined decision or action is a deterministic system. However, this deterministic system is affected by various uncontrollable and unobservable factors, i.e. $m < n$, so f is very hard to identify. Previously in data-based decisions of both business and science (natural and social), decisions were often based on the data of a given number of variables from a number of samples. This corresponded to introducing the approximation $m = n$ for identifying f. Such an approximation might be valid if f is supported by a large sample set, because individual-specific features of samples can be ignored or cancel one another's influences. However, with the changes in the social and the natural environment, e.g. the growth of the Internet, the sudden price movements of stocks, the appearance of unknown diseases, or signs of great earthquakes, we are confronted with unexpected fatal impacts of new specific features of rare events. This chapter is dedicated to developing process models of discovery, for aiding the human process of *chance discovery*.

1.2 Is Chance Discovery Itself an Event or a Process?

The word 'chance', in this book, means information about an event or a situation significant for making decisions. Here arises the following question.

'Can a chance be discovered in a moment, if the chance occurs at a moment?'

'Yes', one may answer, 'when I hear a new message, I can immediately distinguish good news from bad news. Just think of weather news — you do not take long before feeling happy after you hear it will be fine all day'.

We have reason to reject this answer. Why does this person feel happy at the moment he catches the news? He feels so because he has been desiring a fine day, for several possible reasons, e.g. he was planning to go out or he likes the natural blue color of the sky. We can hardly show an example where one can feel that an opportunistic event occurred without one's preparatory state of mind. In other words, as the saying goes, *a chance favors a prepared mind*.

All in all, the discovery of a chance by a human is realized by a process including his/her *involvement* in a dynamic environment. That is, one's *concern* with chance events increases by being involved in the dynamic changes in the surroundings. For example, one will pay more attention to weather news during a season of changing weather. In this phase, one comes to be ready to catch the information about a new occurrence of an event as being one's chance. Being *aware* of the (occurrence of a) chance, i.e. paying attention to the chance strongly enough, one can begin to understand its meaning in detail. Finally, one can decide to take advantage of the chance, and make or test actions for evaluating the utility of the chance. Thus, chance discovery is a process rather than a momentary event.

Many problems remain. For example, does the awareness of a chance always precede the understanding of it? We can not simply say 'yes', because we usually can not pay attention to a new event and feel its significance if we do not have the slightest understanding about the event. Chance discovery is a complex process in which the occurrence, the awareness, and the understanding of chances interleave with each other. This chapter aims to illustrate the process of chance discovery, where a human and a computer collaborate for mining chances from a dynamic environment.

1.3 From Knowledge Discovery to Chance Discovery

Complex rules have been learned for predicting a predefined rare event, such as a break in signal transmission along a certain line, by the extension of genetic algorithms, cost-sensitive learning, etc. [1.19, 1.18]. These work for automated prediction where the candidates of the prediction target, i.e. significant rare events, are not *selected* online by the prediction system but are *predefined* offline. That is, the obtained rules are optimized according to a certain predefined object function corresponding to the likelihood of the occurrence of the target event under the observed conditions. All in all, using such existing data-analysis and data-mining methods we can not explain the significance of (1) unnoticed features not described in the given data, or (2) rare events (values of existing features) people seldom count as worthwhile predicting.

In explaining significant financial anomalies, the latent popularity of new products, and the signs of fatal hazards, we always suffer from both difficulties (1) and

(2) above. For example, the emotion of customers such as anger, happiness, stress, etc. may cause the growth of their desire for a certain rare product. However, these emotions of customers are hardly given in existing data.

1.3.1 The Human as a Main Component of Chance Discovery

For evaluating the significance of a new factor (a feature or its value corresponding to (1) or (2) above) unfamiliar to us, it becomes necessary to shift our mental context from the usual situation to concern ourselves with unusual situations where the factor becomes influential. Otherwise we cannot smoothly accept a new situation, because of the disturbance caused by cognitive dissonance [1.4] due to an unfamiliarity with the situation. On the notion of 'context shifting', the process of chance perception has been modeled [1.8]. It is essential to shift the mental context of a human who is not concerned with hidden factors potentially causing new events to accepting the new situation and to reasoning whether the event is significant for making a decision.

By comparing with the prevalent process of knowledge discovery from data (KDD) [1.3] in Fig.1.1, we can show that the process of chance discovery focuses strongly on the context shifting of human as in Fig.1.2. For the early stage of context shifting in chance discovery, the essential information is a *peripheral* clue [1.13], as in Fig. 1.2, i.e. information not directly explaining the significance of the chance but attracting humans to chances. For example, a persuasive authority (e.g. a renowned professor of seismology) showing a visualized simulation where a certain event becomes a meaningful chance (a small quake preceding a big earthquake, displayed on TV) typically works as a peripheral clue. If the human accepts the new context, a *central clue* [1.13] directly explaining the significance of the chance becomes acceptable and pushes the human forward to a real *action*. This central information is sometimes acquired from recent data of event occurrences. Possibly, a familiar part of the data may work as a 'peripheral clue 2' in Fig. 1.2 and attract the human's attention to the environment where the chance may occur. During real actions, she can evaluate the current chance and so trigger herself to desire the next chance if necessary. It is noteworthy, however, that spiral models of chance perception [1.15, 1.10] commonly point out that an action may concentrate the attention of the human too deeply on the problem in hand for her to become aware of other new chances. Chapter 6 by Shoji will be dedicated to showing how human-to-human communication aids this awareness of chances.

This evolution of the process model directs the studies on chance discovery to answering the question 'How can we humans discover chances?', not leaving chance discovery only to the power of computers. A chance is here evaluated according to the value of decisions or actions, interacting with the external environment, based on the chance. In this interaction, new essential features are detected and reflected in the 'observation' of the value of those features.

Let us compare Fig. 1.1 and Fig. 1.2 in detail. The 'interpretation' in Fig. 1.2 is separated from 'evaluation', reflecting the discussion above. The spiral model of human perception of chances is integrated with the KDD process and a human is

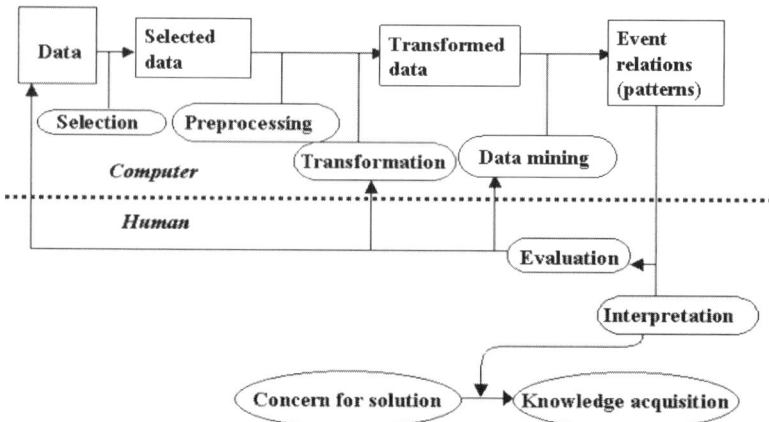

Fig. 1.1. Fayyad's model of the knowledge-discovery process

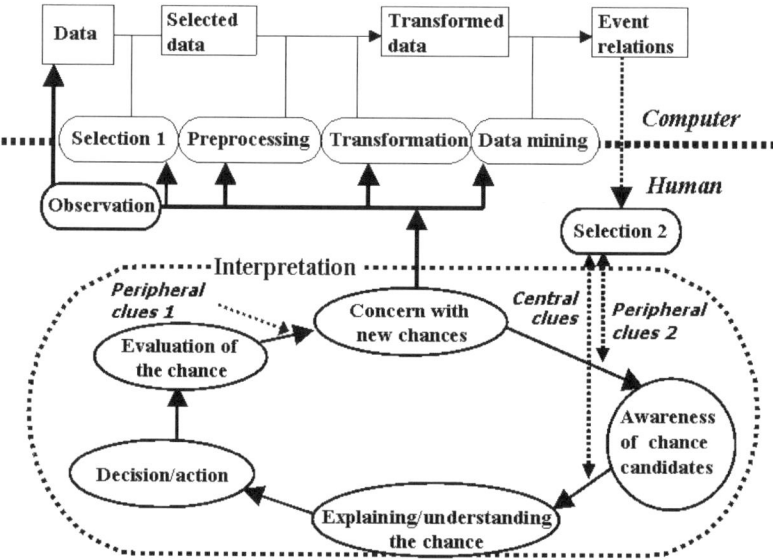

Fig. 1.2. The process of chance discovery

positioned as a stronger engine than a machine to select, interpret, and evaluate the value of a chance in Fig. 1.2.

The contribution of a human to the 'selection' step, where rare events should be picked up if they may really be chances, is in the interpretation cycles where the interpretations of significant chances are recognized as 'central clues' ('selection 2'). The 'selection 1' in Fig. 1.2, corresponding to the 'selection' in Fig. 1.1, plays a rather weaker role in chance discovery, discarding only obvious noise (e.g. 'is', 'are', etc. in the case of text data). In the interpretation cycles, as in the cycles of SECI [1.9] going through implicit and explicit communications among a group of co-workers, humans can reach an explicit understanding of chances by selecting and diffusing the opinions of a few participants who understands the significance of the chances.

Real data can be a useful external information source to guide a human to the explicit awareness and explanation of a chance, if presented in a suitable manner. For example, using a data-mining method showing causalities including a rare event, a human in a 'concerned' state is likely to understand the central explanation about the significance of rare events. Hereby, we noticed three keys to chance discovery:

Key 1: Communications about the central information (i.e. the interpretations) of the significance of an event.

Key 2: Peripheral information (e.g. high-impact scenes visualizing the chances, words from authoritative people, etc.) for enhancing a user's concern with (or imagination of) the context where a chance is desired.

Key 3: New methods of modeling and mining chances from data, for obtaining and showing the causalities involving rare events.

Let us now discuss basic theories and approaches to Keys 1 through 3.

1.3.2 Communications, Key 1 for Chance Discovery

People share some common ideas about daily affairs, but can have different knowledge about new situations. External information sources such as books and Web pages may supply new knowledge, but a human is rather accustomed to consult suitable people for advice on what to do in confronted situations. In particular, when the situation is new and uncertain, human–human communication becomes essential for noticing, diffusing, and establishing useful knowledge socially [1.1].

Let us look briefly at the decision process of a buyer (including a potential buyer who may become a customer, i.e. a frequent buyer) in a market. She becomes involved in buyer-to-buyer communications and the flood of external information. If a peripherally important (mostly authorized) piece of information is found, she becomes concerned with its significance as a chance. This process can be made outside the store by simply chatting about desirable products. Through further communications, either buyer-to-buyer or with a sales clerk, she becomes concerned with central information, i.e. the explanation for why a certain product is beneficial for her, and moves closer to a decision to buy it.

In social psychology, it is known that a simple rumor can be diffused to influence people who have some uncertainty in their own decisions, which is here reflected in the step where peripheral information prevails before central information works [1.1], i.e. when a human is no more confident of the chance sought after in the last cycle of interpretation and the examination of the new chance is not fixed yet. The model of diffusion [1.14], where people differ in their reaction speed to innovative ideas, also corresponds to the step-wise growth of popular concern in communications about any new appearance of chance.

It has been pointed out that a dialog, where chance discovery can be made by a group of humans or agents, includes an *inquiry*, where participants collaborate to find a new solution to a problem (because it is unknown how one can manage a new chance), and a *persuasion*, where one participant seeks to persuade another/others to accept a new proposal (McBurney and Parsons [1.7]). According to the formalization by McBurney and Parsons, an essential locution of participants in the communication involved in chance discovery introduces sharable new value criteria, i.e. the recognized relevance of an event to benefit or loss. Such a creation of value criteria sometimes unifies multiple communities. McBurney and Parsons will present further details in Chap. 10.

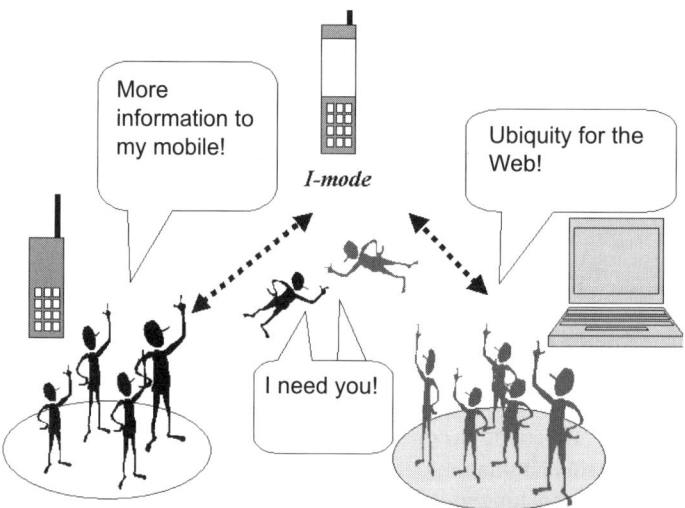

Fig. 1.3. The diffusion of i-mode, a Japanese mobile phone with an interface to the Internet. i-mode appeared as a new value attracting the two communities of cellular and PC users, and made them attract each other. This was a case of fusion into a large market.

If the new value criteria are about products attracting multiple communities, leaders in each community will be unified first because they are ready to touch new products and contact each other to talk about how their communities can cooperate with each other to take full advantage of the product. This can merge the communities to form a new market sharing a new value criterion, by diffusion from leaders, as illustrated in Fig. 1.3.

From the analogy between the human–human interaction of ideas and the genetic process of evolution, we can regard a new combination of ideas as a breakthrough leading to innovation. For a detailed discussion about this lesson from genetic algorithms theory and its possible implementation as a real communication, the reader is referred to Chap. 19 by Goldberg *et al.* By building well-defined formalization frameworks, as these chapters propose, the capability for chance discovery may, in the long term, be transplanted from human to software agents, as proclaimed in Chap. 10. However, regarding a human as a system-learning schema from such logical descriptions (even if some details are missed), we can aim in the near future at acquiring the capability to make chance discoveries. We focus attention on communications, as a main engine of chance discovery. Several chapters in this book are dedicated to showing embodied communications for chance discovery: Chap. 14 by Prendinger and Ishizuka on agent(s)–human communications; Chap. 20 by Sumi and Mase on agent-mediated human–human interaction; and Chap. 8 by McArthur and Bruza on creative human–human online communications.

1.3.3 Scenic Information for Imagination, Key 2 for Chance Discovery

A human becomes aware of the relevance of his or her implicit knowledge (i.e. one knows, but is not conscious of the knowledge) or implicit experience to a confronted situation, using mental imagery [1.6], i.e. scenes imaginable in conjunction with the confronted situation. It has been shown that talking about scenes where a product is used can be an effective explanation of the utility of the product [1.16]. Scenic information has also been proposed as an aid to smooth context shifting [1.8]. In other words, scenic information shifts a human forward in the phase where one is weakly concerned with a chance, working as a peripheral clue. For real situations where scenic information and oral communication interplay with one another to create new aspects of products, e.g. in sales communication in a clothing store, refer to Chap. 6. Also, Chap. 24 by Mizuno introduces a chance navigator with a human interface carrying scenic information. In Chaps. 12 and 13 by Takama and Ohsawa, the theoretical framework for generalizing these effects of scenic information is presented as a basis for an interactive process among machines, humans, and their environment.

For dealing formally with the mechanism of finding relevance between the current and memorized experiences, formulations based on *analogy* seem to be a promising approach. The current state of studies in this direction is presented in Chap. 15 by Dietrich *et al.* and Chap. 16 by Abe. These chapters will also give the reader an opportunity to consider the relations among discovery *of* chances, discovery *by* chance, and inference *for* chances extending from Chap. 2 by Yoshikawa.

1.3.4 Data Modeling and Mining, Key 3 for Chance Discovery

As mentioned at the beginning of Sect. 1.3, the essential reason why existing data-mining methods have never realized chance discovery is that a human and his or her surroundings form a complex system involving unnoticed factors. We can consider at least three types of human vs. environment relations where:

(1) The environment has a one-way effect on the human(s).

(2) The human(s) has a one-way effect on his/her/their environment.

(3) The human(s) and the environment interact with and affect each other.

For example, consider earthquakes occurring from the extremely complex and unknown pressures among land crusts. This system is of type (1), i.e. a pure risk in that we humans can not avoid (with the current state of the art in seismology or disaster-prevention technologies) the risks involved in the future occurrence of great earthquakes. On the other hand, the same system is of type (2) because humans strike the earth to investigate the mechanism underground, and then look at the response to the shock from the land crust on the surface and the inner part of the earth. Also, in trenching surveys, we humans mine the earth to find 100 000 years of the history of underground events. To do so effectively, however, humans must choose the correct part of the land surface to act on, considering the past attacks from the land — the chance discovery in this system is of type (3).

For a full understanding of chance discovery as a system, we need to discuss and model complex systems where a small event may have a major result in the future. Chapter 4 by Jensen is presented from the aspect of the physics of self-organizing complex systems. Other chapters in Part IV show data-mining methods applicable as Key 3 for chance discovery, based on particular models of the real world as a complex system. Part V gives examples of these methods interacting with humans and the real environment of humans.

1.3.5 Criteria for Evaluating a Discovered Chance

Summarizing communication for chance discovery: first, a few leaders notice a chance in spite of its rareness, and an acceptable interpretation is proposed and spread to a number of other people who might share the new value, i.e. the significance of the chance, and the value may grow to be a common value. Thus, the significance of a chance can be measured as $P, U,$ and G below. As in [1.11], the three criteria are measurable and give an intuitively sound evaluation of chances.

Proposability: how reasonable can a proposal of a decision be made based on the chance. This can be measured by the rate, P, of people who agree to a decision proposal based on the chance.

Unnoticeability: how difficult it is to notice the chance without it being proposed by other people. This differs from pure *novelty* [1.3] because an unnoticeable chance might have existed without being noticed. This can be measured by U, the inverse of the rate of people who noticed the significance of the chance within a short constant time, τ, after its first appearance.

Growablility: the speed of increase in the number of people who agree with the significance of a chance. This can be measured by the growth rate, G, i.e. the ratio of [*the number of people who agree with the significance of the chance*] to [*the number of people who noticed the significance of the chance at the beginning of a time period of length τ*].

To date, studies on chance discovery have aided human awareness of chances of high P, U, and G. For example, in [1.11], relationships between breakfast foods were obtained from the data of meal consumptions by residents of Tokyo and visualized using *KeyGraph*, a visual data-mining tool. A group of housewives, being inspired by the graph and exchanging their ideas, enabled the discovery of hidden but influential value criteria of housewives when designing breakfast. In this approach, a few discussants' subjective awareness of the significance of rare events spread throughout the group. The details of this case appear in Chap. 25 by Fukuda.

This success is owed to a revision from the model of the knowledge discovery process in Fig. 1.1. A key point was that it came to be considered that a chance can be something unnoticed, rare, or new. The model in [1.10] was thus proposed and revised to Fig. 1.2, carrying a human to *concern, awareness, understanding of the chance*, and to *decision/actions*, returning to a new concern with another chance. The group of humans stimulated each other through these phases, and data mining helped the externalization of each discussant's ideas. Because the consideration of latent factors helps in the explanation of the significance of new chances, this process model enabled the discovery, in [1.11], of rare but significant food-consumption patterns arising out of deep-level motivations of consumers.

1.4 Double-Helical Model of Chance Discovery Process

In the process of chance discovery, humans and automated data-mining systems work together, each processing spirally toward creative reconstruction of ideas, as in Fig. 1.2. In this section I present the *double helix* [1.5, 1.12], a refined process model of *chance discovery*. This new model has two helical sub-processes, as illustrated in Fig. 1.4. One is the process undertaken by a human(s), who substantially forms the interpretation cycles in Fig. 1.2, progressing to deeper awareness of chances. The other helix is a process undertaken by a computer(s), receiving and mining the data (DMs, standing for data mining i.e. steps employing tools of data mining, in Fig. 1.4). In Fig. 1.4, each step is numbered with '-h' ('-c'), i.e. 'k-h' ('k-c') for integer k means the information process of the human (computer) in increasing order of k. That is, the road (dotted curve) in Fig. 1.4 depicts the processes undertaken by the human(s) (computer). The thin arrows show the interactions between humans and computers.

Hereafter, let me call a human, or a group of humans who co-work, aiming to discover chances a *subject*. The name 'double-helix' means parallel processing, i.e. there are simultaneous runs of these pairs of helices (spiral processes) due to the input of 'the subject data', monitoring the mind of the subject. For example, the words in the thoughts of discovery subject form subject data. This data is input when the subject discusses or thinks — in previous data mining, the computer took a rest while humans were discussing the latest output of the data mining for 'the object data', that is the data from the environment dealt with by the target problem. Between the two helices, the following interactions occur:

- The subject data, taken from the subject's thoughts, is put into the computer.
- The object data, showing the record of the environment, is collected, reflecting the subject's concern with the target domain.
- The mining result from the subject data (DM-a in Fig. 1.4) is shown to the subject her-/himself, or themselves if they make a group, to be reflected in the understanding of the subject's own concerns, awareness, and understanding of chances.
- The mining result for the object data (DM-b in Fig. 1.4) is reflected in the understanding of the chance in the environment.

This new model is useful for realizing an evolutionary process of chance discovery: the thoughts of the subject recorded as a text file and processed by a text miner externalize the subject's concerns with factors which have been unnoticed but which exist in the real world.

Showing this to her-/himself or themselves stimulates the subject to be aware of what has been lying in her/his/their own mind. For example, the latent interests of people using the WWW, revealing the rare behaviors, could be discovered by showing the text of their discussions with themselves [1.12]. This process, to *look at one's own mind*, was not covered in Fig. 1.2. In this section, let me show the double-helix model as a new method for hypothesis creation. For detailed descriptions showing that this process can lead to successful chance discovery, the reader is referred to Chap. 23 by Nara and Ohsawa. Other successful cases so far include real business applications, as in marketing [1.5].

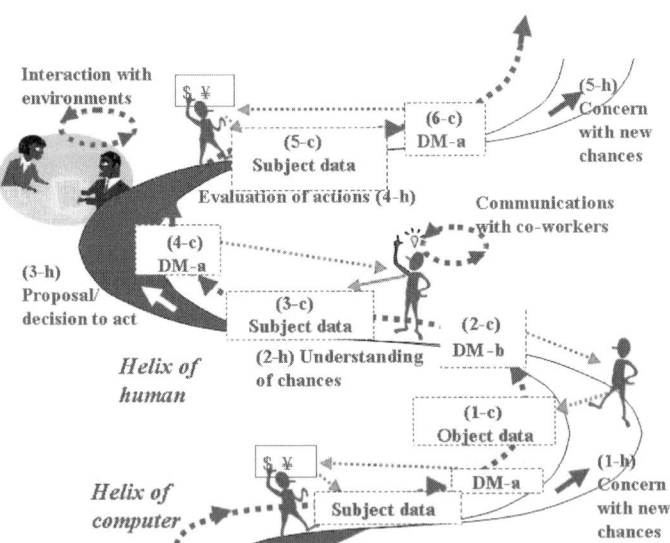

Fig. 1.4. The double-helical process of chance discovery (the numbers such as '1-h' correspond to each step, e.g. in the case of a social survey)

1.5 Subsumption Architecture for Robust Helices

If you desire that a robot should operate in a very dynamic environment, such as in a fire, the robot should be robust, i.e. it should achieve its goals in the face of various effects caused by unexpected changes. Let us take the example of a robot for rescuing survivors in a fire: it must find a path around obstacles blown by the hot wind, move on the path, and reach people who are alive. To achieve this, the robot should not waste time optimizing the walking plan. Instead, it should look at the scene, decide on the direction, and walk on the path all at the same time, and without disturbing people on this path. Furthermore, its path to survivors must not be disturbed by winds coming from the raging fire.

As a good choice for designing such a robot with quick and robust actions, *subsumption* architectures have been applied since first proposed by Brooks [1.2]. The intelligence of the robot is realized in layers, each layer representing a subprocess. Each layer runs constantly without waiting for the result of previous subprocesses, but exchanges messages (sending commands to sublayers and reports to metalayers) as if the sublayer is its slave and the metalayer is its client. If some obstacle comes up suddenly, the robot can react quickly, referring to the newest messages from other layers, instead of waiting for a revised plan to be obtained after some time spent in planning. The result of a subsumption architecture is the efficient spiral (i.e. helical) behavior of the robot, i.e. the action comes after each parallel subprocess, and before the new sensing step which feeds back to the subprocesses of the next action.

Borrowing the idea of subsumption, we can apply a subsumption-based design to chance discovery by humans and assistance in this by data mining using computers, as in Fig. 1.5. Here, 'evaluation' is the metamost layer, subsuming 'concern' which in turn subsumes 'understanding'. And 'understanding' subsumes 'proposal' which finally affects the real actions. Each of these subprocesses is recorded as subject data, and processed using data-mining tools while actions involve some observations — 'sensing' in the terminology of robotics — acquiring object data from the environment. As well as the robot in the subsumption architecture, the discoverer is involved in a helical process, hand in hand with the computer's data mining.

The subsumption architecture is a new design approach for a chance discovery system, and so this book does not include applications of this model. However, this approach is more efficient than following sequentially the stages in the double-helix process because humans have considerable capabilities for parallel processing of thought subprocesses. For example, the human interface of a chance discovery system having the following windows will support the user's subsumption architecture to run in the double-helix process:

The window for understanding and evaluation: The result of data mining, using methods appropriate for visualizing the meaning of rare events, is displayed. A user can click on an interesting part of the result to call essential parts of the real data relevant to the user's interest, to be displayed in the following window for concerns. The clicks here can be regarded as the evaluation of understandable chances. Here the part 'understanding -> proposal -> evaluation', or more directly 'understanding

-> evaluation of understood chances', occurs, and proposals, if generated, can be entered in the following window for proposal.

The window for concerns: The part of the real world is shown by image, text, etc., invoked upon request by the user, i.e. called by a click on the window for understanding and evaluation above. The result of data mining for the data shown here is returned to the window above.

The window for proposal: The window in which the user can input the discovered chances, or new proposals based on these chances. The user can simulate or make actions, based on the proposals entered here. The result of text mining for the proposals entered here can be returned to the window above for understanding and evaluation.

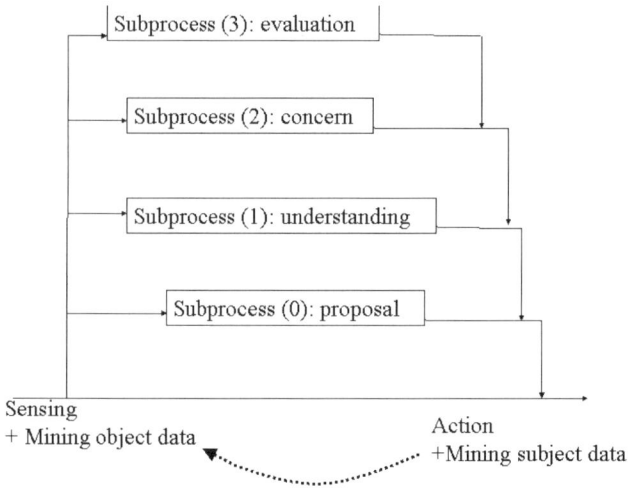

Fig. 1.5. The subsumption model of chance discovery

1.6 A Tentative Conclusion

In this chapter, the overall process of chance discovery has been sketched, and each step in the process is linked with studies on chance discovery presented so far and in this book. The double-helix model, for the process in which a human discovers what we call chances, has been presented, along with its extension to '*subsumption*', a robust and fast architecture. Insights into deep-level factors in the real world can be obtained, and proposals of actions based on the new awareness of rare but significant data items become acceptable.

This chapter focused chiefly on the role of human–computer interactions. In the following chapters, we go into details of both machine-centric and human-centric approaches to chance discovery.

References

1.1 Allport GW, Postman L (1947) *The Psychology of Rumor*, Holt, Rinehart & Winston, New York, NY
1.2 Brooks R (1986) A Robust Layered Control System for a Mobile Robot, *IEEE Transactions on Robotics and Automation*, 2(1): 14–23
1.3 Fayyad U, Shapiro GP and Smyth P (1996) From Data Mining to Knowledge Discovery in Databases, *AI Magazine*, 17(3): 37–54
1.4 Festinger L (1957) *A Theory of Cognitive Dissonance*, Row, Peterson, Evanson, IL
1.5 Goldberg DE, Ohsawa Y (2002) *AI for Chances in Business, Tutorial Note TU-1 in Pacific Rim International Conference on Artificial Intelligence (PRICAI 2002)*
1.6 Kosslyn SM (1980) *Image and Mind*, Harvard University Press
1.7 McBurney P, Parsons S (2001) Chance Discovery Using Dialectical Argumentation, *New Frontiers in Artificial Intelligence LNAI 2253*, Springer-Verlag, Heidelberg, Germany, pp. 414–424
1.8 Nara Y, Ohsawa Y (2000) Tools for Shifting Human Context into Disasters, in *Proceedings of IEEE Conference of KES2000*, pp. 655–658
1.9 Nonaka I, Takeuchi H (1995) *The Knowledge Creating Company*, Oxford University Press
1.10 Ohsawa Y (2002) Chance Discoveries for Making Decisions in Complex Real World, *New Generation Computing*, 20(2): 143–163
1.11 Ohsawa Y, Fukuda H (2002) Chance Discovery by Stimulated Groups of People, Application to Understanding Consumption of Rare Food, *Journal of Contingencies and Crisis Management*, 10(3):129–138
1.12 Ohsawa Y, Nara Y (2003) Decision Process Modeling across Internet and Real World by Double Helical Model of Chance Discovery, *New Generation Computing*, 21(2):109–121
1.13 Petty RE, Cacioppo JT (1981) *Attitudes and Persuasions: Classic and Contemporary Approaches*, WC Brown, Dubuqute, IA
1.14 Rogers EM (1962) *Diffusion of Innovations*, The Free Press, New York, NY
1.15 Shibata H, Hori K (2001) A Creativity Support System for Long-Term Use in Everyday Life - IdeaManager: Its Concepts and Utilization, *New Frontiers in Artificial Intelligence LNAI 2253*, Springer-Verlag, Heidelberg, Germany, pp. 455-460
1.16 Shoji H, Hori K (2001) Chance Discovery by Creative Communications Observed in Real Shopping Behaviors, *New Frontiers in Artificial Intelligence LNAI 2253*, Springer-Verlag, Heidelberg, Germany, pp. 462–410
1.17 Simon H (1945) *Administrative Behavior: a Study of Decision-Making Processes in Administrative Organizations*, 1st edition 1945, 4th edition 1997, Simon & Schuster, Inc., New York, NY
1.18 Turney PD (1995) Cost-sensitive Classification: Empirical Evaluation of a Hybrid Genetic Decision Tree Induction Algorithm, *Journal of Artificial Intelligence Research*, 2: 369–409
1.19 Weiss GM, Hirsh H (1998) *Learning to Predict Rare Events in Event Sequences*, Proc. of KDD-98, pp. 359–363

2. Decisions by Chance and on Chance: Meanings of Chance in Recent News Stories

Fumiko Yoshikawa[1]

Hiroshima Shudo University, 1-1-1 Ozukahigashi, Asaminami-ku, Hiroshima 731-3195, Japan
email: fumiko-y@shudo-u.ac.jp

Summary.

   Recognition of chance can be influenced by a person's situations or surroundings at a specific time. It is useful to know the situations and contexts in which people use the word *chance* because, by considering the relationships between a sentence including the word *chance* and the context, we can construct the logic of our recognition of chance. In this paper we have collected texts which include the word from recent articles in newspapers and magazines, and, taking its etymology into consideration, have explored the use of the word and its implications. The sentences and the contexts show the ways we recognize some events as chances.

2.1 Introduction

What kind of events do we recognize as chances? Recognition of chance can be influenced by a person's situations or surroundings at a specific time. It is useful for us to know the situations and contexts in which people use the word *chance* because, by considering the relationships between a sentence including the word *chance*, and the context, we can construct the logic of our recognition of chance. In order to observe the context including the sentences with the word *chance*, we utilize recent articles from newspapers and magazines that include the word *chance*, which can mirror the modern recognition of chance quite faithfully. The articles are collected mainly through the Internet. Some are randomly selected from newspapers.

   Before we observe the concrete examples, we should grasp the major meanings of the word *chance*. Referring to some dictionaries, we notice that the meanings of chance are quite extensive. Some are close to *opportunity*, some to *probability*, some to *fortune*, and others to *accidental events*. These meanings appear quite independent and are unlikely to be integrated into a single, certain concept, if one just considered them. However, if one looks more closely at the original meaning of the word, one can work out the connection between the meanings.

   In this chapter, we analyze the uses of *chance* in recent news stories and through this analysis we comprehend the logic of our recognition of chance, referring to its etymology. We will give details about the etymology and the meanings of *chance* defined in some dictionaries in the following section, and actual examples of the word quoted from recent articles in later sections.

2.2 The Etymology of *Chance*

2.2.1 Meanings of *Chance* in ME and Its Synonyms in OE

According to the 2_{nd} edition of *The Oxford English Dictionary*, (hereafter *OED2*) [2.5], the word *chance* originated in the present participle of Latin *cadere* meaning 'falling'. English adopted the word from Old French *cheance* in the Middle English period (see also *The Oxford Dictionary of English Etymology*[2.6]).

The Middle English Dictionary [2.4] supplies quotations under the following definitions. The first definition is '1 (a) Something that happens or takes place; an occurrence or event, esp. one that is unexpected, unforeseen, beyond human control, or attributed to providence or destiny'; the earliest citation is from *Arthour and Merlin* (c 1300). The definition 1 (b) is the idiomatic use *what chance so bitide*, meaning 'whatever may happen'. The second definition is 'something that happens (to sb.) and affects (his) circumstances for better or worse; a stroke of (good or bad) luck'. The earliest citation is from the Becket legend of the *South English Legendary* (c 1300). The third definition is 'one's luck, lot, or fate (whether good or bad)', whose earliest citation is from *The Land of Cokaygne* (a 1300). The fourth is '4 (a) Something that may or may not come about or be realized; a fortuitous event or circumstance, an eventuality; vicissitude', whose earliest citation is again from *Arthour and Merlin*, and its specific use is '4 (b) An incidental donation'. The fifth definition is '5 (a) A situation or circumstance; a case; (b) a favorable circumstance, an opportunity' and the sixth 'An adventure or exploit in arms'. Their earliest citations are both from Robert of Gloucester (c 1300), the *Metrical Chronicle*. Other definitions are '7 (a) A force that shapes man's life on Earth; (b) influence upon events, control of man's destiny, dominance', '8 (a) The falling of the dice, the number turning up at a throw, the number thrown; (b) *fig.* one's fortune as predicted by the fall of the dice', and other idiomatic uses including *bi chaunce* and *per chance* meaning 'by chance, accidentally'.

As mentioned above, the word *chance* is a loanword from Old French in the Middle English period. Some concepts which the word *chance* bore later, however, were expressed by other words from the native English tongue before *chance* was borrowed from French. In the Old English period, the native words such as *(ge)byre, gedafenlicnes, inca, intinga, gelimplicnes, sæl, tid, tidlicnes, tima, towyrd* (see *A Thesaurus of Old English* [2.7], Section 05.11.08) had functioned like *chance* meaning *opportunity*.

According to *An Anglo-Saxon Dictionary* edited by Bosworth and Toller [2.2], *(ge)byre* meant 'the time at which anything happens'. The second word appearing in the thesaurus, *gedafenlicnes*, had such senses as 'decency, convenience, opportunity'; this word seemingly means an appropriate situation rather than appropriate timing. *Inca* has rather a negative sense of *occasion, opportunity*, for the other senses of the word are 'a cause of complaint' and 'a scruple, doubt'. *Intinga* had such senses as 'case, occasion'. The original meaning of *gelimplicnes (gelymplicnys)*, which meant 'opportunity, occasion', is almost the same as that of *chance*. The verb *gelimpan* meant 'to happen, occur, befall, come to pass, take place'. *Sæl* is the original form of *sele*, meaning 'good fortune or opportunity' (see *sele* in the *OED2* [2.5]), *tid* is *tide* (see *tide* in [2.5]), and *tima* is *time* (see *time* in [2.5]). Now *time* has a more general meaning than *chance*, but, at that time, their meanings were very similar. The verb *getimian* meant 'to happen, befall'. *Tidlicnes* is a nominalization of the adverb of *tid*, which is *tidlice*. *(To)wyrd* also had meanings close to *chance*. *An Anglo-Saxon Dictionary* [2.2] supplies the definitions 'what happens, fate, fortune, chance' to *wyrd* and 'occasion' to *towyrd*.

Taking a survey of the native words which had meanings close to *chance*, one may say that some of the native English words which have senses close to *opportunity* or *chance* have a similar original meaning of 'what happens'. Especially, the last word *wyrd* seems to have followed a course similar to *chance* on its semantic development. The rivalry between *chance* and its synonyms compels interest, but it lies outside the scope of this chapter; thus we have concentrated on the subject of the original meaning of *chance* in this chapter (the following references are available for Old and Middle English word studies: Cameron, Kingsmill and Amos (1983) and Sylvester and Roberts (2000) [2.3, 2.9]).

2.2.2 Definitions of *Chance* in the *OED2*

Having glanced at the meanings in the earlier periods, we may now turn to the main subject of this section: the integration of senses of present-day English *chance* into the etymology. We find in some etymological dictionaries (e.g. *The Kenkyusha Dictionary of English Etymology* [2.11]) that the original meaning is connected particularly with the falling of dice. *The Chambers Dictionary of Etymology* [2.1] also refers to the meaning of falling of dice, as the meanings of OF *cheance* are 'accident, falling of dice, dice game'. And the same dictionary also makes mention of the meanings of *cheance* in Middle English 'something that takes place, especially unexpectedly, fortune, luck'.

The dice image of *chance* explains the reason for the polysemy of the word a great deal; the etymology connects those meanings which are apparently independent. Hereafter, we verify the meanings of *chance* referring to the definitions of the *OED2*.

The first definition of the noun *chance* in the dictionary is:

> 1. a. **The falling out or happening of events; the way in which things fall out; fortune; case.**
> b. **A happening or occurrence of things in a particular way; a casual or fortuitous circumstance;** = <u>accident</u> *n.* **1 b.**

The earliest citation for Definition 1a is from Robert of Gloucester, which is also one of the earliest citations for this word.

> 1297 <u>R. Glouc.</u> (1724) 465 To come . . . to helpe is moder, that was her ofte in feble chaunce.

The second definition is as follows:

2. (with *pl.***) A matter which falls out or happens; a fortuitous event or occurrence; often, an unfortunate event, mishap, mischance;** = <u>accident</u> *n.* **1 a, c.** *arch.*

The earliest citation for the second definition is from *Cursor Mundi* (a 1300). Both Definition 1 and Definition 2 are extremely close to the meanings of *cheance* in the Middle English period. The difference between Definition 1 and Definition 2 is in their foci: the former focuses on the event itself and the latter focuses on the matter, what happens.

The third definition is marked by a dagger symbol, which illustrates that the word is scarcely used at present.

†3. a. **That which befalls a person; (one's) hap, fortune, luck, lot.** *Obs.* **or** *arch.*

b. **in the game of Hazard.**

The earliest citation for this definition is also from Robert of Gloucester:

1297 <u>R. Glouc.</u> (1724) 14 Hym þouȝte þe ymage in hys slep tolde hym hys cheance.

There is no citation for this definition after the late seventeenth century. The use of *chance* in this sense might have been declining in those days and gone out of use in the course of time. As listed in the definition, there are not a few words meaning 'fortune'. The sources of those words may be worth mentioning (see the *OED2* [2.5]). The word *fortune* originates in Latin *fortūna* and English borrowed it from French in the twelfth century; and *luck* originates in Middle Dutch *geluk* and the shortened form was borrowed from Low German in the Late Middle English period. *Lot* is a native English word. The form in OE is *hlot*. *Hap* was borrowed from Old Norse in the Early Middle English period (there is an OE word which was derived from the same origin as *hap*. It is *ȝehæp*).

This word *hap* also has the meaning of 'an event or occurrence which befalls one' (see the second definition of *hap* in the *OED2*), which is extremely close to the second definition of *chance* in the *OED2*. At present, the meaning is mainly borne by such words as 'happening', 'accident', 'event', and 'occurrence', but once *chance* was a rival in that meaning category. The adverb *perchance*, which used to be synonymous with *perhaps*, is a good example. Now the former is supplanted by the latter.

We shall now return to the subject of the *OED2*. This sense of *chance* meaning 'fortune' has a specific use for a dice game. Under Definition 3b 'in the game of Hazard', a line from Chaucer's 'Pardoner's Tale' is cited.

c 1386 <u>Chaucer</u> *Pard. T.* 325 Seuene is my chaunce, and thyn is cynk and treye.

The word *hazard* is also a rival of *chance* in meaning. The word *chance* has some rivals from other sources and *hazard* is one which shows the same kind of semantic development. The original source of the word is probably an Arabic word meaning 'die' (see the *OED2* [2.5]). *Hazard* also has the senses of 'fortuity' and 'risk'. The phrase *at hazard(s)* is equivalent to *by chance*.

Next, we shall examine the fourth definition:

> 4. a. **An opportunity that comes in any one's way. Often const. of. Also** *pregnantly* = **chance or opportunity of escape, acquittal, or the like. (Often passing into sense 5.)** b. **A quantity or number; used with adjs., as** *fine, nice, smart. U.S. dial.* c. *Cricket.* **An opportunity of dismissing a batsman, given to a fieldsman by the batsman's faulty play; chiefly in phr.** *to give a chance.*

Another citation from Robert of Gloucester is for this Definition 4a.

> 1297 R. Glouc. (1724) 468 The king let Henri is sone, as God ʒef the cheaunce, Lowis doʒter spousi.

After this, there are no citations in this sense for the fourteenth and fifteenth centuries. The following citation for Definition 4a is from Shakespeare's *Cymbeline*.

> 1611 Shakes. *Cymb.* v. iv. 132, I That haue this Golden chance, and know not why.

Today, this meaning of *chance* is frequently used.

The meanings described with the fifth and sixth definitions appeared much later than the other meanings.

> 5. a. **A possibility or probability of anything happening: as distinct from a certainty: often in plural, with a number expressed.** b. *Math.* = probability; **so also** *theory or doctrine of chances.*
> 6. **Absence of design or assignable cause, fortuity; often itself spoken of as the cause or determiner of events, which appear to happen without the intervention of law, ordinary causation, or providence;** = accident *n.* 2.

The fifth one is a mathematical term. The sixth one is interesting in that it hints at English speakers' philosophy of life, a kind of fatalism. Fate is one of our greatest concerns even in classical antiquity. Philosophers have discussed it from time immemorial and it is no exaggeration to say that every religious artist drew personified fates. In classic paintings, fortune is spinning the wheel of fortune and swaying our fates. I infer from Definition 6 that, in our history over the centuries, a sort of deterministic fatalism, which is quite different from the classical fatalism, might have stolen people's hearts. People might have come to believe that every event has its reason to happen, even in the case that it seems fortuitous for them.

Other definitions are for idiomatic uses of *chance* such as *by chance*; *in, through, with chance*; *take one's chance*; *take a chance* or *chances*. There is no space for an extended discussion of these phrases. The formation consisted of a preposition and *chance* is the same kind as *perchance* mentioned above. *Hap* also adopts the same kind of formation. The last two phrases *take one's chance* and *take a chance (chances)* are rather important in this chapter. In those phrases, the word *chance* always involves a certain degree of risk as well as a certain degree of benefit. The definition for *take one's chance* is as follows:

11. a. **to take what may befall one, submit to whatever may happen; to 'risk it'.** So † *to stand to one chance (obs.).*
b. **To seize one's opportunity (see 4).**
c. *To take a chance or chances***: to take a risk or risks. orig.** *U.S.*

We now have surveyed the definitions of the word *chance. The Oxford English Dictionary* was compiled with the aim of showing the vicissitude of English words on a diachronic principle. We must also consider the definitions of *chance* in dictionaries on synchronic principles. *The Longman Language Activator* [2.8] classified the meanings of *chance* roughly in three categories: one is 'chance/opportunity', the second is 'chance/by chance', and the third is 'lucky person'. However, the dictionary tells us to refer to the headword 'lucky' about the third category. *Kenkyusha's New College English-Japanese Dictionary* [2.10] divides the meanings of *chance* into five categories: (1) opportunity; (2) probability, estimate; (3) accidental event, fortune; (4) risk, adventure; (5) lottery. In the following section, we shall be examining the sentences including the word *chance* excerpted from articles in newspapers and magazines, giving careful consideration to their context.

2.3 Meanings of *Chance* in Recent News Stories

In order to understand the situations and contexts where people use the word *chance*, in this section we collect texts which include the word from recent articles in newspapers and magazines in English, and explore the uses of the word and its implications, taking its etymology into consideration. The sentences and the contexts show the ways we recognize some events as chances.

2.3.1 The Classificatory Criterion

Here we discuss the actual examples of the word *chance* according to the four categories, which were broadly divided because a lexical database called WordNet (provided by Princeton University, http://www.cogsci.princeton.edu/~wn/) defines those categories and some learners' dictionaries (e.g. *Kenkyusha's New College English-Japanese Dictionary* [2.10]) also classify the meanings into the four categories. The first category is quite close to *opportunity*, the second to *probability*, the third to *for tune*, which seems to be quite close to the original meaning 'something which takes place', and the final category is 'something dangerous from some angles, something which has some dangerous sides'. The last one can be regarded as part of the third one, *fortune*; that is, it is also possible to classify the word into three basic groups. The logical classification of the meanings is our ultimate goal, but such semantic discussion is beyond the scope of this brief chapter. Here, for the sake of convenience, we adopt the four categories to sort out the articles treated in this chapter.

The instances are collected chiefly from the following web-sites:

– *The Japan Times Online* (http://www.japantimes.co.jp)

– Asahi.com (http://www.asahi.com/english/)
– *The Newsweek Archives* (http://archives1.newsbank.com/newsweek/)

The first two are newspaper sites, and the the the third one, *The Newsweek Archives*, is a news journal site. The site of *The Newsweek Archives* has a search engine which arranges the articles in order of word frequency, and, using this service, we can pick up the articles which include the word *chance* with high frequency.

2.3.2 *Chance* Meaning 'Opportunity'

The most prominent use of *chance* in the recent articles is the one meaning 'opportunity'. A large number of instances can be found in every news site. The following quotations are from the articles appearing in Asahi.com (hereafter underline added to the word *chance*).

– (1) Soccer players switch nationalities for a <u>chance</u> to play.

(Asahi.com: January 4, 2002)

This quotation implies that chances can be intentionally demanded.

– (2) Weekly Japanese language lessons conducted in their mother tongues provide the women with a <u>chance</u> to meet and talk with their compatriots.   (*The Japan Times*: January 8, 2002)

This sentence was quoted from an article with the heading 'Foreign brides fill the gap in rural Japan — bachelors looked abroad after eligible local girls fled village life for city.' In opposition to the first article about soccer players, the foreign brides in this article don't seem to have attended the Japanese language lessons so as to meet other brides from the same country. The attendance at the language lessons for foreign brides inevitably led them to meeting their fellow countrywomen.

The following quotation is not from any kind of news story but from an advance billing of a movie in the magazine, and is a sentence which briefly sketches the plot of a movie called *Training Day* starring Denzel Washington, who plays the role of a corrupt sergeant and Ethan Hawk, a rookie:

– (3) Given a <u>chance</u> to be a hero and rescue civilians, Det. Sgt. Alonzo Harris would rather steal someone's drug money.

(*Newsweek Web Exclusive*: October 4, 2001)

This example shows that the subject of this sentence, Det. Sgt. Alonzo Harris, might not notice the chance or not have interests in seizing upon the chance even if he notices it. Examining the above three quotations, we find that the speaker (in Example (3), the writer of the article) takes for a chance the situation that the speaker expresses in a sentence with the word *chance*, and that not every man can regard the same situation as a chance.

2.3.3 *Chance* Meaning 'Probability'

The following examples belong to chance in the sense of 'probability'. Example (4) is a quotation from an article on BSE:

– (4) Experts say that they might have a better <u>chance</u> of determining the source of mad cow disease — and thereby prescribe effective countermeasures — if they have a bigger sample of mad cow diseases to work with, but they caution that numbers alone are no sure bet.

(The Japan Times: January 7, 2002)

The next is from an article on anthrax:

– (5) And patients who start treatment before symptoms set in have excellent <u>chances</u> of surviving.   (*Newsweek*, US edition, p. 36: October 22, 2001)

In other words, the sooner one who was infected with anthrax has treatment, the better possibility he/she has of a cure. This kind of *chance* frequently accompanies adjectives of degree. In Example (4), it is modified by an adjective *better* and in example (5) by *excellent*. In the same article about anthrax, there are other two instances of *chance* in the same sense:

– (6) Despite recent events in Florida, New York, and Nevada, a typical civilian's <u>chances</u> of getting infected are still vanishingly small.   (*Newsweek*, US edition, p. 36: October 22, 2001)
– (7) As word of the NBC case spread late last week, many New Yorkers rushed to get themselves tested — "Our physician referral service is swamped with calls," says Dr. Robert Holzman of Bellevue Hospital and the NYU School of Medicine — but experts stressed the slim *chances* of contracting the infection without warning.   (ibid.)

The same kind of *chance* can be seen in the following headline:

– (8) Sri Lanka's *chance* of ending conflict is bigger than ever

(The Japan Times: December 14, 2001)

In the body of the article, we can see another *chance* with the same meaning:

– (9) If the general mood is one of conciliation now, there are other reasons why peace has a greater *chance* today.   (ibid.)

2.3.4 *Chance* Meaning 'Fortune'

The third type of *chance* could not be found as often as the above-mentioned two meanings. The following article was found in *The Asian Wall Street Journal*:

She Got Laid Off,
He Missed the Train;
Such Lucky Breaks
* *

For Some Americans, Avoiding Tragedy Seemed Like Pure <u>Chance</u>

Monica O'Leary thought her luck had taken a turn for the worse Monday afternoon when she got laid off from her job.

But the fact that she didn't go to work Tuesday turned out to be nothing short of miraculous for Ms. O'Leary. She had worked as a software saleswoman for eSpeed Inc., a technology company with offices on the 105th floor of the World Trade Center.

Ms. O'Leary, 23, is still grappling with memories of her last visit with co-workers Monday afternoon. "I worked with a lot of guys, so I kissed them on the cheek and said 'good-bye'", she says. "Little did I know that it was really good-bye".

For hundreds of people in New York, Washington and other cities affected by the deadly terrorist assault, Tuesday morning turned out to be an incredibly lucky time to oversleep, reschedule a meeting or take time off to sneak in a haircut. By doing so, they managed to sidestep the almost unimaginable fate that befell their co-workers and friends in the devastating attacks on the World Trade Center and the Pentagon.

Greer Epstein, who works at Morgan Stanley & Co.'s offices on the 67th floor of the World Trade Center, escaped possible injury by slipping out for a cigarette just before a 9 a.m. staff meeting. Bill Trinkle, of Westfield, New Jersey, had planned to get an early start on his job as sales manager for Trading Technologies Inc., a software concern with offices on the 86th floor of the World Trade Center's Tower One. But after fussing with his two-year-old daughter and hanging curtains in her bedroom, he missed the train that would have taken him into the office about a half-hour before the attack. Instead, he took a later train directly to visit a client company, where workers hugged him as soon as he walked through the door.

Joe Andrew, a Washington lawyer and former chairman of the Democratic National Committee, had a ticket for seat 6-C on the ill-fated American Airlines flight 77 from Dulles International Airport to Los Angeles, but switched to a latter flight at the last minute. "I happen to be a person of faith", says Mr. Andrew, 'but even if you aren't, anybody who holds a ticket for a flight that went down . . . will become a person of faith'.

In some cases, it was simply a good day to sustain seemingly bad luck. Nicholas Reihner was upset when he twisted his ankle while hiking during a vacation to Bar Harbor, Maine. But it was the reason he missed his Tuesday morning trip home to Los Angeles from Boston on the American Airlines flight that was hijacked and crashed into the World Trade Center.

'After I sprained my ankle, I was bellyaching to my hiking companion about how life sucks', says the 33-year-old legal assistant. "I feel now that life has never been sweeter. It's great to be alive".

Then there is George Keith, a Pelham, New York, investment banker who had a meeting at 9 a.m. Tuesday on the 79th floor of the World Trade Center. While he was driving through Central Park the night before, however, the transmission of Mr. Keith's brand-new BMW sport-utility vehicle got stuck in first gear. The breakdown forced him to cancel the morning meeting. By the time he called the BMW dealer Tuesday, he was anything but furious. "I told them it was the best transmission problem I'll ever have".

David Gray, a compliance officer for Washington Square Securities, lives in Princeton, New Jersey, and was due to arrive by commuter train at the Trade Center for a meeting with one of the firm's brokers just at the time the first plane hit. However, a few days earlier, Mr. Gray, who is married to New York City Ballet principal ballerina Kyra Nichols, broke his foot while jumping rope at home. He said he had been feeling very 'sheepish' about the nature of the accident, but now says "Thank God I was a lousy jump-roper".

After he broke his foot, Mr. Gray rescheduled the meeting for later in the day so that he could drive into Manhattan instead of taking the commuter train. "So I was on the New Jersey Turnpike watching the World Trade Center go up in flames, instead of being in it".

In some cases, a chain of unlikely circumstances added up to a collective near-miss. For Irshad Ahmed and the employees of his Pure Energy Corp., the circumstances were these: a postponed meeting, a delay at a child's school, and a quick stop at the video store. Mr. Ahmed, president of the motor-fuels maker, had been set to attend a 9 a.m. meeting in the company's 53rd-floor conference room inside Tower One.

But last week, the participants decided to push the meeting back. As a result, none of Pure Energy's nine employees were at work when the terrorists struck. Some were at a New Jersey lab. Others were out at appointments. Mr Ahmed's secretary was running late at her child's school. As for Mr. Ahmed, he decided to stop off and return a couple of Blockbuster videos. "It's one of those little decisions you make that lead up to big events in life", he says.

For others, a decision to defy orders proved lifesaving. Michael Moy, a software engineer for IQ Financial Inc., was at his workstation getting ready to write software on the 83rd floor of World Trade Center Tower Two when the first jetliner struck Tower One. A few minutes later, he says building security came on the speaker and instructed occupants to remain in their offices, saying that it would be more dangerous in the streets due to falling debris from the other building.

Disobeying those instructions, Mr. Moy and his boss told the 15 or so employees in their wing to start heading down the stairs, Mr. Moy says. Once again an announcement came over the speaker system, instructing employees to return to their respective floors. A few employees decided to do so and headed toward the lobby's elevators. Just then, the doors of several elevators exploded, apparently because the second hijacked airplane had slammed into the building just a few floors above them.

Pandemonium followed, but being familiar with the stairway systems in the building, Mr. Moy and his boss directed co-workers to a little-used stairway that was relatively empty. As a result, dozens of people were able to hurry downstairs and escape into the street.

'I'm glad we acted the way we did', says Mr. Moy, 'otherwise I wouldn't be having this conversation with you'.

By Robert Tomsho, Barbara Carton and Jerry Guidera.

The Asian Wall Street Journal. Vol. XXVI, No. 10, p. 1; September 14 – 16, 2001

In this article, each paragraph consists of sentences (clauses) whose themes are usually persons who happened to avoid being involved in the terrorism on September 11. Monica O'Leary was fired from a company whose offices were located in the World Trade Center on the day before. Greer Epstein went out of the building to smoke just before the attack. Bill Trinkle missed his usual train. Joe Andrew canceled his seat on the American Airlines flight 77 and got a seat on another plane. Nicholas Reihner missed the flight because he twisted his ankle. George Keith canceled a meeting in the World Trade Center because of a car breakdown. David Gray postponed a meeting in the building because he broke his foot. These are about personal experiences, and as for groups, first, Irshad Ahmed, the president of a company, and his employees all avoided the disaster. The president also put off a meeting, which had been scheduled in that morning. His employees, therefore, were not in the office then, and his secretary was not there yet because she took her child to school. Secondly, Michael Moy, his boss, and his colleagues, who were working in Tower Two when the first airplane crashed into Tower One, neglected the instructions to stay in their offices over the public-address system, ran down the stairs to the ground, and took refuge from the tower.

In all the cases except for the last episode, they took actions independently of the terrorism, or events, most of which were seemingly unfortunate ones, occurred just before the terrorism, and consequently those actions and unfortunate events saved their lives; so called, bad luck often brings good luck. As for the last episode, after the first airplane crash, they disregarded the instruction over the speaker; that is usually considered as a dangerous action because such stampedes to stairs, in many cases, increase the dangers of falling down one after another. As a result, they survived the disaster because they did not obey the instruction. The phrase included in the headline, *pure chance*, means that they knew that they were lucky after the terrorism. It was by chance.

Even in the first sense, which is quite close to *opportunity*, *chance* is sometimes said to have an adventitious nuance; while *opportunity* does not have such a nuance (see *Kenkyusha's New College English–Japanese Dictionary* [2.10]). The article in *The Asian Wall Street Journal* shows that this word *chance* in the third sense can be used in the situations where people could not predict the influence of their actions or of events. In situations which can be recognized as *chance*, understanding the cause-and-effect relationship between the action (or event) and the later event in advance is beyond human ability.

2.3.5 *Chance* Involving a Certain Degree of Risk

The final sense of *chance* is the one which involves a certain degree of risk. The following concerns Dale Earnhardt, a racing driver who died in last year's Daytona 500:

> When the '78 season unfolded, with all of this attention and hope, Osterlund and Wlodyka quickly decided they had made one major miscalculation. They didn't like their choice of driver. They didn't think [Dave] Marcis was testing the car in races enough. They felt he was solid, a fine guy, certainly competent, stayed away from trouble, finished most of the time in a respectable third or fourth or fifth ... but he never won. He didn't take the chances they wanted.
>
> (*Newsweek Web Exclusive*: October 15, 2001)

– (10) He didn't take the *chances* they wanted.   (ibid.)

The subject of the last sentence in the quotation, Example (10), *He* refers to another racing driver Dave Marcis and *they* in the relative clause refers to his racing team owners, Osterlund and Wlodyka. This sentence says that Dave Marcis always held the third, fourth, or fifth places, but that he never won. The phrase *take chances* can be replaced by another phrase *take risks*; that is to say, the actions the drivers take trying to win the championships might not necessarily turn out as they want them to. The results caused by those actions are unpredictable matters for human beings.

2.4 Discussion and Conclusion

The original meaning of the word *chance* is 'falling of dice'. The sense of 'happening' was derived from the original meaning. The logic of the semantic development is exactly analogous to Caesar's famous words on the Rubicon, "the die is cast". The word *chance* particularly expresses an event unpredictable for us as if we could not predict the number on the dice, though each number would be shown at a fixed rate. We easily understand that this kind of problem is probabilistic. This is a plausible logic of the reason why the word *chance* has a sense meaning 'probability'.

In a fatalistic view, predicting the number is beyond human-beings' ability. We only realize the number when the dice has rolled over. We take that kind of event as accidental because we realize the relation of cause and effect after we have been affected by the effect. The unexpected nature of the word *chance* might have been connected with classic fatalism, *moira*, for one is likely to attribute such accidental events to supernatural strength. It is little wonder that *chance* has the sense of 'one's fortune'.

We do not know the way to make a particular face of the dice appear even though we know how many faces the dice has. We have similar situations to this in our lives. The particular face of the dice one wants is similar to the particular situation we are waiting for. This must be the logic of cognition of *chance* meaning 'opportunity'. When we find the face of the dice we want, we recognize that the chance has come, but we do not know the way to get the face. The chance is still an accidental situation. There is another thing to note. As we saw in Example (1) about soccer players, sometimes we know the way to enhance our chance, or probability of success. Even in that case, we cannot assure our success.

Sometimes we cast the dice being prepared for the bad faces, as we find risky faces on a dice as well as beneficial faces. This is our psychological mechanism in cases where we use the phrase *take a chance (take chances)*. We can recognize from the example of *take chances* introduced in the preceding section that one is fully conscious of bad aspects as well as good aspects in his/her situation when he/she uses the phrase.

From the observations in the last section, one general point is very clear. Our cognition of chance varies according to the speakers' interpretations of their situations, though we are not conscious of the variety. In some cases, the chances were not obvious until the events had occurred; in other cases, they were obvious even before the events, but people did not know how to achieve success.

In ancient days, people heard sibylline predictions through spritualistic media such as shamans and oracles. It is a decided advantage to foresee what will happen in the future. To know the future will make it possible for us to avoid risk and certainly bring us economic success. That is why people have always showed untiring zeal in prophecies or oracles. Now computers are performing the equivalent social function on scientific lines. However, we see that the implementation of the algorithms for providing chances should be demanding because we have to take every situation developing around people who want to get chances into consideration. The observation of the use of the word *chance* in each context and comprehension of our cognition of chance offer the key to successful imprementation.

Acknowledgements. First, I gratefully acknowledge helpful discussions with Prof. Yukio Ohsawa, Dr. Yutaka Matsuo, and Mr. Naohiro Matsumura. Next, I wish to express my gratitude to Prof. Malcolm J. Benson for reading the entire text and making a number of helpful suggestions. I would also like to thank Prof. Hideki Watanabe and Mr. Yoshitaka Kozuka for permission to use their books and for their words of valuable advice, and Prof. Clive H. Pemberton for helpful comments on an earlier draft. Needless to say, however, responsibility for any errors rests entirely with the author. Finally, I wish to thank my family members for their great encouragement.

References

2.1 Barnhart RK, Steinmetz S (eds) (2000) *Chambers Dictionary of Etymology*. (Previously published as the *Barnhart Dictionary of Etymology* by The HWWilson Company in 1988.) Chambers, Edinburgh, UK

2.2 Bosworth J, Toller TN (eds) (1972) *An Anglo-Saxon Dictionary*, Oxford University Press, London, UK

2.3 Cameron A., Kingsmill A, Amos AC (eds) (1983) *Old English Word Studies*, University of Toronto Press, Toronto, Canada

2.4 Kurath H, Kuhn SM and Reidy J (eds) (1952) *Middle English Dictionary*, University of Michigan Press, Ann Arbor, MI

2.5 Murray JAH, Bradley H, Craigie WA, Onions CT (eds) (1933) *The Oxford English Dictionary, the 2nd edition*. Oxford University Press, Oxford, UK

2.6 Onions CT (ed) (1966) *The Oxford Dictionary of English Etymology*. Oxford University Press, Oxford, UK

2.7 Roberts J, Kay C, Grundy L (eds) (1995) *A Thesaurus of Old English*. King's College London, London, UK

2.8 Summers D, Rundell M, et al (eds) (1993) *The Longman Language Activator*. Longman, Harlow Essex, UK

2.9 Sylvester L, Roberts J (2000) *Middle English Word Studies—A word and Author Index*. Brewer, Rochester, NY

2.10 Takebayaship S, Yoshikawa M and Ogawa S (eds.) (1994) *Kenkyusha's New College English-Japanese Dictionary, the 6th edition*. Kenkyusha, Tokyo, Japan

2.11 Terasawa Y (ed) (1997) *The Kenkyusha Dictionary of English Etymology*. Kenkyusha, Tokyo, Japan

3. Prediction, Forecasting, and Chance Discovery

Yutaka Matsuo

Cyber Assist Research Center, National Institute of Advanced Industrial Science and Technology, Aomi 2-41-6, Tokyo 135-0064, Japan
email: y.matsuo@aist.go.jp

Summary.

This chapter addresses the relation and difference between prediction, forecasting, and chance discovery. Prediction and forecasting have a long history. So far, many studies have been devoted to prediction and forecasting. However, in complex real-world systems, contrary to scientific laws, it is sometimes very difficult to predict the future. In such situations, model creation, model selection, and parameter fitting are all important in the complex changing real world. Chance discovery targets three aspects that prediction and forecasting methods have not shed light on, i.e. emphasis on model and variable creation and discovery, emphasis on rare events, and emphasis on human and computer interaction.

3.1 Introduction

This chapter addresses the relation and difference between prediction, forecasting, and chance discovery. Prediction and forecasting have a long history. From remote history, such as in ancient Greece, man demonstrated the desire to predict the future and understand the past; these desires motivated the search for laws that explain behavior of observed phenomena.

Scientific discoveries are sometimes verified through prediction: prediction of the planet Neptune's existence by Leverrier, prediction of deviation of light by Einstein, prediction of the helical structure of DNA by Watson and Crick, etc. [3.23]. Prediction has a very strong force of argument. So far, many studies have been devoted to prediction and forecasting. However, in complex real-world systems, contrary to scientific laws, it is sometimes very difficult to predict the future. The difficulty of prediction depends on the degree of freedom and complexity of the system; if too many parameters should be fixed, it is impossible to make a precise prediction. If the evolution law amplifies initial uncertainty too rapidly, one can not make long-term predictions.

In such situations, choice of a prediction model strongly affects the prediction performance. A model which works well in one case might not work well in other cases. Therefore, model creation, model selection, and parameter fitting are all important in the complex changing real world.

In contrast to the long history of prediction and forecasting, chance discovery is a brand-new research field; formally it began in 2000 (although many essential pieces of research had already begun in the late 1990s). Chance discovery targets aspects that prediction and forecasting methods have not shed light on: rather, those aspects that prediction and forecasting had considered as given. In this section, differences are classified into three categories: emphasis on model and variable creation and discovery, emphasis on rare events , and emphasis on human and computer interaction. Conventional prediction and forecasting methods presume that a user (of the method) knows already which variables to predict, and which variables should be cast into the methods. For example, an investor wants to know the trend of a certain stock price based on the history of the price or data of other stock prices and economic indices; a marketer wants to predict sales of a product based on the sales history; a traveler wants to know tomorrow's weather based on the history of weather changes and current weather. However, sometimes one does not know which variable to predict: one can imagine a woman who is not aware of the risk of great earthquakes living in a quake-prone area, or a man who is not aware of the potential chance of developing a new hit product. These people do not know which variable to predict. In the real world, often in very important situations, we are not aware of which variables to predict, and which variables to cast into prediction methods.

Furthermore, ordinal prediction and forecasting methods postulate the existence of a coherent model behind data. If we assume coherence, many prediction and forecasting methods work very well. Certainly, scientific laws are very coherent. However, in the real world, sometimes it is not reasonable to assume coherence. Social and economic relationships are constantly changing. New products appear day by day. The Internet emerged globally, completely changing our way of life and business activities. Greenhouse gases have become a problem on a world-wide scale, resulting in the regulation of greenhouse-effect gas emissions, and leading to a new market for ecological hybrid cars. In such a real world, the assumption of a coherent model sometimes does not hold. Rather, we should develop methodology in the structurally changing world in which we live.

The following section makes a brief survey of prediction and forecasting methods. Knowing that prediction and forecasting constitutes a long-studied area, we cover only limited aspects of that field. Further information can be found, for example, in [3.28, 3.15, 3.7, 3.5]. Recent advances in data-mining methods open a new direction to prediction and forecasting. After overviews presented here, we will discuss the difference and relevance between prediction/forecasting and chance discovery in Sect.3.3. Section 3.4 is devoted to one model which we think captures the changing world: the small world. Some surveys and discussions are made there with regard to the small world .

3.2 Existing Method of Prediction and Forecasting

3.2.1 Time-Series Prediction

Weigend and Gershenfeld indicate that time-series analysis has three goals: forecasting, modeling, and characterization [3.28]. Forecasting is also called predicting; it aims at accurately predicting the short-term evolution of a system. (Prediction is also referred to as estimating unobservables, for example of an RNA structure or of a VLSI circuit.) The goal of modeling is to find a description that accurately captures features of the system's long-term behavior. The third goal, system characterization, attempts with little or no a priori knowledge to determine fundamental properties, such as the number of degrees of freedom of a system or the amount of randomness.

Before the 1920s, forecasting was done by simply extrapolating the series through a global fit in the time domain. The beginning of 'modern' time-series prediction might be set at 1927 when Yule invented the autoregressive technique in order to predict the annual number of sunspots. His model predicted the next value as a weighted sum of previous observations of the series [3.31].

According to [3.28], two crucial developments occurred around 1980 due to general availability of powerful computers. The first development was state-space reconstruction by time-delay embedding. The second development was emergence of the field of machine learning; it was able to adaptively explore a large space of potential models. With the shift in artificial intelligence from rule-based methods toward data-driven methods, the field was ready to apply itself to time-series.

3.2.2 ARMA Model

Linear time-series models are one of the most simple predictive models; they can be understood in great detail and are straightforward to implement. ARMA models have dominated all areas of time-series analysis and discrete-time signal processing for more than half a century [3.28]. Two crucial assumptions will be made: the system is assumed to be both linear and stationary.

Assume that we are given an external input series $\{e_t\}$ and seek to modify it to produce another series $\{x_t\}$. In the MA (moving average) model, the present value of x is influenced by the present and N past values of the input series e:

$$x_t = \sum_{n=0}^{N} b_n e_{t-n} = b_0 e_t + b_1 e_{t-1} + \ldots + b_N e_{t-N}.$$

In the AR (autoregressive) model, some feedback is considered:

$$x_t = \sum_{m=1}^{M} a_m x_{t-m} + e_t.$$

Depending on the application, e_t can represent either a controlled input to the system or noise.

The ARMA model is a combination of the AR and MA models; the ARMA(M, N) model is stated as

$$x_t = \sum_{m=1}^{M} a_m x_{t-m} + \sum_{n=0}^{N} b_n e_{t-n}.$$

We can estimate coefficients of the AR(M) model from the observed correlational structure of a signal. Estimation of the coefficients can be viewed as a regression problem: expressing the next value as a function of M previous values. This can be done by minimizing squared errors: the parameters are determined such that the squared difference between the model output and the observed value, summed over all time steps in the fitting region, is as small as possible. Standard techniques exist, often expressed as efficient recursive procedures, for finding MA and ARMA coeficients from observed data.

Historically, an important step beyond linear models for prediction was taken 20 years ago; it used two linear functions instead of one globally linear function. This threshold autoregressive model (TAR) is globally non-linear. Such non-linear models significantly expand the scope of possible functional relationships for modeling time series, but this benefit comes at the expense of simplicity. One solution to this is in a connectionist framework.

3.2.3 Pattern Recognition

One new developing method of forecasting is through pattern imitation and recognition [3.22]. Consider the time series as a vector

$$\boldsymbol{y} = \{y_1, y_2, \ldots, y_n\},$$

where n is the total number of points in the series. The current state is represented as y_n. One possible simple method of prediction is based on identifying the closest neighbor of y_n in the past data, say y_j, and predicting y_{n+1} on the basis of y_{j+1}. This simple approach may be extended by taking an average prediction based on a set of nearest neighbors. The definition of the current state of a time series may be extended to include more than one value. Optimal state size must be determined experimentally on the basis of achieving minimal errors on standard measures.

Consider again the time series $\boldsymbol{y} = \{y_1, y_2, \ldots, y_n\}$. A segment in the series may be defined as a difference vector $\boldsymbol{\sigma} = (\sigma_1, \sigma_2, \ldots, \sigma_{n-1})$, where $\sigma_i = y_{i-1} - y_i$ ($\forall i, \ 1 < i < n - 1$). A pattern contains one or more segments and may be visualized as a string of segments

$$\boldsymbol{\sigma} = (\sigma_i, \sigma_{i+1}, \ldots, \sigma_h)$$

for given values of i and h, where $1 < i < h < n - 1$. If we choose to represent the pattern more simply, we encode the time series $\boldsymbol{y}$ as a vector of change in direction: a value y_i is encoded as 0 if $y_i - 1 < y_i$, as a 1 if $y_i - 1 > y_i$, and as a 2 if $y_i - 1 = y_i$. A pattern in the time series may now be represented as

$$\boldsymbol{\rho} = (b_i, b_{i-1}, \ldots, b_h).$$

In this approach, time-series forecasting refers to the process of matching a current state of the time series with its past state. Success in correctly predicting the series depends directly on the pattern-matching algorithm. Also, the size k has an important impact on error minimization and correct prediction. The match itself is sometimes not exact and can be done by a fuzzy matching algorithm.

Similarly, aside from fuzzy methods, a large number of studies have been done for forecasting using neural networks, genetic algorithms, and Markov models.

3.2.4 Information Between the Past and the Future

In [3.4], Bialek et al. say that the only components of incoming data that present the possibility of being useful are those that are predictive. It makes sense to isolate the predictive information from non-predictive information. Learning a model to describe a data set can be seen as an encoding of that data; the quality of this encoding can be measured using information-theory concepts.

From the information-theory perspective, past data T provides information about future data T'. We can write the average of this predictive information as

$$I_{\text{pred}}(T, T') \leq \left\langle \log_2 \frac{P(x_{\text{future}}|x_{\text{past}})}{P(x_{\text{future}})} \right\rangle \tag{3.1}$$

$$= S(T) + S(T') - S(T + T'), \tag{3.2}$$

where $\langle \cdots \rangle$ denotes an average over the distribution; $S(T) = -\langle \log P(x_{\text{past}}) \rangle$ is the entropy of observations on a window of duration T. From the formula above, we can view $I_{\text{pred}}(T, T')$ as either the information that a data segment of duration T provides about the future of length T', or the information that a data segment of duration T' provides about the immediate past of duration T.

If we have been observing a time series for a long duration T, then the total amount of data we have collected is measured by the entropy $S(T)$. Under some assumptions, we can write $S(T) = S_0 T + S_1(T)$; of the total information we have taken in by observing x_{past}, only a vanishing fraction is relevant to the prediction:

$$\lim_{t \to \infty} \frac{\text{Predictive information}}{\text{Total information}} = \frac{I_{\text{pred}}(T)}{S(T)} \to 0.$$

In this sense, most of what we observe is irrelevant to the problem of predicting the future.

3.2.5 Data-Mining Methods

Time-series data has been recently studied in the context of data mining. Many methods attempt to find frequent patterns in time-series data (e.g. [3.10]). APRIORI is one of the most well-known methods to find association rules

$$X \to Y.$$

Agrawal and Srikant introduced the sequential pattern-mining problem in [3.24]. Many methods which are based on the APRIORI property [3.1] have been proposed for mining sequential patterns (e.g. [3.2, 3.24, 3.9]).

Han *et al.* studied periodicity search, that is, search for cyclicity in time-related databases [3.12]. They found segment-wise periodicity in the sense that only some of the segments in a time sequence have cyclic behavior. For example, Laura may read a newspaper at 7:00 to 7:30 every weekday morning, but may do all sorts of things afterwards.

In contrast to mining frequent patterns or periodical patterns, several studies focus on rare events. Weiss proposed a method to predict extremely rare events such as hardware-component failures in the AT&T network [3.29, 3.30]. Their system, called Timewaver, is a genetic-based machine learning system for predicting events. Following their description of the event-prediction problem, a prediction occurring at time t, Pt, is said to be correct if a target event occurs within its prediction period. The system searches the solution space using a genetic algorithm. Prediction rules are encoded into each individual. The rule is for example: if two (or more) A events and three (or more) B events occur within an hour, then predict the target event. They use precision and recall to evaluate a solution. Recall is the percentage of target events correctly predicted and precision is the percentage of times that a target event is predicted and actually occurs. The evaluation function is based on both precision and recall. The F-measure, which is used in information retrieval, is used as the evaluation function:

$$f = \frac{(\beta^2 + 1)\text{precision} \times \text{recall}}{\beta^2 \text{precision} + \text{recall}}.$$

Instead of usual direct association, Tan et al. introduced the concept of indirect association between items [3.26]. They believed that some of the infrequent item sets may provide useful insight about the data. Consider a pair of items, (a, b), that seldom co-occur together in the same transaction. If both items are highly dependent on the presence of another item set, Y, then the pair (a, b) is said to be indirectly associated via Y. In market basket data, this method can be used to perform competitive analysis of products. For text documents, indirect association between a pair of words often corresponds to synonyms, antonyms, or words that are present in the different contexts of another word. This method is also used for mining Web-usage data [3.25].

Domeniconi et al. attempted prediction of significant events from sequences of data with categorical features [3.6]. Co-occurrence analyses of events are done by means of singular value decomposition of examples constructed from data. Starting with an initial rich set of features, they clustered features based on correlation. The resulting classifier was expressed in terms of a reduced number of examples; thereby, predictions can be performed efficiently.

In [3.23], catastrophic events are discussed such as the rupture of composite materials, great earthquakes, turbulence, abrupt changes of weather regimes, financial crashes, and human parturition. A central property of such complex systems is the possible occurrence of coherent large-scale collective behaviors with a very rich structure, resulting from repeated non-linear interactions among their constituents. These systems in natural and social sciences exhibit rare and sudden transitions, which occur over time intervals that are short compared to the characteristic time scales of their posterior evolution. Such extreme events express, more than anything else, underlying forces. In case of the rupture of materials, the fracture process depends strongly on the degree of material heterogeneity: if the disorder is too small, then the precursory signals are essentially absent and prediction is impossible. If heterogeneity is large, rupture is more continuous.

Finally, Last et al. [3.14] introduced new aspects and difficulties of time-series databases (TSDB). The process of knowledge discovery in TSDB includes cleaning and filtering of time-series data, identifying the most important predicting attributes, and extracting a set of association rules that can be used to predict future time-series behavior. They used a fuzzy approach to express extracted rules in natural language.

3.3 Difference between Prediction/Forecasting and Chance Discovery

As seen above, myriad frameworks have been developed for predicting the future, including statistical, pattern-recognition, and data-mining algorithms. The major concern of chance discovery is also in the future, e.g. predicting earthquake occurrence, developing new merchandise, and planning new strategies. However, chance discovery targets those aspects that prediction and forecasting methods have not shed light on. Rather, on what in prediction/forecasting had been considered as given. We will discuss three aspects of chance discovery: model and variable creation and discovery, rare events, and human–computer interaction.

3.3.1 Emphasis on Model/Variable Creation and Discovery

Conventional prediction/forecasting methods postulate the existence of a coherent model behind the data. If we assume the coherence, many prediction/forecasting methods work very well. Certainly, scientific laws are very coherent. However, in the real world, is it reasonable to assume coherence? In the real world, the assumption of a coherent model often does not hold. Rather, we should develop methodology for a structurally changing world that resembles the one in which we live.

In real life, such as in the business world, human networks, social development, and so on, it happens very often that the structure of the system changes at some points. In [3.11] the way in which a little thing can cause a big structural change is discussed. Not only does a system evolve gradually as time passes, but the system may also completely change its structure at some points. This is due to large-scale collective behaviors with a very rich structure and repeated non-linear interactions among its constituents. In such a situation, conventional prediction and forecasting methods are not as effective as in a stable situation. In fact, when we face dramatic structural change, we may not be able to predict the future. It is very important to grab what happens, and find which variables to focus on.

Therefore, chance discovery is not concerned so much with predicting the precise values of some variables in the future. Although such prediction is very important in a stable situation, it is not effective in dynamic situations. Knowing what is happening, determining which variables to monitor, and creating a new model are of great importance.

Model selection is a key issue in prediction/forecasting. There are some heuristics to find the proper model, such as Akaike information criteria (AIC) or minimal description length criteria. In the context of data mining, feature selection is also an important process, which selects informative attributes. However, what we mention here includes a big change of the model based on complex dynamics.

3.3.2 Emphasis on Rare Events

Rare events sometimes have a very large impact on social, economic, and business worlds. It is relatively easy to obtain knowledge about a frequent pattern, and thus it can be understood well. In this sense, a frequent pattern does not have large information if we assume that the a priori probability is modeled by common awareness of the event: if all competitors of a company know about an event, the information can not be a powerful strategic card.

On the other hand, rare events are not easy to recognize and use for decision making. Events with low frequency are sometimes neglected; thus they have much information. If most competitors of a company do not know about an event, it can present opportunities for the company.

Ordinary statistical methods are very useful if a model is assumed and the number of samples is large. However, these methods are not proper for rare events. (Note that there are some techniques to analyze rare events statistically [3.8].) If the number of samples is small, it is generally not statistically supported. Chance discovery focuses on the tail of the distribution. Even if a large number of samples are collected, the tail exists and sometimes the tail is a good source of information.

The above discussion is based on information in Shannon's sense [3.21]. That is, we discard the meaning of the event and only focus on the probability of an event. However, in the real world, we must also consider the impact of the event. Prediction of a big earthquake with low probability is important, but prediction of an event with low probability and a low impact has no merit. Therefore, whether an event has an impact or not is an essential aspect.

When we deal with rare events, it is not practical to consider the meaning and impact of every rare event beforehand because such rare events can emerge in a variety of ways. It is essential to use computer calculation to reduce the number of rare events which *might be* important.

3.3.3 Emphasis on Human and Computer Interaction

The third point of difference is that chance discovery is thus exploiting the future with the aid of humans.

Some rare events are simply noise, while others indicate great impact. It is completely impossible to fully automate the judgement of rare events. To understand the rare event, it is necessary to have a large amount of background knowledge. To implement a computer with a large amount of background knowledge is virtually impossible, as much artificial intelligence research has shown. Therefore, human and computer interaction is essential, which is discussed below.

In prediction and forecasting methods, it is assumed that a user (of the method) knows already which variables are to be predicted, and which variables are to be used (including the case where a part of a large number of variables are used). For example, an investor wants to know the trend of a certain stock price; a marketer wants to know how the sales will be; a traveler wants to know tomorrow's weather. However, how about those who are not aware of the risk of great earthquakes living in a quake-prone area? How about those who are not aware of the potential chance of developing a new hit product? In the real world (and often in very important situations), we are not aware of which variables to predict and which variables should be used.

Therefore, it is important to suggest new variables to humans. Textual information is a good source of information to provide humans with new aspects of targeting data because natural language has an extremely large number of dimensions. It is very often the case that humans can discover a new variable to predict through the stimuli of language. In addition, visualization and communication are both very important aspects for aiding humans' creativity, which is described in detail in Chap. 6.

3.3.4 Relevance of Prediction/Forecasting and Chance Discovery

Although prediction and forecasting and chance discovery have different aspects based on different presuppositions, they are not exclusive. Rather, they are complementary. To predict the future, it is very important to understand the events; sometimes we must invent the model and variables. Chance discovery focuses on the process of understanding data and model-creation. Model selection, parameter fitting, and hypothesis verification follow this understanding and model-creation stage.

Actually, commonly used methods are the combination of *KeyGraph* and a statistical hypothesis test: *KeyGraph* is first used to understand the data and to create a hypothesis. Then, statistical prediction methods are used to evaluate the hypothesis.

3.4 Importance of Structural Information for Rare Events

Though it is very difficult to predict the future with structural change, some recent research shows promising results. One method for addressing structural changes is to concentrate on the network structure of data, and discover which node might cause a great structural change. *KeyGraph* is an algorithm to visualize the data and provide an insight to the future, especially on the rare events if they co-occur with multiple frequent clusters. The details of *KeyGraph* are given in Chap. 18.

The same idea can be grasped in other structural analysis: small worlds. Strength of weak ties is known in social psychological science. Centrality in a network is another example of measuring what is important and what is not.

3.4.1 Small Worlds

Graphs that occur in many biological, social, and man-made systems are often neither completely regular nor completely random, but have instead a 'small world' topology in which nodes are highly clustered yet the path length between them is small [3.27]. For instance, if one is introduced to someone at a party in a small world, one can usually find a short chain of mutual acquaintances that connect. In the 1960s, Stanley Milgram's pioneering work on the small-world problem showed that two randomly chosen individuals in the USA are linked by a chain of six or fewer first-name acquaintances (in the scope of their experiments), known as 'six degrees of separation'[3.19]. Watts have shown that a social graph (a collaboration graph of actors in feature films), a biological graph (a neural network of the nematode worm C. Elegans), and a man-made graph (the electrical power grid of the western USA) all have a small-world topology [3.27]. The World Wide Web also forms a small-world network [3.3].

To formalize the notion of a small world, Watts define the clustering coefficient and the characteristic path length [3.27]:

– The characteristic path length, L, is the path length averaged over all pairs of nodes. The path length $d(i,j)$ is the number of edges in the shortest path between nodes i and j.
– The clustering coefficient, C, is a measure of the cliqueness of the local neighborhoods. For a node with k neighbors, then at most $_kC_2 = k(k-1)/2$ edges can exist between them. The clustering of a node is the fraction of these allowable edges that occurs. The clustering coefficient, C, is the average clustering over all nodes in the graph.

Watts define a small-world graph as one in which $L \geq L_{\text{rand}}$ (or $L \sim L_{\text{rand}}$) and $C \gg C_{\text{rand}}$, where L_{rand} and C_{rand} are the characteristic path length and clustering coefficient of a random graph with the same number of nodes and edges.

They propose several models of graphs, one of which is called β-graphs. Starting from a regular graph, they introduce disorder into the graph by randomly rewiring each edge with probability p as shown in Fig.3.1. If $p = 0$, then the graph is completely regular and ordered. If $p = 1$ then the graph is completely random and disordered. Intermediate values of p give graphs that are neither completely regular nor completely disordered. They are small worlds.

For example, Fig.3.2 is a graph constructed from a document as follows[1]: first the document is preprocessed by stemming and removing *stop words* as in [3.20], and extracting an n-gram. Then, each sentence of the document is considered to be in a basket, each of which consists of words (or phrases). After the preprocess, nodes are settled by selecting a word which appears over a user-given threshold number of times (e.g. three times). For every pair of nodes, the co-occurrence for every sentence is counted; an edge is added if the Jaccard coefficient exceeds a threshold, J_{thre}. The Jaccard coefficient is simply the number of sentences that contain both terms divided by the number of sentences that contain either term. This idea is also used in constructing a referral network from WWW pages [3.13]. Figure 3.2 shows a graphical visualization of the world of a document. Nodes are clustered, yet the whole graph is connected loosely. The co-occurrence graph of a technical paper comprises a small world.

Recently, many studies have revealed small-world characteristics. Mathias and Gopal investigated small-world networks from the point of view of their origin

---

[1] Note that *KeyGraph* was also invented as a document-processing algorithm.

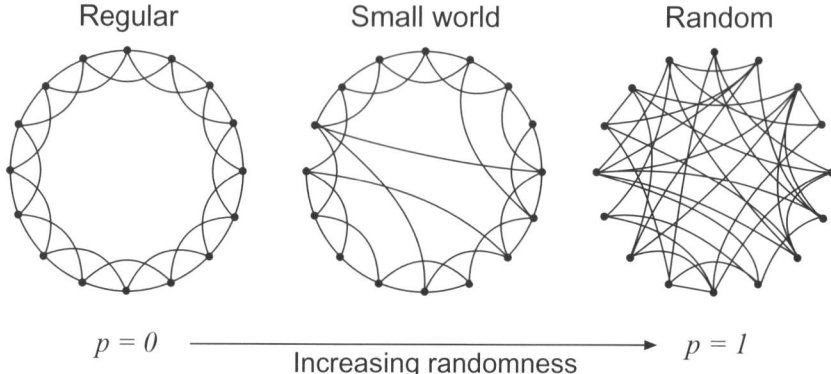

Fig. 3.1. Random rewiring of a regular ring lattice

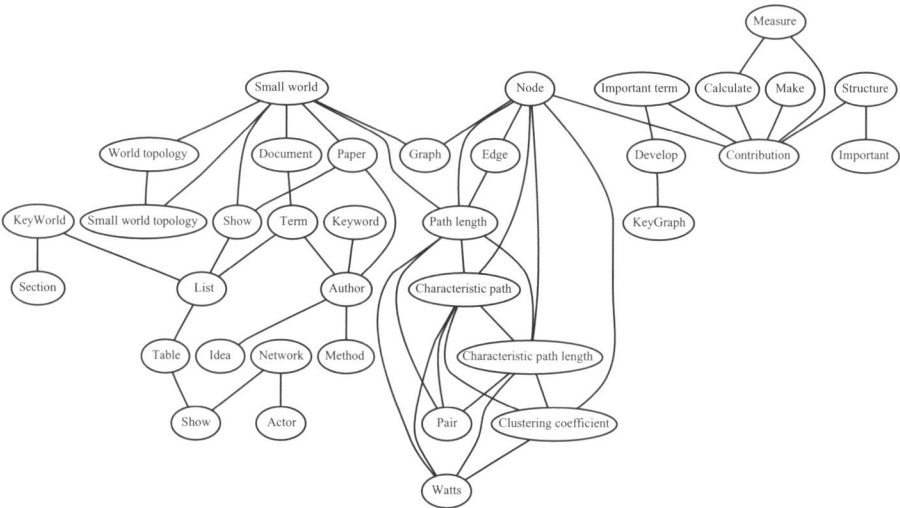

Fig. 3.2. Small world of a document

[3.16]. They showed that small-world topology arises as a consequence of a tradeoff between maximal connectivity and minimal wiring.

3.4.2 Structural Importance

In [3.18], node contribution is considered in the context of a small-world: if a node is to be deleted, at what point will the small-world topology break? The contribution of node v, CB_v, is measured by

$$\mathrm{CB}_v = L_{G_v} - L_v, \tag{3.3}$$

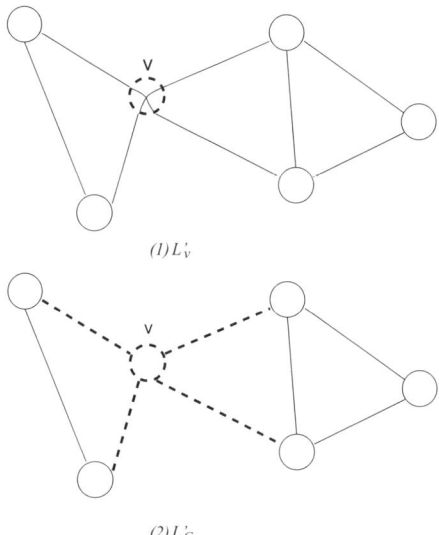

(1) $L_v^{'}$

(2) $L_{G_v}^{'}$

Fig. 3.3. L_v and L_{G_v}

where L_v is the characteristic path length averaged over all pairs of nodes except node v and L_{G_v} is the characteristic path length of the graph without node v 3.3.

The larger CB_v is, the greater its contribution to a small world. We can detect which nodes are structurally important from the viewpoint of a small world, that is, those that contribute to the efficiency of network flow and efficiency of network cost.

This method is based on the same idea as *KeyGraph*: if a node (an event) shares an important position in a graph, it might have an impact even if the frequency of the event is low. Importance is defined, in the *KeyGraph* case, by co-occurrence of two or more big clusters, and in the small-world case by the contribution for the graph to be highly connected. The method in [3.17] employs another importance criterion: if the flow on the graph is through a certain node, the node is important.

There can be many other ways to define importance on the network. However, this direction seems promising because the structure (or context) is considered to evaluate the importance of events. Certainly, such methods will not detect important rare events by themselves, but by being used in combination with human understanding, they have great potential for data analysis and prediction (or even invention) of the future.

3.5 Conclusion

This chapter gives an overview of prediction methods including the ARMA model, pattern recognition, and data mining; differences and relevance between prediction and forecasting and chance discovery are discussed. Although the presuppositions of prediction and forecasting and chance discovery differ, both will be useful in different stages of data analysis and decision making.

References

3.1 Agrawal R, Srikant R (1994) Fast algorithms for mining association rules. In Bocca,J.B, Jarke,M, and Zaniolo, C, editors, *Proc. 20th Int. Conf. Very Large Data Bases, VLDB*, Morgan Kaufmann, San Francisco, CA, pp. 487–499

3.2 Agrawal R, Srikant R (1995) Mining sequential patterns. In Philip S. Yu and Arbee S. P. Chen, editors, *Proceedings of the Eleventh International Conference on Data Engineering*, IEEE Computer Society Press, Los Alamoitos, CA pp. 3–14,

3.3 Albert A, Jeong H, Barabási A (1999) Diameter of the World-Wide Web. *Nature* 401(6749)

3.4 Bialek W, Nemenman I, and Tishby N (2001) Predictability, complexity, and learning. *Neural Computation*, 13:2409–2463

3.5 Brockwell PJ, Davis R (1996) *Introduction to Time-Series and Forecasting*. Springer Verlag, Heldelberg, Germany

3.6 Domeniconi C, Perng C, Vilalta R, and Ma S (2002) A classification approach for prediction of target events in temporal sequences. In *Proc. 6th European Conference on Principles and Practice of Knowledge Discovery in Databases (PKDD'02)*, pp.125 – 137

3.7 Chatfield C (1996) *The Analysis of Time Series*. Chapman and Hall, London, UK, 5th edition

3.8 Embrechts P, Klüppelberg C, and Mikosch T (1991) *Modelling Extremal Events for Insurance and Finance*. Springer Verlag, Heidelberg, Germany

3.9 Garofalakis MN, Rastogi R, and Shim K (1999) SPIRIT: Sequential pattern mining with regular expression constraints, In *Proc. 25th International Conference on Very Large Data Bases (VLDB'99)*, pp.223–234

3.10 Geurts P (2001) Pattern extraction for time series classification. In *Proc. PKDD 2001*, pp.115–127

3.11 Gladwell M (2000) *The Tipping Point: How Little Things Can Make a Big Difference*. Little Brown & Co, Boston, MA

3.12 Han J, Gong W, Yin Y (1998) Mining segment-wise periodic patterns in time-related databases. In *Fourth International Conference on Knowledge Discovery and Data Mining*, AAAI Press, Menlo Park, CA, pp.214–218

3.13 Kautz H, Selman B, Shah M (1997) The hidden Web. *AI magazine*, 18(2):27–35

3.14 Last M, Klein Y, Kandel A (2001) Knowledge discovery in time series databases. *IEEE Transactions on Systems, Man, and Cybernetics*, 31(1): 160 – 169

3.15 Mannila H, Toivonen H, Verkamo AI (1995) Discovering frequent episodes in sequences. In *Proc. 1st International Conference on Knowledge Discovery and Data Mining (KDD'95)* pp.210 – 215

3.16 Mathias N, Gopal V (2001) Small worlds: How and why. *Physical Review E*, 63(2):021117 – 021128

3.17 Matsumura N, Ohsawa Y, Ishizuka M (2002) Pai: Automatic indexing for extracting asserted keywords from a document. In *Proc. AAAI Fall Symposium on Chance Discovery* pp.28 – 32

3.18 Matsuo Y, Ohsawa Y, Ishizuka M (2001) A document as a small world. In *Proceedings the 5th World Multi-Conference on Systemics, Cybenertics and Informatics (SCI2001)*, 8: 410–414

3.19 Milgram S (1967) The small-world problem, *Psychology Today*, 2:60–67

3.20 Salton G (1989) *Automatic Text Processing*. Addison-Wesley, Boston, MA

3.21 Shannon CE (1948) A mathematical theory of communication. *Bell System Technical Journal*, 27:379–423 and 623–656

3.22 Singh S (2000) Pattern modelling in time-series forecasting. *Cybernetics and Systems - An International Journal*, 31(1): 49 – 66

3.23 Sornette D (2002) Predictability of catastrophic events: material rupture, earthquakes, turbulence, financial crashes and human birth. In *Proc. National Academy of Sciences USA* pp.60 – 67

3.24 Srikant R, Agrawal R (1996) Mining sequential patterns: Generalizations and performance improvements. In Peter M. G. Apers, Mokrane Bouzeghoub, and Georges Gardarin, editors, *Proc. 5th Int. Conf. Extending Database Technology, EDBT*, volume 1057, pp.3–17, Springer Verlag, Heidelberg, Germany

3.25 Tan N, Kumar V (2001) Mining indirect associations in web data. *Proc. of WebKDD 2001: Mining Log Data Across All Customer TouchPoints*, pp.145 – 166

3.26 Tan PN, Kumar V, Srivastava J (2000) Indirect association: Mining higher order dependencies in data. In *Proc. the 6th European Conference on Principles and Practice of Knowledge Discovery in Databases*, pp.632–637

3.27 Watts D (1999) *Small worlds: the dynamics of networks between order and randomness*. Princeton University Press, Princeton, NJ

3.28 Weigend AS, Gershenfeld NA (1993) *Time Series Prediction*. Addison-Wesley, Boston, MA

3.29 Weiss GM, Hirsh H (1998) Learning to predict rare events in event sequences. In R. Agrawal, P. Stolorz, and G. Piatetsky-Shapiro, editors, *Fourth International Conference on Knowledge Discovery and Data Mining (KDD'98)*, New York, NY, 1998. AAAI Press, Menlo Park, CA, pp.359–363

3.30 Weiss GM (1999) Timeweaver: a genetic algorithm for identifying predictive patterns in sequences of events. In *Proc. the Genetic and Evolutionary Computation Conference (GECCO-99)*, pp.718–725

3.31 Yule GU (1927) On a method of investigating periodicities in disturbed series with special reference to Wolfer's sunspot numbers. Philosophical Transactions of the Royal Society of London, Series A, 226:267–298

4. Self-organizing Complex Systems

Henrik Jeldtoft Jensen

Department of Mathematics, Imperial College, 180 Queen's Gate,
London SW2BZ, UK
email: h.jensen@ic.ac.uk

Summary.

The present chapter deals with chance discovery from the perspective of complex systems. We will limit ourselves to a type of complex systems known as self-organized critical systems; to discuss these systems we need to make clear what we mean by notions such as complex, scale free, critical, and self-organization. To fix some ideas we will introduce two self-organizing complex models, i.e. a simple cellular automaton model and a model of evolutionary ecology. This highlights how seemingly innocent local perturbations may propagate through the entire system, totally altering the composition of the system. We consider the possibility of prediction in complex systems, and elaborate on the specifics of chance discovery in interconnected and highly sensitive complex systems.

4.1 Introduction

The present chapter deals with chance discovery from the perspective of complex systems . Our main concern will be to discuss what constitutes a chance in a complex system. There is at present no comprehensive well-established theory of complex systems in the same way as we have, for instance, for equilibrium thermodynamic systems. Despite decades of *complexity* research we still lack a definition of what precisely to consider as a complex system. In any case the aspects of complex systems we will emphasise are lack of scale, global sensitivity, and emergence of properties at the collective level which are generated by the interaction and collective dynamics of the components of the system.

In systems where the global state depends in a crucial way on the action of the individual components, it is far from clear what can be considered a 'chance'. A possible action which might appear at a certain moment in time as a chance for a single member of the system may turn out to alter the entire configuration of that system, leaving the individual members in a very different situation than anticipated.

We will limit ourselves to a type of complex systems which are known as self-organized critical systems; to discuss these systems we need to make clear what we mean by notions such as complex, scale free, critical, and self-organization. Section 4.2 will attempt to specify the meaning of these concepts in somewhat general terms. To fix some ideas we will introduce two self-organizing complex models in Sect. 4.3. One model is the celebrated sandpile model introduced by Bak, Tang, and Wiesenfeld [4.3] in 1987. This is a simple cellular automaton model which has become paradigmatic for the field of self-organizing critical systems. The dynamics of these systems typically involve bursts of activity or avalanches. The other model is more recent and somewhat more complex than the sandpile model. The model is thought of as a model of evolutionary ecology and is called the tangled nature model [4.7, 4.9]. It is useful for a discussion of how intermittent dynamics can occur in complex systems even if the micro-dynamics always ticks along at the same constant rate. We shall also use the tangled nature model to illustrate dynamical consequences of the strong interconnectedness frequently associated with complex systems. The model highlights how seemingly innocent local perturbations may propagate through the entire system, totally altering the composition of the system. The technicalities of the presentation of the sandpile and the tangled nature models can probably be jumped over by readers interested in the qualitative aspects only. We then move on to a discussion of what a chance may look like in a scale-free complex system in Sect. 4.4. Although chance discovery is supposed to be different from predictability, we will in Sect. 4.5 briefly consider the possibility of prediction in complex systems, since this may also be related to possible different behaviors between events for which the finite size of a system is relevant and events which do not probe the boundaries of the system. In Sect. 4.6 we elaborate on the specifics of chance discovery in interconnected and highly sensitive complex systems. Finally we present some concluding remarks in Sect. 4.7.

4.2 What Is a Self-Organizing Complex System?

As an example of one type of complex systems we will focus our discussion on what has become known as self-organized critical systems and we shall introduce these systems through a number of steps. First we will mention what is meant by a critical system in thermodynamics. Thereafter we will discuss the notion of scale invariance or scale-free systems; this will allow us to mention the idea of self-organized critical systems and we will discuss in what sense these systems are complex.

First a few words concerning complex systems. What is a complex system? Though no simple and precise definition exists of what a complex system is, I believe most would agree that complex systems have the following features in common. They consist of many interacting components, it is possible to identify various levels of hierarchies, and many different scales are simultaneously relevant.

We will introduce complex systems and self-organizing complex systems by means of a description of critical thermodynamic systems.

4.2.1 What Is a Critical System?

We would not call a gas of non-interacting particles a complex system although the gas may contain 10^{23} particles. This system may be complicated but it is not complex. It is impossible to follow or predict the precise microscopic motion of the particles; nevertheless the overall macroscopic behavior, as well as the statistics, of the system are obtained from the physics of a single particle. No collective effects exist. The situation is different when interaction between particles is included. Interaction can give rise to new emergent properties, properties which plainly do not exist at the level of individual particles. One can think of elasticity of a material. When we bend a solid material, we need to overcome elastic forces which originate in the interaction between the particles. A simple solid already exhibits a hierarchy of order: namely the individual particles, which possess their own microscopic properties or order. The next level of order is the collective arrangement of the particles into the solid crystal. Despite the existence of these two levels of order a simple solid is not really to be considered as a complex system. The reason for this is that if we perturb the system at a given point, the perturbation will remain local. The effect of the perturbation will decay exponentially as a function of distance x according to the functional form $\exp(-x/\xi)$, where the rate of decay with distance is determined by the so-called correlation length ξ. Thus what happens at a given position in the crystal is by and large only influenced by what happens in the vicinity, i.e. basically within the distance ξ, from that position. There are situations where the correlation length diverges. This happens for systems in thermodynamic equilibrium when the temperature is tuned to the critical temperature.

A thermodynamic equilibrium system at the critical temperature, where ξ is infinite, is often considered an example of a complex system. The reason is that the entire system is intimately connected and fluctuations are very strong. A famous example of these strong fluctuations is the critical opalescence of liquids [4.5]; here the density fluctuates so strongly that an otherwise transparent liquid, like water, becomes opaque because the density fluctuations of the molecules of the liquid are macroscopic and able to scatter the light illuminating the liquid.

We summarize: a thermodynamic equilibrium system is said to be critical when the temperature is tuned to a value such that the correlation length becomes infinite and all parts of the system become interconnected in an essential way. This is related to all length scales between the microscopic, say the distance between the molecules, up to the macroscopic length scale, the size of the system becoming relevant for the behavior of the system [4.5].

4.2.2 Scale-Free Systems

A thermodynamic system at the critical temperature is said to be scale free or scale invariant [4.5]. The point is that since the correlation length is infinite no length scale stands out as being of particular relevance; the system does not possess one particular scale more relevant than other scales. This is because all length scales are to be considered equally relevant. As an effect of this lack of scale, configurations in critical systems do not change their appearance when magnified, or in other words scaled up. A famous example of such a scale-invariant object is the cauliflower. Viewed close up one little bouquet of the cauliflower looks like the head of the entire plant. For our discussion the best way to illustrate the idea of scale invariance is perhaps to leave equilibrium systems and instead describe the metaphor of the sandpile introduced by Bak, Tang, and Wiesenfeld in 1987 [4.3]. Imagine one is building a pile of sand by adding sand grain after grain gradually producing a heap. Once in a while the new grain will fall on a site where the local slope is too steep to support the extra amount of mass. The added grain will tumble down the slope and may, as it moves down, trigger the motion of more grains. These new grains may themselves tumble further down and release even more grains in a chain reaction. In this way an avalanche can be released. Sometimes the avalanche will stop after only a few grains have tumbled, other times the avalanche might grow into a major landslide. The size of the avalanche will depend in a very sensitive way on the local slope at the positions reached by sand released by the new grain of sand we added. Thus we do not expect any typical size of the resulting avalanche. Sometimes the avalanche will be small, sometimes very big. In the original computer model [4.3] the probability density $p(s)$ of the sizes of released avalanches follows a power-law distribution $p(s) \propto s^{-\tau}$. A power-law distribution is free of any characteristic scale in the sense that the relative probability $p(s_2)/p(s_1) = \alpha^{-\tau}$ between events of sizes s_1 and $s_2 = \alpha s_1$, which differs by a factor α, does not depend on the size s_1. So the ratio between the number of avalanches of size 10 and of sizes 100 is the same as the ratio between the number of avalanches of sizes 100 and 1000. This is of course not at all the case if one deals with a quantity distributed according to, say, a Gaussian probability density. Note also that a quantity distributed according to a power law may have no average if the power-law exponent τ is smaller than 2, since the integral $\int_1^\infty sp(s)\mathrm{d}s = \int_1^\infty s^{1-\tau}\mathrm{d}s$ will diverge (see Fig. 4.1).

The above is an idealized picture which applies in detail to the computer model introduced by Bak, Tang, and Wiesenfeld [4.3] and in some cases even to real granular systems (the relation to experiment is discussed in [4.1] and in more detail in [4.11]).

The actual behavior of real physical sandpiles is not our main concern here. The important point is the idea that the result of one and the same type of perturbation of a system may lead to responses of a very broad range of sizes. Many different types of systems have been suggested to possess this property; among them are biological evolution and commercial markets as well as a number of physical systems such as earthquakes [4.1, 4.11].

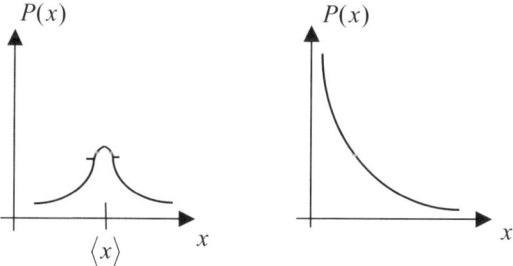

Fig. 4.1. A narrow distribution with a well-defined typical average event size given by $\langle x \rangle$. The right-hand graph indicates the shape of a scale-free power-law distribution. When $p(x)$ decays too slowly for large x, the average value $\langle x \rangle$ may be infinite, i.e. not defined

4.2.3 Self-organizing Critical Systems

Above we have discussed the concept of a critical system. This concept was first developed in the context of thermodynamic phase transitions. We have mentioned that criticality is related to the lack of a spatial characteristic scale or a scale characterizing the distribution of 'events' (e.g. avalanche size or it could be sizes of clusters of interacting components). We will now combine these ideas with the notion of self-organization. Thermodynamic systems only exhibit infinite correlation length and lack of characteristic scale when the temperature is tuned by some external mechanism (by e.g. an experimenter) to its critical value. Can scale-free critical behavior be achieved spontaneously in systems without external fine tuning? The notion of self-organized criticality claims that this in fact happens in a natural way in many dynamical systems consisting of large numbers of interacting components. The thinking behind this idea is most easily described by explaining the metaphor of the sandpile introduced by Bak, Tang, and Wiesenfeld [4.3]. Imagine adding slowly grain by grain more and more sand to the pile. The slope of the pile will tend, all by itself, to organize about some critical value. This is imagined to happen in the following way. When the slope is shallow, it is likely to be stable and additional sand grains will not induce any avalanches. Hence, the continued addition of new grains to a shallow slope will gradually increase the slope. If the slope of the pile becomes too steep, big avalanches will occur, the effect of which will be to decrease the slope of the pile, see the sketch in Fig. 4.2. One may imagine that an intermediate slope separates these two regimes. This could be a critical slope where avalanches of a broad range of sizes are released as new grains are added. The pile should self-organize to this critical slope because no avalanches will occur when the slope is smaller than the critical value and hence the slope will grow in this regime, but as soon as the slope becomes greater than the critical value, avalanches will begin to be released and the slope of the pile will decrease as a result. Thus we can imagine that the gradual addition of more and more grains of sand will drive the pile into a fluctuating state where the *intrinsic* dynamics of the avalanches will keep pushing the slope back towards its critical value. It seems that if we simply drive the sandpile by adding grains slowly, so slow that the induced avalanches can manage to come to a halt before we add another grain, then the avalanche dynamics will automatically bring the system into a critical state. Moreover this critical state will be an attractor for the dynamics: even when the drive takes the system away from the critical state, the intrinsic fluctuations will bring the system back towards criticality.

Do real sandpiles behave in this way? The answer is: yes, sometimes as discussed in [4.11]. We do not need to go into this discussion here. The essential point we want to make is that dynamical systems do exist for which the dynamics (when slowly driven) keeps bringing the system back towards a critical state.

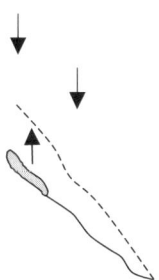

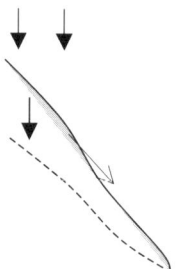

Subcritical: slope growing   Supercritical: slope decreasing

Fig. 4.2. Fluctuations of the slope around the critical slope: to the left; sand is added to a shallow slope and the grains build up locally without inducing flow of sand down along the slope; as a result the local slope increases. To the right; grains hit a steep section of the pile and an avalanche starts to slide, bringing the slope of the pile down

4.3 Details of Two Models

We now turn to a somewhat more detailed discussion of two specific models of complex systems that by their own intrinsic dynamics end up in a critical state. First the paradigmatic sandpile model which has produced a wealth of activity and inspired very many different concrete models, all of which intend to capture certain aspects of complexity. The feature of the sandpile model most important to our present discussion is the ability of equal-size perturbations to lead to very different effects. After introducing the Bak–Tang–Wiesenfeld sandpile we discuss the tangled nature model of ecological evolution. The main point of the description of that model is to highlight that complex systems generically exhibit intermittent dynamics at the macroscopic level, even when the dynamics at the microscopic level flows at a uniform rate. The tangled nature model also highlights the point that complex systems tend to be at the brink of a major reorganization, a point also elaborated by Kauffman; see e.g. his book [4.12]. Moreover the discussion of the tangled nature model will be used to mention that complex systems are likely not to be in a statistical stationary state. This is a point often neglected, but which is of great importance since it means that the emergent statistical properties of such systems change as an effect of the dynamical evolution. It should be emphasized that this can happen even when at a local microscopic level the operation of the system remain unchanged.

4.3.1 The Bak–Tang–Wiesenfeld Sandpile Model

This model was introduced in 1987 in [4.3] as an illustration of how interacting many-component systems may drive themselves into a critical state. The model is very simple to formulate and we go through some of the details as a way to highlight a selection of the ingredients that may be at play when a system more or less by itself develops a critical complex state. Key words are meta-stability, slow driving, and local threshold or rigidity.

We think of adding sand grains, represented by the small rectangular boxes in Fig. 4.3, one by one to a flat surface. The rule is that whenever the heights of two neighboring columns h_i and h_{i+1} differ by more than a certain value $h_i - h_{i+1} > h_c$ the top grain at the position with the tallest column is to be moved to the neighboring shorter column[1]. In Fig. 4.3 we have chosen $h_c = 2$ and a grain from column i must be moved onto column $i + 1$. This is meant to mimic a local slope which is so steep that the grain starts to slide down. After this is done the slope at position $i + 1$ becomes too large and a grain must be moved onto column $i + 2$. In this way the avalanche continues until rearranging the grains does not induce new over-critical slopes. One imagines that the pile is built up against a wall on a table and whenever sand grains reach the right edge, they simply leave the system by falling over the edge.

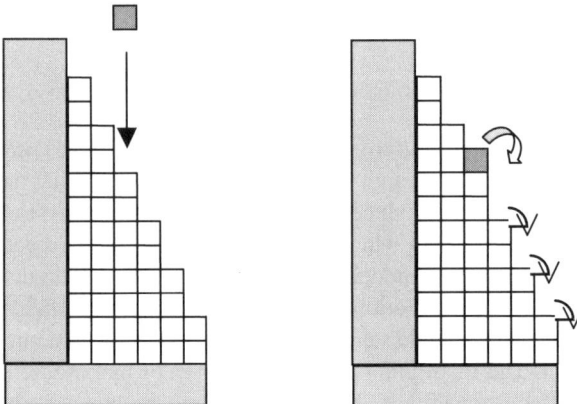

Fig. 4.3. Adding an extra grain of sand; the local slope becomes over-critical and a grain of sand starts tumbling down the slope of the pile. Eventually the grain leaves the system over the table edge at the far right

When the model is defined in two or higher dimensions the sizes of the avalanches, induced by repeated random addition of new grains of sand, follow a power-law probability distribution. Whenever an avalanche has come to a stop the system is in a static state. Activity, in the form of rearrangement of sand grains, only starts to happen again when the local slope somewhere on the pile becomes too steep. However, this static state is only meta-stable. Adding more sand will eventually produce a too-steep slope somewhere and another avalanche will be released. The avalanches take the system from one meta-stable configuration to another. If no height difference between two adjacent columns were allowed $h_c = 0$, that is, if there were no local rigidity or threshold, the sandpile would always relax to the same completely

---

[1] Note that the sandpile model is often formulated in terms of the criterion directly on the height itself and not on the height difference. This is a technical detail mentioned here only to make reading of the original papers less confusing. See e.g. [4.11] for more details.

flat configuration. This would correspond to 'stacking' small spheres, say ball bearings, for which rearrangement only stops when a single layer of spheres is found along the surface of the table and all the other spheres have fallen over the edge. The local threshold is responsible for the existence of many different meta-stable configurations. The system will be trapped temporarily (for periods of varying duration) in these meta-stable configurations while the 'stress' gradually builds up – in the sandpile model the stress is represented by the additional sand grains brought onto the pile from the exterior.

The size of a released avalanche is very difficult to predict even though the avalanche evolves according to completely deterministic rules. The only stochastic element in the model is the random choice of the position at which the grains are added to the pile. Prediction is so difficult because the evolution of the avalanche depends in detail on the local slopes of the region covered by the avalanche. Hence very detailed information about the state of the entire system would be needed in order to know beforehand the precise effect of adding an extra grain of sand. The difficulties in predicting the dynamics of the system are also seen in the lack of correlation between consecutive avalanches. When the system is driven by adding new grains at random positions across the bulk of the pile the total flow of sand is only weakly correlated in time [4.11]. Driving the model by only adding sand at random positions up against the wall induces $1/f$ correlations on a time scale *longer* than the duration of the largest avalanches [4.11], so these correlations do reach across the individual avalanches. Time correlations at time scales long compared to the duration of the individual avalanches are also seen in other models of self-organized criticality [4.10]. However, this is the collective global dynamics of the system over very long time scales, and it is far from clear how knowledge of this aspect of the dynamics might help the discovery of a 'chance' here and now.

Further, we may note that in some circumstances it might be considered a great chance to be able to add one more grain quietly without releasing an avalanche. Whether this is possible or not requires only local knowledge, namely the actual value of the local height difference. The situation is completely different if chance corresponds to the ability to trigger a big avalanche. Then knowledge concerning non-local spatial and temporal aspects is required.

4.3.2 The Tangled Nature Model

Our second example is concerned with a model intended to capture some of the salient features of biological evolution, in particular aspects of co-evolution in ecosystems. The model is somewhat more complicated than the sandpile model, though in the light of the overwhelmingly intricate nature of biological reproduction the model is certainly very schematic. To emphasise the model's focus on co-evolutionary ecological aspects it is called the *tangled nature model* [4.7, 4.9] with a reference to Charles Darwin's notion of the tangled bank [4.8]. The reason to mention the tangled nature model in relation to a discussion of chance discovery is that the model is an example of a highly connected self-organizing system that exhibits a number of emergent properties of general interest to complex systems.

The model consists of individuals characterized solely by their particular genotype . The ability of an individual to reproduce depends on its genotype and on the composition of the population . Mutations can occur during reproduction leading to a difference between parent and offspring. From these simple ingredients develops some very interesting behavior which may be interpreted as species formation, extinctions and creations, intermittency, and a gradual collective adaptation. Here follows a description of the model in more detail.

The model contains $N(t)$ individuals enumerated $\alpha = 1, 2, \ldots, N(t)$, where the total size of the population $N(t)$ is self-adjusting and varies with time according to the microscopic dynamics of the model to be described now. An individual i is characterized by a genome $\mathbf{S}^\alpha = (S_1^\alpha, S_2^\alpha, \ldots, S_L^\alpha)$ where the individual 'genes' can assume the values $S_i^\alpha = \pm 1$. Thus the genotype of an individual is determined by which of the corners of the L-dimensional hypercube the individual belongs to. A time step of the model consists of two actions: annihilation and reproduction. First one picks at random one of the $N(t)$ individuals. That individual is then removed from the system with probability p_{kill}; this killing probability is for simplicity kept constant. It is the same for all individuals at all times. The reproductive event is a little bit more involved. Here we will discuss asexual reproduction only. First one picks again at random an individual. This individual, say number α, is then replaced by two new copies with probability $p_{\text{off}}(\mathbf{S}^\alpha)$, and the original individual is removed from the model. This corresponds to cell division. The offspring probability $p_{\text{off}}(\mathbf{S}^\alpha)$ is calculated in the following way. For the specific genotype $\mathbf{S}^\alpha$ we calculate a weight function $W(\mathbf{S}^\alpha)$ by summing up couplings to other occupied positions $\mathbf{S}$ in genotype space; each coupling is multiplied by the number of individuals, or the occupancy, $n(\mathbf{S}, t)$, of that genotype. In anecdotal terms the idea is that if it is bad for a rabbit that one fox is around, then it is even worse for a rabbit if two foxes are in the neighborhood.

In Fig. 4.4 we sketch the computation of $W(\mathbf{S}^\alpha, t)$, which is given by the expression

$$W(\mathbf{S}^\alpha, t) = \frac{1}{N(t)} \sum_{\mathbf{S}} J(\mathbf{S}^\alpha, \mathbf{S}) n(\mathbf{S}, t) - \mu N(t).$$

Here the couplings $J(\mathbf{S}^\alpha, \mathbf{S})$ assume values in the interval $[-1/c, 1/c]$ and are given as a random deterministic function of $\mathbf{S}^\alpha$ and $\mathbf{S}$. The factor c is a constant introduced to determine the width of the distribution of coupling strengths. The second term in $W(\mathbf{S}^\alpha, t)$ determines the average size of the total population; the factor μ determines the 'ecological' carrying capacity of the system: larger μ corresponds to fewer resources leading to a smaller population. The probability that the individual $\mathbf{S}^\alpha$ produces offspring is obtained by turning $W(\mathbf{S}^\alpha, t)$ into a probability in the following simple way (here the specific functional form is not essential, all what is needed is a function from the entire real axis onto the interval $[0, 1]$):

$$p_{\text{off}}(\mathbf{S}^\alpha, t) = \frac{\exp(W(\mathbf{S}^\alpha, t))}{1 + \exp(W(\mathbf{S}^\alpha, t))}.$$

$n(\mathbf{S}, t)$ = occupancy at the location $\mathbf{S}$

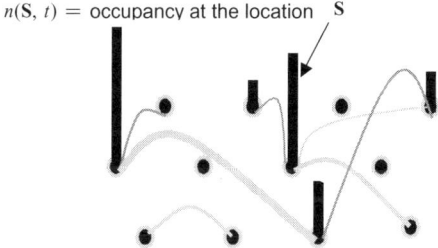

Fig. 4.4. A sketch of how positions in genotype space are coupled together and how the occupancy, indicated by the black columns, differs from position to position(in reality genotype space is a cube in L dimensions and therefore difficult to draw)

Mutations can occur with probability p_{mut} per gene and consist simply in a change of the sign of a gene inherited from the parent: $S_i^\alpha \mapsto -S_i^\alpha$.

The easiest way to describe the way the dynamics of the model operates is to consider the initiation of the system. In Fig. 4.5 the occupancy in genotype space is plotted as function of the first few thousand time steps along the x-axis. All possible genotypes are labeled from 1 to $2^{20} \approx 10^6$ (we use $L = 20$ in this case) up along the y-axis. At each time step every occupied genotype position is indicated by a dot. Initially all individuals, 500 in this case (see Fig. 4.6), are given the same genotype, call it $\mathbf{S_0}$; hence for about the first 2500 time steps we observe a single straight line parallel to the x-axis in Fig. 4.5.

The total size of the population as well as the diversity, defined as the number of different genotypes occupied at a given moment, is plotted for the same time interval in Fig. 4.6. We notice that initially the population size contracts. This is because the first term, the coupling term, in $W(\mathbf{S_0}, t)$ is zero because $n(\mathbf{S}, t) = 0$ for all $\mathbf{S} \neq \mathbf{S_0}$. The resulting offspring probability,

$$p_{\mathrm{off}}(\mathbf{S_0}, t) = \frac{\exp(-\mu N(t))}{1 + \exp(-\mu N(t)),}$$

is too small to counterbalance the killing probability. This corresponds to the carrying capacity being unable to support the size of the population, as least as long as all the individuals are identical. So the population size steadily declines, see Fig. 4.6, which allows $p_{\mathrm{off}}(\mathbf{S_0}, t)$ to gradually increase. After about 2500 time steps reproduction becomes appreciable and therefore mutations also start to occur; this produces new genotypes and, consequently, as seen in Fig. 4.6, the diversity of the system increases. The more diverse population is now able to take advantage of the interactions between individuals of different genetic composition, as mimicked by the interaction term in $W(\mathbf{S}^\alpha, t)$; this allows a further increase in the offspring probabilities and the population can start to grow. It should be stressed that not all configurations with many different genotypes present are able to sustain a bigger population. The system has to adapt in a self-organized way. Among all the different possible ways of distributing a population in genotype space the system has to find

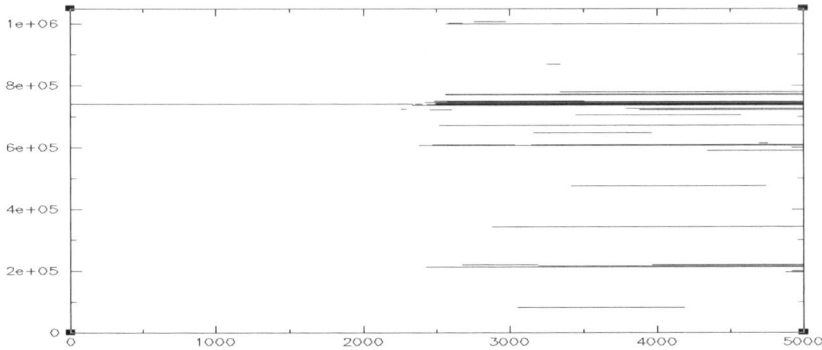

Fig. 4.5. The early-time behavior of the occupancy in genotype space. Time is along the x-axis and genotype label is along the y-axis. A black dot is placed at those labels which happen to be occupied at a certain instant in time

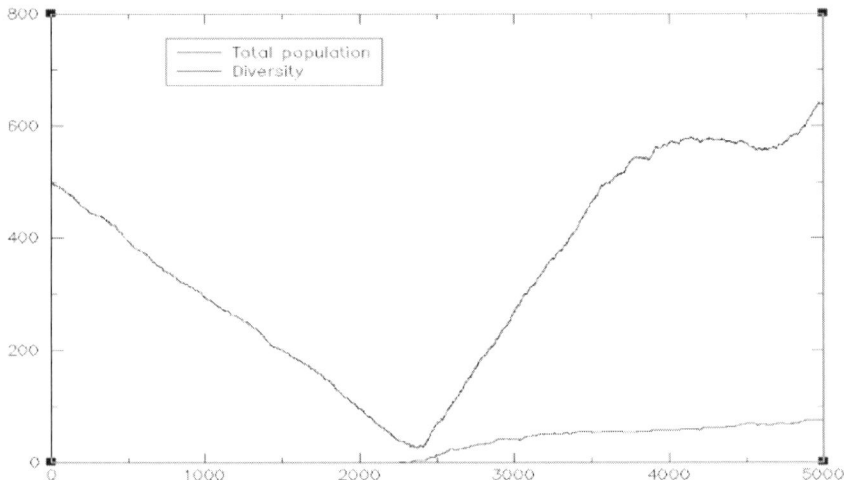

Fig. 4.6. The population size and diversity (number of different genotypes present) as function of time in the very early evolution of the tangled nature model

those subsets of configurations with genotype positions coupled together in a beneficial way. Thus, out of the completely random set of couplings specified at initiation of the model, in the function $J(\mathbf{S}^\alpha, \mathbf{S})$ sets of sites which are coupled together in a way that ensures some collective coherence and stability must be selected.

The model continues now to evolve for as long as one can afford to simulate. After the first few thousand time steps intermittent dynamics emerges where long periods of fairly stable configurations in genotype space all of a sudden are replaced by new configurations. This transition from one relatively stable transition to the next happens through relatively short periods of hectic extinction and creation ac-

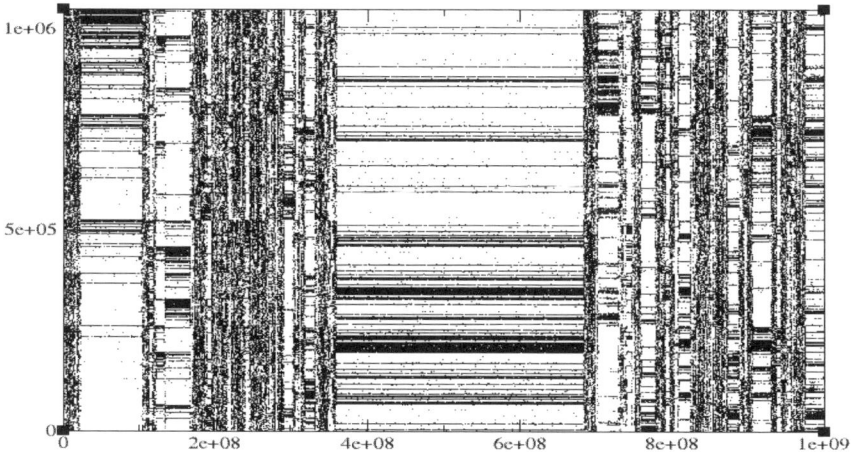

Fig. 4.7. The long-time behavior of the occupancy in genotype space, similar to Fig. 4.5. Time is along the x-axis. The different genotypes are enumerated along the y-axis

tivities. Figure. 4.7 shows the intermittent nature of the dynamics of the tangled nature model when considered at the level of 'species' or genotypes.

One notices the long stretches of co-existing straight lines in Fig. 4.7. These correspond to relatively stable genotype configurations. We call these periods quasi-evolutionary stable strategies (q-ESS) with a reference to Maynard Smith's game-theoretic concept evolutionary stable strategies [4.13]. During these periods the tangled nature model has found a way to arrange the population in genotype space which is able to withstand the perturbations produced by the never-ceasing fluctuations in the selection pressure produced by the intrinsic fluctuations in the mutations and population size. To a certain degree the q-ESS configurations are stable against 'invasions' of different strategies or genotypes. But only to a degree; eventually the intrinsic fluctuations trigger a collapse of the state and the system undergoes a short hectic period of extinction and creation while the system searches for a new q-ESS.

Is the tangled nature model a self-organizing critical system? It is in the following sense. The micro-dynamics consisting of annihilation, reproduction, and mutations involv various uncorrelated stochastic processes from which emerge a number of scale-free quantities at the macroscopic level. The durations, or lifetimes, of the q-ESS periods, the duration of the brief hectic transition periods separating the q-ESS, and the time for which single positions in genotype space are occupied all follow broad power-law-like distributions. It should be noted that the duration of all these different time periods are much longer than-time interval survived by any individual. Accordingly, the lifetime of the occupation of a single position in genotype space (the analog to the lifetime of a species), the q-ESS, and the transition periods all are genuine emergent collective properties of the model.

There are other indications of critical behavior in the tangled nature model. For instance the clusters of interacting occupied positions in genotype space appear to exhibit various scale-free statistical measures, such as the distribution of cluster sizes. Furthermore, similar to real macro-evolution the model is not in a stationary state. The average duration of the q-ESS periods slowly increases with time as the system achieves a higher and higher degree of collective adaptation. This adaptation is truly collective since it corresponds to selecting more efficient configurations of occupied genotype positions and their mutual couplings. Recall that the couplings in genotype space are ascribed at random initially and that the dynamics will need to select subsets of positions with attached couplings which produce the most beneficial $p_{\mathrm{off}}(\mathbf{S}, t)$ for the large number of genotype positions $\mathbf{S}$.

After having presented two concrete models of complex self-organizing systems we return to the discussion of chance discovery.

4.4 What Does a Chance Look Like When There Is No Scale?

Perhaps the most important reason for thinking about chance discovery from the perspective of critical complex systems is that the very notion of what constitutes a chance may be different in such systems. To phrase the problem in a clear way from the onset one may wonder whether a chance is a big event, a big system-spanning avalanche, or is a chance to be identified by the small responses that avoid stirring the entire system. Clearly the answer must depend on context.

The sandpile model highlights that the same action, namely adding one grain of sand, can produce very different effects. So if we believe that we somehow have developed an intuition from previous experience that a certain situation allows us a chance to achieve a desired result by a certain action, we should keep in mind that the state of the system could be somewhat different this time and that even a slight difference could lead to a very different effect if we were to repeat our action. An example of this aspect of complex systems is the often disastrous consequences of solving political problems by a militaristic approach which perhaps worked 'last time'.

The interconnectedness of critical complex systems also raises certain questions. Strictly speaking any action in or on the system will influence the entire system. So a chance must be considered with reference to the collective co-operative nature of the dynamics of the system. The consequences of any action may turn out to be very different from what a local short-time consideration might suggest.

This point can be illustrated by the use of the tangled nature model. The state of a system, which is given by the distribution of genotypes, is completely altered as one passes from one q-ESS to the next. Figure. 4.8 shows an example of such a transition. The branching tree in the top of the figure indicates how the descendants of a given genotype (e.g. the one indicated as 'root' in the figure) may end up far from the original ancestor as the mutations produce branching upon branching event, as indicated by the highly forked horizontal tree structure in Fig. 4.8. One way to trigger the termination of a q-ESS is to remove all individuals of a certain

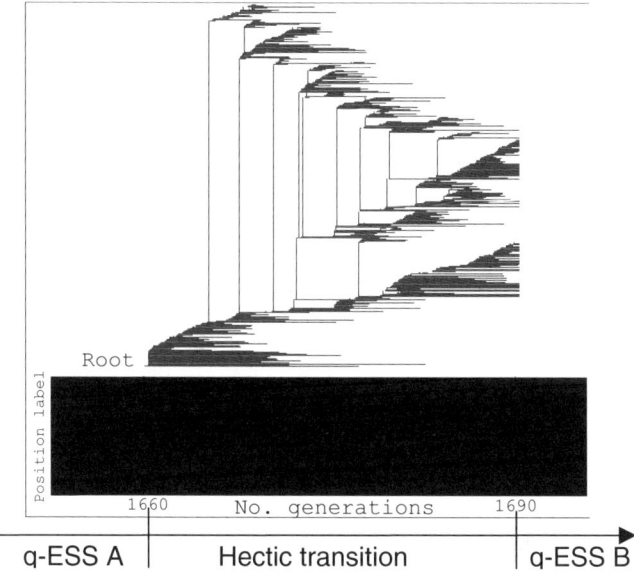

Fig. 4.8. A close-up view of a transition between two q-ESS periods in the tangled nature model. The top of the figure shows the forking out of one particular genotype present in q-ESS A. Mutation after mutation produce a wealth of new branches but the majority of these go extinct after only a few generations. A few, though, survive and become established in the successive q-ESS. The bottom of the figure shows the occupancy in genotype space using the same representation as in Figs. 4.5 and 4.7

genotype. In the model such a perturbation leads typically to the collapse of the present ecological configuration, whereafter the system undergoes a period of hectic fluctuations before another q-ESS is established. In ecology, species which are of special significance to the stability of an ecosystem are called keystone species [4.14]. In the highly interconnected tangled nature model of evolutionary ecology most species, or types, turn out to be 'keystone species'. The reason for this is that the effect of removing one specific type of individual from the system tends to propagate across the system. Even if we happen to eliminate a type, call it S_1, which is only connected to one other type, say S_2, the removal of S_1 may change the conditions necessary for the existence of S_2, which then becomes unstable and could go extinct. In this way the effect of removing species S_1 can propagate through the system and may cause havoc across the entire ecology. This is just to say that the set of components of a system that are crucial to the overall stability of a system may not only be limited to those having *direct* links to many other components of the system. Chain-like processes can in a complex system magnify a small perturbation until it becomes of damaging consequence to the entire system.

In systems that behave in this way it will be very difficult indeed to identify what constitutes a chance. For sake of argument one might imagine that we would consider eliminating some bacteria responsible for an unpleasant contagious decease.

We would probably evaluate this in the first instance as a great chance. But if now it turns out that the eliminated bacteria were preventing some other even more disagreeable bacteria from establishing itself in the environment, we might reconsider what we first found was a chance.

We have gone through details of specific models of self-organizing complex critical systems to be able to substantiate the point that the identification of a chance can be difficult in complex systems. The two models discussed here both point to the fact that what might look at a promising chance in the short run and at some local level, could turn out to have profound, and potentially disastrous consequences on longer time scales and on a more global level. The ability to correctly recognize a chance as a true chance is in complex systems inevitably dependent on an understanding of the global collective nature of the dynamics of the system. Can such an understanding be obtained? This is at least to some extent linked to the issue of predictability in complex systems, which we briefly discuss in the next section.

4.5 Predictability in Complex Systems

This is obviously a complicated problem. The link between complexity and predictability is studied as e.g. described in the reviews by Bialek et al. [4.4] and by Boffetta et al. [4.6]. These studies are more related to the important question as to whether a certain system in principle allows predictions to be made. This may not help us much in a specific situation when we want to judge the effect of a range of different possible actions. Above we discussed, in the sandpile model, as well as in the tangled nature model, the extreme sensitivity and contingency of the dynamical evolution of these models and we therefore expect detailed prediction based on the knowledge of the state of the system at a given time to be difficult. The scale-free properties of the system are related to this question. A scale-free distribution of avalanche sizes would not apply to the model if large avalanches with certainty were followed by a number of small avalanches before yet another large avalanche again was released.

It might of course be that some events influence the system so dramatically that they fall in a category different from the one of the scale-free events. This is Sornette's idea that the really big catastrophic events simply do not follow the same scale-free probability distribution as the one followed by the smaller events [4.15]. Consider for example earthquakes. In general earthquakes of a given fault exhibit scale-free statistics as recent analysis supports [4.2]. However, it may be though that some earthquakes can be so big that they are influenced by the finite size of the system and therefore will be subject to conditions which are different from those smaller earthquakes experience (see e.g. [4.11]). Sornette suggests that certain precursors may precede these giant events permitting predictions to be made.

Nevertheless, prediction in scale-fee complex systems is difficult and will most likely involve some sort of global probing involving a large number of parameters.

4.6 Chance Discovery as an Understanding of the Common Good

The above discussion can be summed up as pointing out that chance has to be considered from the global collective point of view in complex critical systems. The interconnectedness and the sensitivity of such systems can turn what first appeared as a beneficial chance into an act of fatal consequences for the system as a whole. One may argue that scale invariance implies that there is really no essential difference between small and big events except their size and that the ability of the system to self-organize will ensure that recovery occurs. This is true but only in a statistical sense. Let us imagine that we induce an extinction of some species in an area because we saw a great chance in, say, producing genetically modified crops on some fields. A complex self-organizing system probably bounces back in some way or another. But when the system recovers it can have changed its nature completely and we can not be sure that the new 'species composition' is one we like. That is, if we assume we did not directly trigger a major ecological catastrophe with immediate and direct fatal consequences.

Recently, we have been studying models of interacting agents who play two-person games using information they receive from a small group of other agents. One can think of this either as 'business partners' or as interacting biological organisms. The games are zero-sum games. When an agent in a game wins a score of value $+1$, the partner wins a score of value -1, i.e. loses a score. After a certain number of test games the agent with the lowest score is replaced by a new agent with random properties and a new round of test games is performed. One can study the properties of those agents that survive longest in the system. Recall that agents with low score are in danger of being removed after each round of test games. Perhaps one would for a moment think that the agents with the greatest longevity are those that manage to make the largest score. This turns out to be incorrect. Agents survive many test rounds by tuning their strategy towards an overall 0 score. The reason is that when an agent is too greedy and produces a large score he is inevitably connected to agents who run up a large negative score. This means that partners of agents with a big score are in particular danger of being removed from the system. When that happens the big-scoring agent all of a sudden finds himself completely out of tune. His strategy had been adapted to interact with a partner who is no longer around and it is most unlikely that this strategy is efficient in the new environment encountered when the badly scoring agent is renewed. In contrast those agents that are able to pair up with partners in neutral games do not run the risk of removing their own foundation of existence. This scenario appears rather familiar. If the foxes manage successfully to eat all the rabbits they may feel very satisfied but only for a very short while.

Assume that we now agree that in complex systems chance must be seen from the global collective perspective. Will we then be able to spot when the right conditions arise? Above we have argued that this will be very difficult in systems with a sensitivity comparable to the one found in the sandpile model and in the tangled nature model. So if we by chance have in mind to execute an action that leads to some beneficial outcome predictable by us, then this seems to be hopeless in the type of systems discussed here. Perhaps we by chance sometimes have in mind that we are given the possibility to act but without knowing in detail the effect of our action. If we find this notion of chance relevant, the chance discovery in the language of the sandpile will correspond to identifying the moment when adding an extra grain of sand is possible. In the tangled nature model we might consider chance as being able to remove a certain genotype, say because fluctuations have taken the number of individuals of that specific type down to a level we can manage to eradicate. This is at an anecdotal level what is done when an infectious disease is successfully eliminated from a region. Although we are unable to evaluate the long-term global consequences of our action we might be willing to take the chance, or perhaps run the risk, if the short-term local effect is sufficiently positive. Isn't that what we do every time we undergo treatment with antibiotics?

4.7 Some Concluding Remarks

Our discussions of complex systems have perhaps reminded us that the effect of our action can eventually only be judged from a perspective that goes across the globe and across time. Isn't this self-evident? So was it all in vain? I think not entirely and this for two reasons. Though most people intuitively appreciate that we live in a very complex world where it is difficult to predict the eventual effect of our actions, decisions are frequently made neglecting the complexity of a problem. I believe a prime example of this is the issue of genetically modified (GM) food. From a short-time narrow perspective GM food promises a solution to many difficulties and may be considered as a major chance. The situation is more involved when one considers the potential perils of the production and spread of GM food in a wider context. What are the long-term consequences for biological evolution and ecosystem function and stability? Is it at all possible to judge the value of GM food neglecting that the most dangerous effects could easily be indirect and only show up only after a long time?

The other and related reason why I think the present chapter might be worthwhile is that the study of specific models may help to quantify the discussion of *chance discovery*. People may be sympathetic to the qualitative argument that we live in an interconnected world, but what can we do about it and what are the consequences anyway? Even simplified and not very realistic model systems may be able to help clarify just how sensitive a given system might be. It could be possible to gain a better understanding of the different levels of dynamics involved. In the tangled nature model the dynamics of the individuals is very fast compared to the dynamics at the levels of species of genotypes. Perhaps understanding the time of response between the different levels can make it possible to know which types of actions to treat with care because they tend to induce major system-spanning perturbations and which actions are less problematic. In the tangled nature model removing one genotype entirely from the system is much more likely to induce a major transition in the composition of the population than depleting the number of individuals across all existing genotypes by the same factor.

Models of complex systems suggest that one should distinguish between global and local chances and between short-term and long-term chances. Moreover, in a complex system one is probably more likely to differentiate between a chance and a catastrophe-triggering action if one employs a multi-parameter approach involving as many aspects of the properties of the system as possible.

Acknowledgements. I am grateful to Peter McBurney for inviting me to contribute this chapter and his comments on the manuscript.

References

4.1 Bak P (1997) How Nature Works, Oxford University Press
4.2 Bak P, Christensen K, Danon L and Scanlon T (2002) Unified Scaling Law of Earthquakes, Physical Review Letters 88(178501):1-4
4.3 Bak P, Tang C, and Wiesenfeld K (1987) Self-Organized Criticality An Explanation of 1/F Noise Physical Review Letters 59:381
4.4 Bialek W, Nemenman I, Tishby N (2001) Predictability, Complexity and Learning, Neural Computation, 13:2409-2463
4.5 Binney JJ, Dowrick, NJ, Fisher, AJ, and Newman MEJ (1992) The Theory of Critical Phenomena, Oxford University Press
4.6 Boffetta G, Lencini M, Falcioni M, Vulpiani A (2002) Predictability: a way to characterize complexity Physics Report, 356:367-474
4.7 Christensen K, Collombiano SA, Hall M, Jensen, HJ (2002) Tangled Nature: A Model of Evolutionary, Ecology, J. of Theoretical Biology 216:73–84
4.8 Darwin C (1859) The Origin of Species by Means of Natural Selection First published by John Murray, London, UK, reprinted by Penguin Books, London, UK, in 1968
4.9 Hall M, Christensen K, Collombiano SA and Jensen, HJ (2002) Time-depenent extinction rate and species abundance in a Tangled-nature model of biological evolution Physical Review E 66(011904) 1-20
4.10 Hamon D, Nicodemi M, Jensen HJ (2002) Continuously Driven OFC: A Simple model of Solar Flare Statistics, Astronomy and Astrophysics, 387:326
4.11 Jensen HJ (1998) Self-Organized Criticality, Cambridge University Press
4.12 Kauffman S (1995) At Home In The Universe, Viking, New York, NY
4.13 Maynard Smith J (1982) Evolution and the Theory of Games, Cambridge University Press
4.14 Pimm SL (1991) The Balance of Nature, University of Chicago Press
4.15 Sornette D (2002) Predictability of catastrophic events: material rupture, earthquakes, turbulence, financial crashes and human birth, Proc. of National Academy of Sciences of USA, 99:2522-2529

5. Anatomy of Rare Events in a Complex Adaptive System

Paul Jefferies[1], David Lamper[2], and Neil F. Johnson[1]

[1] Physics Department, Oxford University, Oxford OX1 3PU, UK
[2] Oxford Centre for Industrial and Applied Mathematics,
   Oxford University, Oxford OX1 3LB, UK
   email: linc0227@herald.ox.ac.uk, lamper@maths.ox.ac.uk,
   n.johnson@physics.ox.ac.uk

Summary.

   Here we provide an analytic, microscopic analysis of rare and extreme events in an adaptive population comprising competing agents (e.g. species, cells, traders, data packets). Such large changes represent a form of *chance discovery* and tend to dictate the long-term dynamical behavior of many real-world systems in both the natural and social sciences. Our results reveal a taxonomy of these infrequent yet extreme events, and provide a microscopic understanding as to their build-up and likely duration.

5.1 Can We Deal with Large Changes

Large unexpected changes – such as crashes in information networks or financial markets , and punctuated equilibria in evolution – happen infrequently, yet tend to dictate the long-term dynamical behavior of real-world systems in disciplines as diverse as biology and economics, through to ecology and evolution. The task of understanding, forecasting, and possibly even controlling such extreme, rare, or 'chance' events represents a major challenge for both science and society. To what extent are they predictable? Even if we could identify their imminent arrival, what could we do to minimize the potential risk or possible hazards?

   The intrinsic ability to generate large, endogenous yet unexpected changes is a defining characteristic of complex systems, and arguably of nature and life itself [5.14, 5.16, 5.13, 5.4, 5.5, 5.15, 5.12, 5.8, 5.2]. Such changes are manifestations of subtle, short-term temporal correlations resulting from internal collective behavior. They seem to appear out of nowhere and have long-lasting consequences. To what extent can they ever be 'understood'? Followers of the self-organized criticality view [5.2] would claim this question is naïve because of an inherent self-similarity in nature: any large changes are simply magnified versions of smaller changes, which are in turn magnified versions of even smaller changes, and so on. Such self-similarity is presumed to underlie the power-law scaling observed in natural, social, and economic phenomena [5.2]. However, there are reasons for believing that the largest changes may be 'special' in a microscopic sense [5.8]. Power-law scaling is only approximately true, and does not apply over an infinite range of scales. Apart from being atomistic at the smallest scale, a population of competing agents can not cause any effect larger than the population size itself: in short, the largest changes will tend to 'scrape the barrel' in some way. Jahansen and Sornette [5.8] quote Bacon from Novum Organum: "Whoever knows the ways of Nature will more easily notice her deviations; and, on the other hand, whoever knows her deviations will more accurately describe her ways.".

This paper addresses the task of understanding, and eventually controlling, the large endogenous changes arising in a complex adaptive system comprising competing agents (e.g. species, cells, traders, data packets). Our work reveals a taxonomy of large changes, and provides a quantitative microscopic description of their build-up and duration. Our results also provide insight into how a 'complex-systems manager' might contain or control such extreme events .

5.2 A Complex Adaptive System of Competing Agents

We consider a generic complex system in which a population of N_{tot} heterogeneous agents with limited capabilities and information repeatedly compete for a limited global resource. Our model was introduced in [5.9, 5.7], and is a generalization of the El Farol bar problem and the minority game, concerning a population of people deciding whether to attend a popular bar with limited seating [5.1, 5.10]. At time step t, each agent (e.g. a bar customer or a market trader) decides whether to enter a game where the choices are option 1 (e.g. attend the bar or buy) and option 0 (e.g. go home, or sell): N_0 agents choose 0 while N_1 choose 1. The 'excess demand' $D[t] = N_1 - N_0$ (which mimics price change in a market) and number $V[t] = N_1 + N_0$ of active agents (which mimics volume of market orders) represent output variables. These two quantities fluctuate with time, and can be combined to construct other global quantities of interest for the complex system studied (e.g. summing the price changes gives the current price). This model can reproduce statistical and dynamical features similar to those in a real-world complex adaptive system, namely a financial market [5.9, 5.7], and exhibits the crucial feature of seemingly spontaneous large changes of variable duration [5.9, 5.7, 5.11]. The resulting time series appears 'random' yet is non-Markovian, with subtle temporal correlations which put it beyond any random-walk-based description. The temporal correlations of price changes and volume, and their cross correlation, are of intense interest in financial markets where so-called chartists offer a wide range of rules-of-thumb [5.3] such as 'volume goes with price trend'. Although such rules are unreliable, the intriguing question remains as to whether there could *in principle* be a 'science of charting'.

A subset $V[t] \leq N_{tot}$ of the population, who are sufficiently confident of winning, are active at each time step. For $N_1 < N_0$ the winning decision is 1 and vice versa, i.e. the winning decision is given by $H[-D[t]]$, where $H[x]$ is the Heaviside function. The global resource level is so limited, or equivalently the game is so competitive, that at least half the active population lose at each time step [5.1, 5.10]. The only global information available to the agents is a common bit string 'memory' of the m most recent outcomes. Consider $m = 2$; the $P = 2^m = 4$ possible history bit-strings are 00, 01, 10, and 11, which can also be represented in decimal form: $\mu \in \{0, 1, \ldots, P - 1\}$. A strategy consists of a response, $a^\mu \in \{-1, 1\}$, to each possible bit string μ: $a^\mu = 1 \Rightarrow$ option 1 and $a^\mu = -1 \Rightarrow$ option 0. Hence there are $2^P = 16$ possible strategies. The heterogeneous agents randomly pick s strategies each at the outset, and update the scores of their strategies after each time step with the reward function $\chi[D] = [-D]$, i.e. $+1$ for choosing the minority action, -1 for choosing the majority action. Agents have a time horizon T over which strategy points are collected, and a threshold level r which mimics a 'confidence'. Only strategies having $\geq r$ points are used, with agents playing their highest-scoring strategy. Agents with no such strategy become temporarily inactive [5.9, 5.7]. We focus on the regime where the number of strategies in play is comparable to the total number available, since this yields seemingly random dynamics with occasional large movements [5.9, 5.7, 5.11]. The coin tosses used to resolve ties in decisions (i.e. $N_0 = N_1$) and active-strategy scores inject stochasticity into the game's evolution. Jefferies et al.[5.6] showed that a simplified version of this system in the limit $r \to -\infty$ and $T \to \infty$ can be usefully described as a stochastically disturbed deterministic system. We are interested in the dynamics of large changes, and adopt the approach and terminology of [5.6]. Averaging over our model's stochasticity yields a description of the game's deterministic dynamics via mapping equations for the strategy score vector $\underline{S}[t]$ and global information $\mu[t]$. For $s = 2$ the deterministic dynamics is given exactly by the following equations:

$$\underline{S}[t] = \underline{S}[0] - \sum_{i=t-T}^{t-1} \underline{a}^{\mu[i]} \text{sgn}\left[D[i]\right], \tag{5.1}$$

$$\mu[t] = 2\mu[t-1] - PH\left[\mu[t-1] - P/2\right] + H\left[-D[t-1]\right].$$

The corresponding demand function is given by

$$D[t] = \sum_{R=1}^{2P} a_R^{\mu[t]} H[S_R - r] \sum_{R'=1}^{2P} \left(1 + \text{sgn}\left[S_R[t] - S_{R'}[t]\right]\right) \Psi_{R,R'},$$

where Elements $\Psi_{R,R'}$ enumerates the number of agents holding both strategies R and R'. Each $\Psi_{R,R'}$ becomes an element of $\underline{\underline{\Psi}}$, the symmetrized strategy allocation matrix which constitutes the *quenched disorder* present during the system's evolution [5.6]. The volume $V[t]$ is given by the same expression as $D[t]$ replacing $a_R^{\mu[t]}$ by unity.

5.3 Large Changes as Crashes in a Bruijn Graph

Large changes such as financial market crashes seem to exhibit a wide range of possible durations and magnitudes, making them difficult to capture using traditional statistical techniques based on one- or two-point probability distributions [5.8]. A common feature, however, is an obvious trend (i.e. to the eye) in one direction over a reasonably short time window: we use this as a working definition of a large change. In fact, all the large changes discussed here represent $> 3\sigma$ events. In both our model and the real-world system, these large changes arise more frequently than would be expected from a random-walk model [5.14, 5.16, 5.13, 5.4, 5.5, 5.15, 5.12, 5.8]. Our model's dynamics can be described by trajectories on a de Bruijn graph [5.6]: see Fig. 5.1 for $m = 3$, with a transition incurring an increment to the score vector $\underline{S}$.

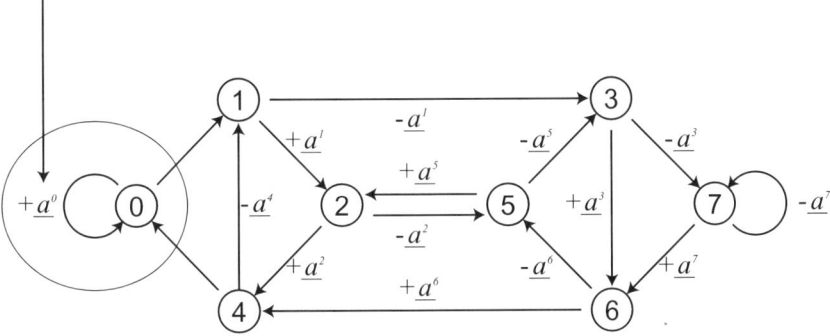

Stable behavior: path with all transitions equally visited

    e.g. $0 \to 0 \to 1 \to 3 \to 6 \to 5 \to 3 \to 7 \to 7 \to 6 \to 4 \to 1 \to 2 \to 5 \to 2 \to 4 \to \ldots$

Crash: path with many negative (position) return transitions

    e.g. $0 \to 0 \to 0 \to 0 \to \ldots$ or $2 \to 4 \to 0 \to 0 \to 1 \to 2 \to 4 \to 0 \to \ldots$

Fig. 5.1. Dynamical behavior of the global information is described by transitions on the de Bruijn graph. Graph for population of $m = 3$ agents. Blue transitions represent positive demand D, red transitions represent negative demand

There are P orthogonal increment vectors $\underline{a}^\mu$, one for each node μ. Setting the initial scores $\underline{S}[0] = \underline{0}$, the strategy score vector in (5.1) can be written exactly as:

$$\underline{S}[t] = c_0 \underline{a}^0 + c_1 \underline{a}^1 + \ldots + c_{P-1} \underline{a}^{P-1} = \sum_{j=0}^{P-1} c_j \underline{a}^j,$$

where c_j represents the *nodal weights* for history node $\mu = j$. The nodal weights enumerate the number of negative-return transitions from node μ minus the number of positive-return transitions, in the time window $t - T \to t - 1$. High absolute nodal weight implies persistence in transitions from that node, i.e. persistence in $D|\mu$. Large changes will occur when connected nodes become persistent. The simplest type of large movement exhibiting perfect nodal persistence would be $\mu = 0, 0, 0, 0, \ldots$ in which all successive price changes are in the *same* direction. We call this a 'fixed-node crash' (or rally). However, there are many other possibilities reflecting the wide range of forms and durations of the large change. For example, on the $m = 3$ de Bruijn graph in Fig. 5.1 the cycle $\mu = 0, 0, 1, 2, 4, 0, \ldots$ has four

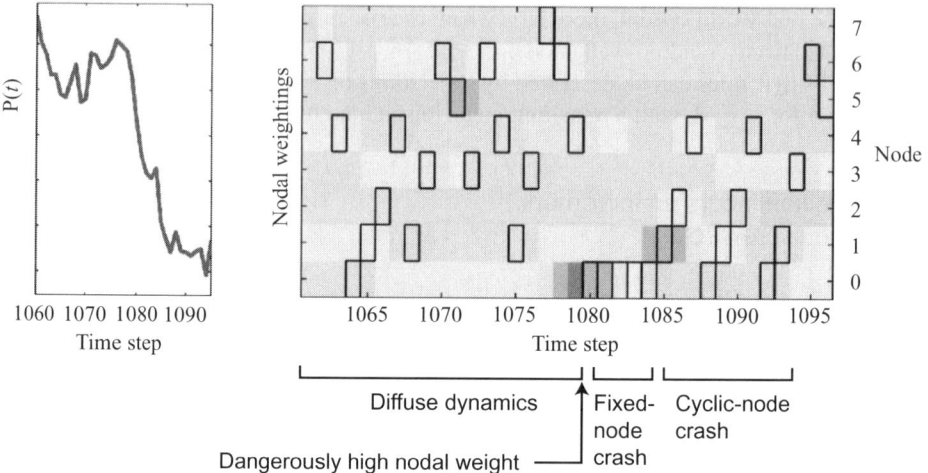

Fig. 5.2. Dynamical behavior of complex system (e.g. price $P[t]$ in financial market) described by evolution of nodal weights c_μ. History at each time step indicated by a black square. Large change preceded by abnormally high nodal weight. Large change incorporates fixed-node and cyclic-node crashes

out of the five transitions producing price changes of the same sign (it is persistent on nodes 1, 2, 4 and antipersistent on node 0). We call this a 'cyclic-node crash' (or rally).

Figure 5.2 illustrates a large change which starts as a fixed-node crash and then subsequently becomes a cyclic-node crash. Cyclic-node crashes can be treated simply as interlocking fixed-node crashes; hence for clarity we focus here on a single fixed-node crash (or rally). For the parameter ranges of interest, the choice about whether a strategy is played by an agent is more determined by whether that strategy's score is above the threshold, than whether it is their highest-scoring strategy. This is because agents are only likely to have at most one strategy whose score lies above the threshold for confidence levels $r \geq 0$. Making the additional numerically justified approximation of small quenched disorder (i.e. the variance of the entries in the strategy allocation matrix $\underline{\underline{\Psi}}$ is smaller than their mean for the parameter range of interest [5.6]), the demand and volume become:

$$D[t] = \frac{N}{4P} \sum_{R=1}^{2P} a_R^{\mu[t]} \text{sgn} \left[S_R[t] - r \right], \tag{5.2}$$

$$V[t] = \frac{N}{2} + \frac{N}{4P} \sum_{R=1}^{2P} \text{sgn} \left[S_R[t] - r \right]. \tag{5.3}$$

Suppose persistence on node $\mu = 0$ starts at time t_0. How long will the resulting crash last? To answer this, we decompose (5.2) into strategies which predict 1 at $\mu = 0$, and those that predict 0. We first consider the particular case where the node

$\mu = 0$ was *not* visited during the previous T time steps; hence the loss of score increment from time step $t - T$ will not affect $\underline{S}[t]$ on average. At any later time $t_0 + \tau$ during the crash (i.e. $\mu = 0$), (5.2) and (5.3) are hence given by:

$$D[t_0 + \tau] = -\frac{N}{4P} \left\{ \sum_{R \ni a_R^\mu = -1} \text{sgn}\,[S_R[t_0] - r - \tau] - \sum_{R \ni a_R^\mu = 1} \text{sgn}\,[S_R[t_0] - r + \tau] \right\} \quad (5.4)$$

$$V[t_0 + \tau] = \frac{N}{2} + \frac{N}{4P} \left\{ \sum_{R \ni a_R^\mu = -1} \text{sgn}\,[S_R[t_0] - r - \tau] + \sum_{R \ni a_R^\mu = 1} \text{sgn}\,[S_R[t_0] - r + \tau] \right\}.$$

$|D[t_0 + \tau]|$ decreases as the persistence time τ increases, and hence the crash ends at time $t_0 + \tau_c$ when the right-hand side of (5.4) changes sign. The persistence time or 'crash length' τ_c is thus given by the mean of the scores of the strategies predicting 0, i.e. $\tau_c = \overline{S}_{R \ni a_R^\mu = -1}[t_0] = -c_0[t_0]$. In the more general case where the node $\mu = 0$ *was* visited during the previous T time steps, τ_c is given by the largest τ value which satisfies:

$$\tau = -(c_0[t_0] + \sum_{\{t'\}} \text{sgn}\,[D[t']]),$$

where $\{t'\} \ni (\mu[t'] = 0 \cap t_0 - T \le t' \le t_0 + \tau - T)$. Assume that the scores have a near-normal distribution, i.e. $S_{R \ni a_R^\mu = -1}[t_0] \sim N[\overline{S}_{-1}, \sigma]$ as in Fig. 5.3(a). For each strategy R there exists an anticorrelated strategy $\overline{R}$ and hence $S_R[t] = -S_{\overline{R}}[t]$ for all t. Consequently, prior to a crash, the score distribution tends to split into two halves as indicated schematically in Fig. 5.3(a). The expected demand (and volume) during the crash are then:

$$\langle D[t_0 + \tau] \rangle \propto \left(\text{erf}\left[\frac{c_0[t_0] + r + \tau}{\sqrt{2}\sigma} \right] - \text{erf}\left[\frac{-c_0[t_0] + r - \tau}{\sqrt{2}\sigma} \right] \right),$$

$$\langle V[t_0 + \tau] \rangle \propto \left(2 - \text{erf}\left[\frac{c_0[t_0] + r + \tau}{\sqrt{2}\sigma} \right] - \text{erf}\left[\frac{-c_0[t_0] + r - \tau}{\sqrt{2}\sigma} \right] \right).$$

These forms are illustrated in Fig. 5.3(b). As the spread in the strategy score distribution is increased, the dependence of $\langle D \rangle$ and $\langle V \rangle$ on the parameters τ and r becomes weaker and the surfaces flatten out leading to a smoother drawdown, as opposed to a sudden severe crash. As the parameters $\overline{S}_{-1}, \sigma, r$ are varied, it can be seen that the behavior of the demand and volume during the crash can exhibit markedly different qualitative forms, yielding a *taxonomy* of *different species* of large change *even within the same single-node family*. This result could explain why financial market chartists' rules-of-thumb [5.3], such as 'volume goes with price trend', are far too simplistic.

We now turn to the important practical question of whether history will repeat itself, i.e. given that a crash has recently happened, is it likely to happen again? If so, is it likely to be even bigger? Suppose that the system has built up a negative nodal

(a)

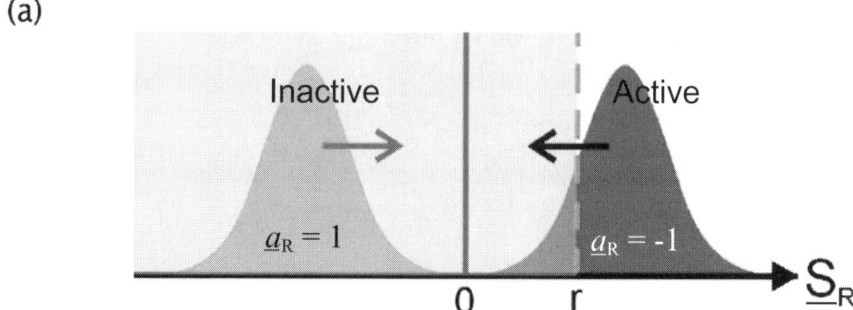

(b)

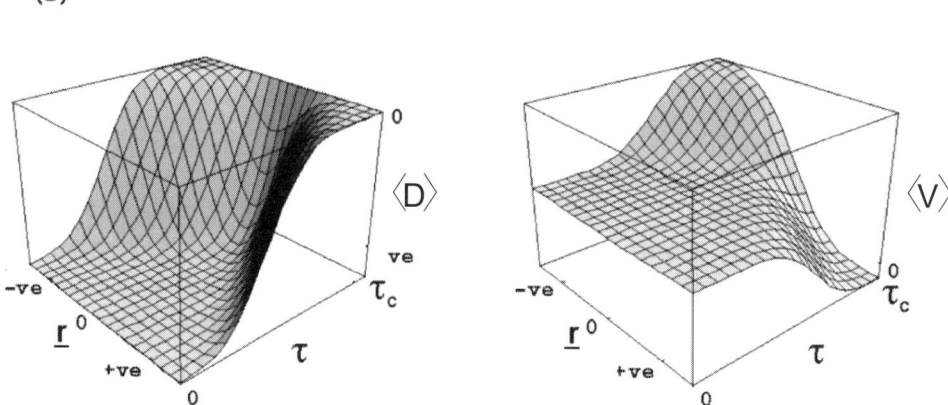

Fig. 5.3. (a) Schematic representation of strategy score distribution prior to crash. Arrows indicate subsequent motion during crash period. (b) Plots of expected demand and volume during crash period showing range of different possible behavior as system parameters are varied

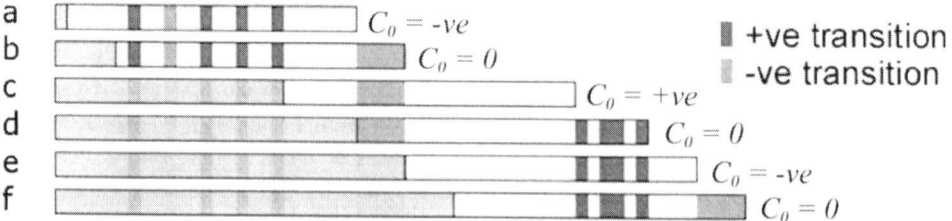

Fig. 5.4. Representation of how large changes can recur due to finite memory of agents. Gray area shows history period outside agents' memory. Example shows recurring fixed-node crash at node $\mu = 0$

weight for $\mu = 0$ at some point in the game (see Fig. 5.4(a)). It then hits node $\mu = 0$ at time t_0 producing a crash (Fig. 5.4(b)). The nodal weight c_0 is hence restored to zero (Fig. 5.4(c)). In this model the previous build-up is then forgotten because of the finite T score window, hence c_0 becomes positive (Fig. 5.4(d)). The system then corrects this imbalance (Fig. 5.4(e)), restoring c_0 to 0. The crash is then forgotten, hence c_0 becomes negative (Fig. 5.4(f)). The system should therefore crash again – however, a crash will *only* re-appear if the system's trajectory subsequently returns to node $\mu = 0$. Interestingly, we find that the *disorder* in the initial distribution of strategies among agents (i.e. the quenched disorder in $\underline{\Psi}$) can play a deciding role in the issue of crash 'births and revivals' since it leads to a slight bias in the outcome, and hence the subsequent transition, at each node. When $c_{\mu[t]} = 0$ (see Fig. 5.4(c)), it follows that $\mathrm{sgn}[D[t]]$ is more likely to be equal to $\mathrm{sgn}\left[\underline{a}^{\mu[t]} \cdot \underline{x}\right]$, where $\underline{x} = \sum_{R'=1}^{2P} \underline{\Psi}_{R'}$ is a strategy weight vector with x_R corresponding to the number of agents who hold strategy R. The quenched disorder therefore provides a crucial bias for determining the future trajectory on the de Bruijn graph when the nodal weight is small, and hence can decide whether a given crash recurs or simply disappears. The quenched disorder also provides a *catalyst* for building up a very large crash.

5.4 Conclusions

Our work opens up the study of how a 'complex-systems manager' might use this information to control the long-term evolution of a complex system by introducing, or manipulating, such large changes. As an example, we give a quick three-step solution to prevent large changes: (1) use the past history of outcomes to build up an estimate of the score vector $\underline{S}[t]$ and the nodal weights $\{c_{\mu[t]}\}$ on the various critical nodes, such as $\mu = 0$ in the case of the fixed-node crash. (2) Monitor these weights to check for any large build-up. (3) If such a build-up occurs, step in to prevent the system hitting that node until the weights have decreased.

References

5.1 Arthur WB (1994) Inductive Reasoning and Bounded Retionality, American Economic Review, 84:406–411
5.2 Bak P (1998) *How Nature Works, Cambridge University Press*, Cambridge, UK
5.3 Blair A (1996) *Guide to Charting* Pitman Publishing, London
5.4 Bouchaud J and Potters M (2000) *Theory of Financial Risks* Cambridge University Press, Cambridge, UK
5.5 Huberman B, Pirolli P, Pitkow J, Lukose R (1997) Strong regularities in World Wide Web surng, Science, 280:95-97
5.6 Jefferies P, Hart M, Johnson NF (2002) Deterministic Dynamics in the Minority Game Physical Reviews E 65:016105
5.7 Jefferies P, Johnson NF, Hart M (2001) From market games to real-world markets, European Physical Journal B 20:493

5.8 Johansen A, Sornette D (2001) Large Stock Market Price Drawdowns Are Outliers, Journal of Risk 4(2): 69-110

5.9 Johnson NF, Hart M, Hui PM, Zheng D (2000) Trader Dynamics in a Model Market In. J. Theoreical and Applied Finance, 3:443-450

5.10 Johnson NF, Hui PM, Zheng D, Tai CW (1999) Minority game with arbitrary cutoffs, Physica A 269:493–502

5.11 Lamper D, Howison D, Johnson NF (2002) Predictability of large future changes in a competitive evolving population Physical Review Letters, 88:017902

5.12 Lux T, Marchesi M (1999) Scaling and Criticality in a Stochastic Multi-Agent Model of a Financial Market, Nature 397:498–500 (1999)

5.13 Mantegna R, Stanley H (2000) *Econophysics*, Cambridge University Press, Cambridge, UK

5.14 McClintock PVE (1999) Unsolved Problems of Noise, Nature 401:23–25

5.15 Nowak MA, May RM (1992) Evolutionarygames and spatial chaos, Nature, 359:826-829

5.16 Palmer RG, Arthur WB, Holland JH, LeBaron B, Tayler P (1994) Artificial economic life: a simple model of a stockmarket, Physica D 75: 264–274

Part II
Key 1 – Communications for Chance Discovery

The first key to chance discovery, given in Part I, is communication. In Part II, we look at human–human, human–agent, and agent–agent communications both in the real and the virtual (online) communities. The emergence of ideas in each of these communications activates the process of chance discovery. Chance discovery is a new challenge, but here you see that it is within the reach of daily chats.

6. Human-to-Human Communication for Chance Discovery in Business

Hiroko Shoji

Department. of Information and Communication Sciences, Faculty of Education, Kawamura Gakuen Women's University, Abiko, Chiba 275-1138, Japan
email: hiroko@da2.so-net.ne.jp

Summary.

Human-to-human communication can provide an opportunity for chance discovery. We tend to think 'information to be communicated' is one its sender wants to convey or one its receiver needs, and others are noise. However, information apt to be overlooked as noise may have potential business opportunities or be a latent pitfall to failure. Although people rarely notice such 'noise information', a person behaving as an agent of communication can bring this subtle sign to a business opportunity without passing it up, if he/she has a high level of awareness. This chapter describes chance discovery in the human-to-human communication in retail business and emails in R & D projects.

6.1 Face-to-Face Communication for Chance Discovery in Retail Business

Capable sales clerks use their strategic knowledge to make an appropriate reaction to their customers, achieving a high sales performance. Such strategic knowledge has been so far considered to be tacit; however, the modern era of highly fluid human resources requires this tacit knowledge to be identified. In addition, the purchase process (i.e. how to buy) as well as products themselves (i.e. what to buy) brings about a value; therefore, sellers need to systematize their knowledge of how to sell. With specific examples presented, this paper describes how capable sales clerks use appropriate communication patterns for their customer's pattern of thinking, that is, how they discover chances through the communication with their customer. Examples presented herein explain only a part of tacit knowledge within capable sales clerks; however, a collection and analysis of many cases from a similar viewpoint is expected to allow for the systematization of sales clerks' tacit knowledge.

6.1.1 Importance of Sales Communication in a Society with Oversupply of Merchandise

Japan and other advanced countries have been referred to as 'a society with oversupply of merchandise' for a long time. With this historical background, manufacturers are required not only to have necessity or functionality of their products as a feature but also to develop products appealing to consumers' sensibility. In addition to products such as apparel and interior goods for which emotional elements have been previously emphasized, industries such as automobiles and computers are also putting emphasis on fashionability. In this world with a wide variety of products, however, the continuous development of products that match the diverse sensibility of various consumers may involve high risk for manufacturers, distributors, or retailers. It is not a good idea in terms of the resource issue and environmental conservation as well. Even limited products could give a better success rate of sales if they match the values for consumers, and could be sold at high prices by adding value to products in the process of sales communication. In the society with oversupply of merchandise, stores are no longer the mere place for shopping, and shopping centers based on an attractive design concept are appearing and trying to explore the new consumption by producing 'the shopping as a process'. Since not all customers initially have a clear image of their needs, it is an important challenge for sellers to convince their customers of their exact wants.

Considering the present state of the society, people in the areas of marketing or consumer praxiology started to indicate the importance of producing 'the purchasing as a process' in order to grab customers with a vague image of desired products. Underhill concluded through his detailed long-term analysis of customer behavior in actual shops that purchasing doesn't mean shopping for their predetermined wants and 'their wants are determined at the shop' [6.11]. He stated that the economy would collapse if people were to buy only necessary things. Schmitt stated that we are currently in the middle of the drastic reform to change the traditional marketing based on features and benefits to the experiential marketing, and indicated that decision-making patterns of consumers are evolving toward the pursuit of experiential values which are sensuous, emotional, and cognitive [6.8]. Peppers and Rogers pointed out that a paradigm shift from 'mass marketing' to 'one-to-one marketing' is taking place based on IT including Internet technology in the domain of marketing, while they stated that the era of one-to-one marketing is the time to re-evaluate the relationship marketing named 'order-taking' that bricks-and-mortar florists or greengrocers on the corner were practicing in the past [6.5]. According to Wind and Rangaswamy, marketing in the future will evolve from customization into 'customerization' [6.12].

The society with oversupply of merchandise is a society with emphasis on experiential values and also a society based on customerization. In this society, customers want to think that they considered and determined everything; however, they actually can not in many cases, because their wants are not always determined previously. This society requires sellers to act as a co-star who shares with their customer the process to articulate the customer's wants. If sales clerks, through communication with their customer, see what information is necessary for the customer's concept articulation and then provide him/her with appropriate information, the customer will probably make a satisfactory purchase, resulting in better sales performance of the sales clerks themselves. In this case, a chance discovery for the sales clerk is to find useful information for the customer's concept articulation .

6.1.2 Purchasing Types and Communication Patterns

Our detailed observation of actual purchase activities has shown that purchasing can be roughly divided into a problem-solving type and a concept-articulation type[6.9].

1. Purchasing as problem solving
   With purchasing as problem solving, the customer initially has a clear idea of what a desired product is like and/or what functionality it requires, and searches for the items which meet his/her requirements. That is, purchasing as problem solving means that the customer has previously determined what to buy. The customer who follows purchasing as problem solving searches for the products meeting his/her requirements to discover solution candidates, balances between them (if there are more than one), evaluates them, and then decides whether to buy them.

2. Purchasing as concept articulation
   With purchasing as concept articulation, on the other hand, the customer initially has unclear requirements for his/her needs and gradually builds up a concrete image of target products through the interaction with a sales clerk. That is, purchasing as concept articulation means that the customer determines what to buy after due consideration in the shop. Customers who follow purchasing as concept articulation start with vague requirements of their own, become aware of their underlying requirements with a trigger of some information provided while looking around various products, understand what their true requirements are, convince themselves of the fidelity of some of the products to the requirements, and then make a decision on whether to buy those products. They don't conceive their true requirements until they actually look at products.

Our analysis also showed sales clerks' communication patterns could be classified into two types, that is, expected reaction and unexpected reaction .

1. Expected reaction
   In regular purchasing, a customer reaches a more satisfactory solution or product through the conversation with a sales clerk. In such a situation, the sales clerk provides the customer with another solution or product that better fits his/her requirements. The sales clerk's role is considered to be presenting the solutions (products) to meet the customer's requirements. The reaction to fill this role is called expected reaction. These kinds of reactions from the sales clerk confirm the customer's requirements or thinking, and present candidates that better fits the requirements. It is often useful for purchasing as problem solving.

2. Unexpected reaction
   On the other hand, the authors observed reactions that promoted customers' decision-making by giving opinions that provided customers with a different viewpoint. The reaction which presents information from a different viewpoint than the customer's current thought is called unexpected reaction. These kinds of reactions from a sales clerk are unexpected, in the sense that they divert from

the usual reaction, which present solutions that better fit the requirements of customers. It is often useful for purchasing as concept articulation.

6.1.3 Creative Communication for Chance Discovery

Which of expected and unexpected reactions is an appropriate interaction depends on the current context. Our analysis also showed that capable sales clerks can successfully grasp values and potential wishes of customers to use an appropriate interaction pattern for the occasion, whereas less capable sales clerks tend to adopt an inappropriate pattern. In other words, skillful sales clerks can communicate with their customers appropriately to discover a chance of the customers' concept articulation, which often leads to a successful sale. Typical examples are shown below. Among different purchasing cases of different subjects, five cases happened to have the same situation where they mind that the jacket under consideration is short. Below is an example of an unsuccessful conversation which didn't lead to the purchase. The sales clerk's reaction in this case was the expected one.

Customer: I want a little longer one. This (candidate A) is a bit too short. I want to hide as much of my waist as possible.
Sales clerk: (after considering for a while) If so, how about this one (candidate B)? This is tucked at the waist and designed to have a long hem.
Customer: Well, let me see ... I'm afraid this is not my taste.

Shown next is an example of a successful conversation which led to the purchase. The sales clerk's reaction in this case was the unexpected one.

Customer: This (candidate B) is a little short, isn't it?
Sales clerk: Such a design is popular this year. Almost every shop deals with short ones. Do you prefer a longer one?
Customer: Too short to cover my waist ... .
Sales clerk: It depends on the balance with your skirt or pants. 'cause you're now wearing a shorter tight skirt, you think that way, but if wearing a long skirt, you will feel better.

In the former case where the conversation didn't lead to the purchase, the sales clerk responded straightforwardly to the customer who was reluctant about the short jacket by presenting a longer one, whereas in the latter case where the conversation led to the purchase, the customer's mental world was changed from one where the relevant attribute is *length of jacket* to another where a different attribute called *balance* is relevant. Through the conversation, the capable sales clerk shown in the latter case could grasp the customer's wish that she make herself look as good-shaped as possible, and induce the appropriate goal (*short but well-balanced jacket*) in accordance with it (as shown in Fig. 6.1).

1. The sales clerk points out to the customer that the balance is more important than the length in presenting a good shape. That is, the sales clerk provides the customer with a different viewpoint (balance) than his/her current one (length).
2. The customer's viewpoint is changed into balance (1 in this figure).

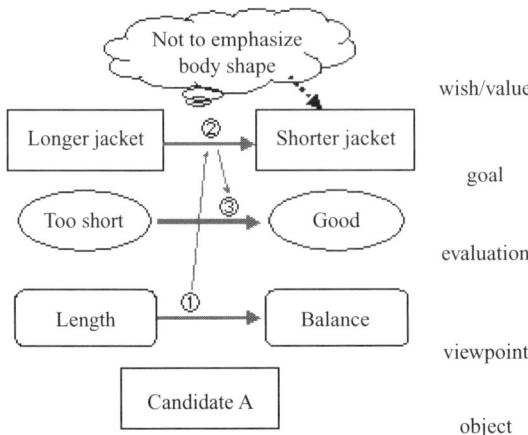

Fig. 6.1. Changes in the customer's mental world caused by the unexpected reaction

3. The customer understands that a short and well-balanced jacket is suitable for presenting a good shape and changes his/her search goal into a short jacket (2 in this figure).
4. The customer, in turn, evaluates the current candidate A to be good (3 in this figure), and resultingly buys it.

Here, let us think what kind of information triggered the customer's concept articulation. How the communication changes the customer's mental world for un-expected reactions is summarized in Fig. 6.1. In case of the unexpected reaction (as shown above), a new focus (*balance*) presented by the sales clerk triggers the cus-tomer's mental leap, and her goal changes accordingly from *long jacket* to *short but well-balanced jacket*, resulting in the change of her evaluation of the current target item (candidate A) to promote her decision making.

The unexpected reaction is useful as a trigger to the concept articulation, mean-ing that it is useful for helping the customer in purchasing as concept articulation. There are two main features useful for facilitating the concept articulation, as fol-lows:

1. Support for conception
   The unexpected reaction sometimes causes the change of the customer's view-point, which in turn triggers the change of the search goal itself, resulting in the promotion of his/her decision making. Skillful sales clerks can use the unex-pected reaction appropriately to facilitate the customer's conception.
2. Support for conviction
   The customer who has become aware of a new viewpoint needs conviction in order to accept the viewpoint smoothly. Skillful sales clerks can facilitate the customer's conviction through their good demonstration of concrete-use scenes and/or usage of possible products.

Our study shows that a creative thinking process is observed in everyday behavior as well as in professional creative activities. Although such *creativity in real life* is a small one compared to those in scientific inventions or discoveries, it is more frequent because of its everyday nature. If you are good at exerting small creativity, or helping others with small creativity, you can gain an advantage in business. For example, a competent sales clerk can help her customers to articulate their concept, which leads to good sales performance. In this sense, creativity in real life, or creative communication for concept articulation, is a kind of *chance discovery*.

6.2 Email-Based Communication for Chance Discovery in an R&D Project

6.2.1 Importance of Communication in Multi-participant Projects

Both complexity and scale of the modern society are steadily increasing. Design/development projects are not an exception to this trend, and the complexity of the development process for various industrial products is increasing according as the progress of science and technology. Therefore, in many cases, many people are required to work together for a single project. The involvement of many people in the same project inevitably means the essentiality of communication and leads to a complex decision making process. Sharing design parameters and/or policies between developers is essential for the design/development process; however, they will be often changed as design proceeds. Insufficient information propagation during the change may bring the project itself to a failure. An example might be the Mars Climate Orbiter (MCO), which is a Mars exploration satellite by NASA. MCO was lost sight of in September, 1999 due to a failed injection into the orbit of Mars. This was reported to be because two navigation teams used different unit systems. Communication between both teams regarding navigation parameters went wrong because both teams didn't know that one team used kilometers while the other used miles. This is a unbelievable rudimentary mistake in one of space-development projects, which are generally supposed to be carefully planned and implemented. Conversely, this case shows the difficulty in smooth communication between members in multi-participant projects.

In this way, smooth communication between participants in a design and development project is certainly a challenge; however, it is even more difficult to properly understand and share the intention behind individual information exchanged in the communication. 'Design rationale' means for what reason an individual decision has been made in the process of design and development. Information on design rationale is important especially when operational trouble occurs and/or in subsequent design activities, but often omitted from regular meeting materials or final reports. Described are mainly resultant design specifications and/or parameters, and design rationale, which shows 'why this decision has been made', is rarely described explicitly. In some cases, even the decision maker themselves forgets the reason. Clarifying decision rationale and sharing it with project members through proper communication can help you prevent operational trouble and failure. It can be also re-used within your organization during subsequent design and development.

As mentioned above, multi-participant design/development projects require the sharing, discovery, and re-use of knowledge. In order to realize them, needless to say, various design information must be properly shared, and, additionally, you need a framework that allows a design decision, which is often left implicit, to be clarified, saved, and shared among everyone in the organization. Furthermore, you need a communication environment to prompt chance discovery that prevents pending issues from being forgotten and overlooked.

6.2.2 Communication Support for Chance Discovery with Mailing List

Currently, one of the most common communication support tools is email. In fact, actual design/development project organizations often use email as a daily communication medium. Therefore, this section describes a co-operative work support with a mailing list .

The study of support for human co-operative work with computers is called computer supported cooperative work (CSCW), and software designed for CSCW is generally called 'groupware'. Co-operative work that CSCW covers is generally grouped into three types by spatial aspect (face-to-face or remote) and temporal aspect (synchronous or asynchronous): (1) face-to-face and synchronous, (2) remote and synchronous, and (3) remote and asynchronous. Co-operative work with mailing lists corresponds to (3). A mailing list was originally designed for transmitting the same information to multiple persons, and, therefore, is excellent at broadcasting. Supposing that it is applied to large-scale complex projects that many people are involved in, however, there are two problems as follows:

1. The formal relevance (referential relationship) between emails may not necessarily reflect the actual relevance of the content of discussion.
2. Due to the transition of communications via email over time, it is difficult to say that a mailing list is suited for the accumulation and re-use of knowledge.

The reason why the first issue is important is as follows: generally, proper understanding of design rationale is required in design and development activities. To do so, it is important to grasp correctly the general structure of discussion, i.e. the flow of discussion leading to decision making. In order to grasp the structure of discussion in a mailing list, each project member reads emails raising issues and/or responding to them in orderly sequence. Information to indicate the relevance between emails, however, is present only in the response relationship defined with in-reply-to or references headers. Additionally, specific interpretation of these information on the response relationship is left to mailer software, and, in some cases, a failure to add these headers properly may cause a loss of the response relationship between emails. In this way, given that the formal relevance between emails may not necessarily reflect the actual flow of discussion, you must reread the content of multiple emails to construct the flow of discussion as necessary in order to properly grasp the flow. The workload of this task on each project member can not be small, and consequent overlooking of important issues may cause subsequent trouble. That is, chance discovery here can prevent important issues from being overlooked.

Aiming at resolving this problem, various studies are being pursued to structure and represent the flow of discussion to clarify design rationale. An example is the study of structured email typified by The Coordinator [6.13]. By making writers of emails describe structured content beforehand, the structured email approach allows readers to grasp the structure. IBIS (issue-based information system) [6.7] studies structuring discussions in design and development and clarifying design rationale. IBIS records the issues of discussion as issue, the possible solutions to issue as position, and arguments for position as argument. Similarly to IBIS, the QOC language [6.10] is another study of the clarification of discussion and design rationale. The QOC language records the issues using question, option, and criteria. IBIS and QOC do not necessarily aim only at mailing lists. An example of the application of IBIS to actual software development is gIBIS [6.1]. An example of its application to real-time co-operative work is rIBIS [6.6]. While any of The Coordinator, IBIS, and QOC makes a writer themself enter the structure, Murakoshi et al. pursue their studies to automatically extract the structure of discussions taking place in a mailing list through natural language processing techniques, achieving some positive results [6.4]. The HISHO system developed by the Communications Research Laboratory, Ministry of Public Management, Home Affairs, Posts and Telecommunications implements a feature to detect topical turning points and/or topical analysis points through analyses of the flow of discussion, and was applied to news groups on the Internet [6.2].

Next, the reason why the second issue is important is as follows: because communication via email is transient, it is difficult by nature to re-use the content of discussions there as knowledge. The more frequently a mailing list is used, a larger amount of emails must be processed. Therefore, overlooking is more likely to happen, and, consequently, can cause low-level miscommunication with a higher probability.

In order to resolve this problem, it is desirable to accumulate and effectively re-use data exchanged in a mailing list to bring about chance discovery. The traditional approach to re-using an email data resource was to Web-archive the content of all emails by using a utility such as MhonArc and then make a full text search by mails. As previously mentioned, however, such as for mailing lists for design/development projects, one topic often consists of multiple mails and therefore the traditional approach with keyword search by single mails often has difficulty in obtaining necessary information. In order to address this problem, HISHO tried to measure the similarities of threads and then present previous similar topics.

6.2.3 Applications to Actual Design and Development Projects

Kato et al. proposed the integrated design information management system (IDIMS) as a framework for promoting knowledge re-use and failure prevention in design/development projects [6.3]. IDIMS sets 'issue' tags to issues especially important in design, and 'Decision' tags to decisions on issues. Issue and decision tags are intended to mitigate a writer's burden of manual tagging by simplifying issue, position, and argument tags used in IBIS. IDIMS also introduces another tag named 'notice' in order to collect comments and/or knowledge with no direct relation to the issues. Figure 6.2 shows an overview of the re-use environment for mailing lists provided with IDIMS. Kato et al. applied IDIMS to a developer mailing list for IN-

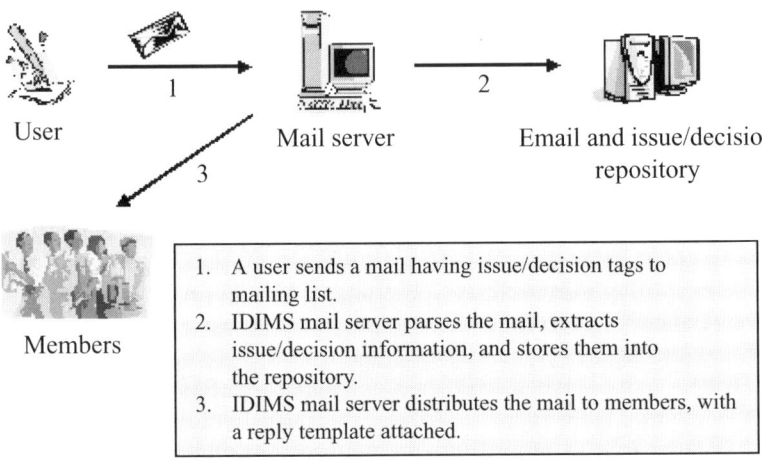

User   Mail server   Email and issue/decision
                                                          repository

Members

1. A user sends a mail having issue/decision tags to mailing list.
2. IDIMS mail server parses the mail, extracts issue/decision information, and stores them into the repository.
3. IDIMS mail server distributes the mail to members, with a reply template attached.

Fig. 6.2. Overview of IDIMS mailing list system

DEX, a small-sized satellite under development in the Space Science Laboratory, to tag and analyze 679 emails posted between April 24, 2001 and October 10, 2001. As a result, 84 issues with no corresponding decisions were found. The majority of them was 56 'no reply' mails with no decisions corresponding to issues. Six 'unknown' mails had no clear description of their issues and therefore the issues indicated by them were indeterminable. As for 'no reply' and 'unkown' mails (62 in total), it is possible that their issues remained unresolved without conclusions. About 10% of all the emails had such unresolved issues; this number is too high to overlook in the satellite-development project which requires exactitude. Unresolved issues are risk factors for the project, and, therefore, discovery of unresolved issues is a chance discovery for the project members. IDIMS could make such a chance discovery from the content of communication via a mailing list.

Figure 6.3 represents a discussion structure for a mailing list by inter mail reply relations (such as in-reply-to and references) implemented in many mailer softwares. In other words, this figure gives you a visualization of formal structure of a mailing list. Figure 6.4, on the other hand, visualizes an actual discussion structure with issue/decision tagging applied to the same mailing list. The node numbers marked in both figures indicate sequence numbers of mails that contain the tag. Comparison of the two figures shows that the referential relationship by mails is certainly different from the actual discussion structure. Especially in case that one mail has multiple issues, even though responses and necessary handling formally

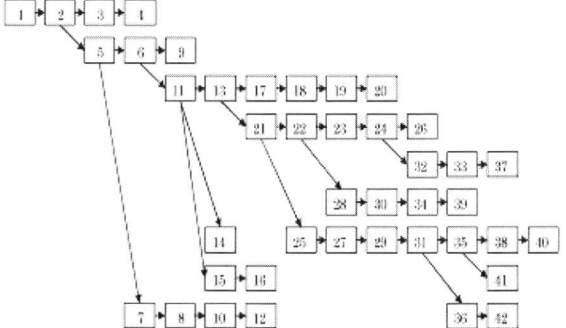

Fig. 6.3. An example of discussion structure by mails

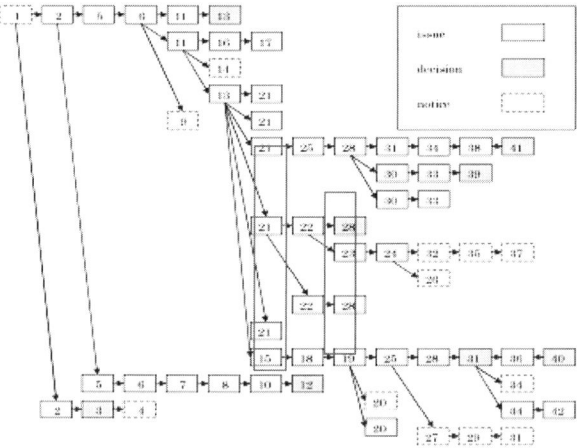

Fig. 6.4. An example of actual discussion structure with the issue/decision tagging

appear to have been made (as shown in Fig. 6.3), some issues, in fact, tend to remain unresolved (as shown in Fig. 6.4). For example, a mail numbered 21 shown in Fig. 6.4 has five issues raised; however, only two of them present their decision and the rest remain unresolved. Kato et al. addressed the reduction of such neglected unresolved issues by implementing additional features of the visualization of issue decision links, automatic search for unresolved issues, and presentation support.

Analysis by Kato et al. also found that discussions already concluded tend to recur. In order to omit these useless discussions, IDIMS introduced search by threads to allow knowledge to be searched by a series of structural units including design knowledge instead of by mails. Evaluation tests by Kato et al. showed that this relevant-thread-searching feature can provide knowledge accumulated in a mailing list also with users who have no special searching know-how.

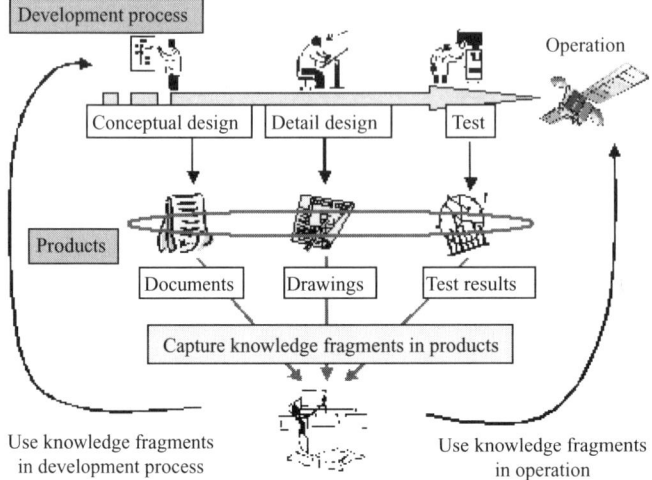

Fig. 6.5. A framework for knowledge recycling

The study by Kato et al. introduced here shows that actual design and development projects do have communication problems such as neglect of unresolved issues regardless of meticulous care by each member and the recurrence of the same discussions. Assuming that communications based on mailing lists that project members daily use are tagged by themselves with minimal workload, IDIMS enables the discussion structure to be clarified with issue and decision tags to help knowledge discovery. This section has only described knowledge discovery from mailing lists and its re-use; however, the actual design/development process involves the occurrence of volumes of "knowledge fragments" which reflect design developers' thinking, knowledge, and experience, and many of them are buried without being leveraged. Kato et al. proposed a knowledge-recycling framework that allows knowledge fragments to be collected, reconstructed, and used from various documents in design development including reports that describe the cause of failures in testing the system, as well as discussions via mailing lists, and put it into practice (as shown in Fig. 6.5).

6.3 Conclusion

Human-to-human communication is an everyday experience that everyone has in his/her daily life. Because it is everyday behavior, we rarely meditate on its purpose every time we do it. Communication in purchasing activities is improperly assumed to be as follows: a customer tells their wants to a sales clerk, who responds to it by providing them with appropriate information. As for discussions within design development projects, we tend to think that every project member should be aware of and share issues to be resolved and proper discussions on how to resolve them should be taking place. Human-to-human communication actually observed, however, is fuzzy and flexible. The communication context is ever-changing. These changes in the context also brings changes in the mental world within persons in communication with each other. Phenomena such as the inconstantness of customers' wants and the unexplainable oblivion of issues intently discussed, as presented in this paper, are never extreme cases. The communication environment is full of risks. However, these risks can be, at the same time, chances. Section 6.2 of this paper showed that persons with high awareness like capable sales clerks can make an effective chance discovery to be at an advantage over their competitors in business. In fact, however, not everyone can be a capable sales clerk, and, therefore, the support for increasing chance discoverers must be provided. In addition, Sect. 6.3 showed that devising the communication environment will probably enable people to discover chances that they would have previously missed.

References

6.1 Conklin J, Begeman ML (1988) gIBIS: A Hypertext Tool for Exploratory Policy Discussion, Proceedings of CSCW'88, ACM, New York, NY, pp.140–152
6.2 Isahara H, Uchimoto K, Ozaku H (1998) Intelligent Network News Reader with Visual User Interface, COLING-ACL'98 Workshop (Content Visualization and Intermedia Representations), pp.12–18
6.3 Kato Y, Shirakawa T, Taketa K and Hori K (2002) Managing Risks in Development of Large and Complex Systems by Knowledge Recycling, Proceedings od AAAI Fall Symposium Series, AAAI Press, Menlo Park, CA, pp.94–98
6.4 Murakoshi H, Shimazu A, and Ochimizu K (2001) Construction of Deliberation Structure Using Mailing List in Cooperative WorkCComputer SoftwareC18(3):19–23 (*in Japanese*)
6.5 Peppers D, Rogers M (1993) The One to One Future, Doubleday, New York, NY
6.6 Rein G, Ellis CA (1991) rIBIS: a real-time group hypertext system, International Journal of Man-Machine Studies, 34:349–367
6.7 Rittel H, Webber M (1973) Dilemmas in a general theory of planning, Policy Sciences, 4:155–169
6.8 Schmitt BH (1999) , Experiential Marketing: How to Get Customers to SENSE, FEEL, THINK, ACT, and RERATE to Your Company and Brands, The Free Press, New York, NY
6.9 Shoji H, Hori K (2002) Creative Communication for Chance Discovery in Shopping, New Generation Computing, 21(1):73–86
6.10 Shum SB, Hammond N (1994) Argumentation-based design rationale - what use at what cost, International Journal of Human-Computer Studies, 40(4):603-652
6.11 Underhill P (1999) Why We Buy: The Science of Shopping, Touchstone Books, New York, NY
6.12 Wind J, Rangaswamy A (2000) Customerization: The Next Revolution in Mass Customization, MSI Report, No00-108
6.13 Winograd T, Flores F (1986) Understanding Computers and Cognition, Addison-Wesley, New York, NY

7. Topic Diffusion in a Community

Naohiro Matsumura[1,2]

[1] PRESTO, Japan Science and Technology Corporation, 2-2-11 Tsutsujigaoka, Miyagino-ku, Sendai, Miyagi 983-0852, Japan
[2] Graduate School of Engineering, the University of Tokyo, 7-3-1 Hongo, Bunkyo-ku, Tokyo 113-8656, Japan
email:matumura@miv.t.u-tokyo.ac.jp

Summary.

People are easily affected by others' comments, especially if they include topics interesting to us. In other words, interesting topics diffuse from person to person in a community. In this chapter, I consider '*influence*' as a unit of diffusion, and propose the influence diffusion model (IDM) to find valuable information such as influential comments, opinion leaders, and interesting terms from the archives of text-based communication. The IDM is applied to the archives stored in the Yahoo!JAPAN Message Boards, and the results of the experimental evaluation are presented.

7.1 Introduction

Business people, especially marketing researchers, are keen to survey consumers to see what type of product or service they wish to receive. Although this sort of analysis is helpful, frequent surveys are needed to follow constantly evolving trends. However, these surveys are becoming increasingly difficult to put into practice, because conducting a survey usually takes a lot of time and is expensive.

Considering the growth of the Internet, it is becoming a feasible idea to survey the Internet. Here, let us focus on the Internet's communication tools such as email, ICQ, chat, or message boards, where we can contact people who may possess valuable information. For example, if we pose a message, 'What do you think of IBM's laptop PCs?' to a message board discussing laptop PCs, we would expect to get a lot of favorable/critical comments. Such comments are extremely valuable for creating fascinating topics stimulating peoples' interest because they reflect users' reputation that is hard to get from the official homepage. In other words, the reputation implies peoples' potential sense of value.

In the field of marketing, those who possess valuable information and have a lot of influence on peoples' decision making are known as *opinion leaders* [7.5]. We can quickly notice new trends before they become established if we can find out what the opinion leaders are saying [7.14].

In this chapter, I aim to find not only opinion leaders, but also influential comments and interesting terms contributing to the discovery of fascinating topics. For this purpose, I propose the influence diffusion model (IDM) in threaded online discussions, where the influence of comments, participants, and terms are defined as the degree of text-based relevance of messages. The IDM is applied to a message board on the Internet, and the results of the experimental evaluation are presented.

7.2 Background

Diffusion research has attracted the attentions of researchers for decades. In the 1940s and 1950s, the media were considered to have a great influence. Katz and Lazarsfeld [7.5] first proposed the 'hypodermic needle model', meaning that the mass media has a powerful influence on a mass audience. Following that, they proposed the 'two-step flow model', where two steps existed in mass-media persuasion: the first step was from the media to opinion leaders, while the second was from the opinions leaders to the followers. Rogers reconstructed the 'two-step flow theory', and proposed the categorization of people by the time they take to adopt new innovations to their own lives [7.10]: innovators, early adopters, early majority, late majority, and laggards. The adopter categories are presently the most widely used in diffusion research. Granovetter argued the importance of weak ties between sparsely knit communities [7.3]. Supposing that opinion leaders play an equivalent role in weak ties, the weak-tie theory can provide a good explanation for the diffusion of influence. Gladwell proposed a new way of understanding why change so often happens as quickly and as unexpectedly as it does [7.2]. He contends that adoption is predicated on the connectors, mavens, and salesmen, and provides explanations of such change.

Shifting our focus into the diffusion on text-based communication, the researches of computer-mediated communications (CMC)[7.6, 7.13] are deeply relevant. Shibanai and Ikeda analyzed the diffusion process of the 'Pentium bug in 1994' by questionnaires [7.12]. Bordia and Rosnow studied rumor-transmission chains by classifying the content of individual messages [7.1]. Kaneko analyzed the comment chain of emails in a mailing list by using network analysis methods to discover central comments and participants [7.4].

These studies mainly focused on the structure of comment chains, not on the contents. In contrast, our approach focuses not only on the structure of comment chains, but also the contents. Therefore, we can analyze the comment chain in detail, and discover specific terms expressing fascinating topics.

7.3 Topic Diffusion in a Community

People are easily affected by others' comments, especially if they include topics interesting to us. In other words, interesting topics diffuse from person to person in a community. Here I consider 'influence' as a unit of diffusion, and show my idea of diffusing influence with some examples.

7.3.1 My Idea of Diffusing Influence

My method for diffusing influence is based on the propagation of terms throughout discussion threads. Hereafter, I explain my idea based on the structure of a message board as a representative example of threaded online discussions.

One feature of a message board is that communication between participants is performed by exchanging comments, i.e. posting new comments or replying to previous comments. My first assumption is that the relation of comments, called the *comment chain*, shows the flow of influence. For example, if comment C_y replies to comment C_x, it is considered that C_y is affected by C_x. Similarly, if person P_y replies to a comment of person P_x, it is considered that P_y is affected by P_x. In these cases, the influence diffuses from C_x to C_y / from P_x to P_y. In this way, the influence diffuses throughout the comment chain.

Another feature of a message board is that comments are written by natural language composed of terms. My second assumption is that a person's idea is expressed and propagated by the medium of terms. Therefore, the process of diffusion of influence is defined as follows.

Definition 1. In text-based communication, influence diffuses along the comment chain by the medium of terms, i.e. words or phrases.

In this study, I define the influence by the degree of terms propagating through the comment chain. For example, if C_y replies to C_x, the influence of C_x onto C_y, $i_{x,y}$, is defined as

$$i_{x,y} = \frac{|w_x \cap w_y|}{|w_y|}, \tag{7.1}$$

where $|w_y|$ denotes the count of terms in C_y and $|w_x \cap w_y|$ denotes the count of propagated terms from C_x and C_y .

In addition, if C_z replies to C_y, the influence of C_x onto C_z via C_y, $i_{x,z}$, is defined as

$$i_{x,z} = \frac{|w_x \cap w_y \cap w_z|}{|w_z|} \cdot i_{x,y}, \tag{7.2}$$

where $|w_z|$ denotes the count of terms in C_z and $|w_x \cap w_y \cap w_z|$ denotes the count of propagated terms from C_x and C_z .

It is considered that the more a comment affects other comments, the more the influence increases. Furthermore, the same can be applied to the influence of participants and terms. Thus, the influence of a comment, participant, or term becomes measurable.

Definition 2. The influence of a comment, participant, or term on the community is measured by the sum of influences diffused from the comment, participant, or term to all other members of the community.

Applying Definition 2 to C_x, the influence (hereafter, let me skip 'to other members of the community') is measured by the sum of the influences diffused from C_x i.e. $i_{x,y} + i_{x,z}$ if the community has three members x, y, and z.

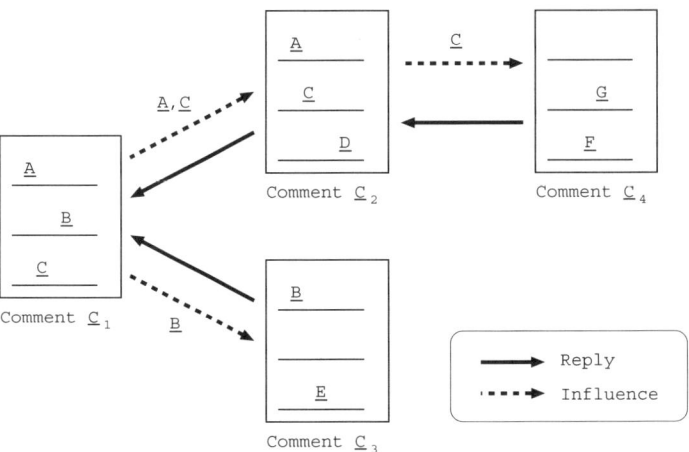

Fig. 7.1. The process of diffusing influence through the comment chain

7.3.2 Measuring the Influence of Comments

Let me measure the influence of comments by using a sample comment chain illustrated in Fig. 7.1, where C_1 is replied to by C_2 and C_3, and C_2 is replied to by C_4. In this case, terms A, C are propagating from C_1 to C_2, term B is propagating from C_1 to C_3, and term C is propagating from C_2 to C_4. Here, the influence of C_1 is calculated as follows.

- **The influence of C_1 on C_2:** the number of propagated terms from C_1 to C_2 is two (A, C), and the number of terms in C_2 is three (A, C, D). Thus, the influence of C_1 on C_2 is $2/3$.
- **The influence of C_1 on C_3:** the number of propagated terms from C_1 to C_3 is one (B), and the number of terms in C_2 is two (B, F). Then, the influence of C_1 on C_3 is $1/2$.
- **The influence of C_1 on C_4 through C_2:** the number of propagated terms from C_1 to C_4 via C_2 is one (C), and the number of terms in C_2 is two (C, F). Considering that the influence of C_1 on C_2 is $2/3$, the influence of C_1 on C_4 via C_2 becomes $2/3 \times 1/2 = 1/3$.

According to Definition 2, the influence of C_1 in Fig. 7.1 is calculated as *(the influence from C_1 to C_2)* + *(the influence from C_1 to C_3)* + *(the influence from C_1 to C_4)* $= 2/3 + 1/2 + 1/3 = 3/2$. Similarly, the influences of C_2, C_3, and C_4 are calculated as $1/2$, 0, and 0 respectively. Therefore, C_1 is selected as the most influential comment in Fig. 7.1.

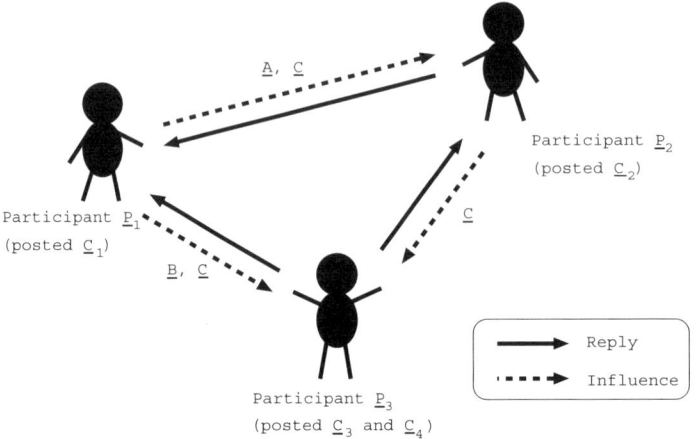

Fig. 7.2. The process of diffusing influence through human network

7.3.3 Measuring the Influence of Participants

Next, let us measure the influence of participants, by assuming that C_1, C_2, C_3, and C_4 in Fig. 7.1 are posted by P_1, P_2, P_3, and P_3 respectively. The relationship between participants, called the *human network*, is illustrated in Fig. 7.2.

Here, the influence of P_1 on P_2 is equal to the influence of C_1 on C_2, while the influence of P_1 on P_3 is the sum of C_1's influence on C_4 via C_2, and of C_1 on C_3. Referring to the above results, the influence of P_1 on P_2 becomes $2/3$, and the influence from P_1 to P_3 becomes $(2/3 \times 1/2) + 1/2 = 5/6$. Therefore, the influence of P_1 is calculated as *(the influence from P_1 to P_2)* + *(the influence from P_1 to P_3)* $= 2/3 + 5/6 = 3/2$. Likewise, the influence of P_2 and P_3 is calculated as $1/2$ and 0, respectively. From these calculations, it is clear that P_1 is the most influential participant in Fig. 7.2.

7.3.4 Measuring the Influence of Terms

Assuming that all terms equally mediate the influence, the influence of each term can be calculated by the sum of influences mediated by the term throughout comment chains. Referring to Fig. 7.1, the influence of A becomes $2/3 \times 1/2 = 1/3$ because A mediates $2/3$ influence together with C. In the same way, the influence of B becomes $1/2$, and the influence of C becomes $(2/3 \times 1/2) + (2/3 \times 1/2) + 1/2 = 7/6$. The influence of other terms (D, E, F) becomes 0. Thus, the most influential term in Fig. 7.1 becomes C.

7.4 Influence Diffusion Model (IDM)

I mathematically formalize the idea of diffusing influence mentioned above as the influence diffusion model (IDM). There are three models included in the IDM: an

IDM for comments, an IDM for participants, and an IDM for terms. In the following sections, I describe each model in detail.

7.4.1 IDM for Comments

Firstly, the influence of a comment C_i is formalized. The influence of C_i diffuses along the comment chain by the medium of terms (Definition 1), and the influence is measured by the sum of influences diffused from C_i (Definition 2). Here, let ξ_i be the comment chain that starts from C_i, i.e. $\xi_i = \{C_i, C_j, C_k \cdots, C_q, C_r \cdots, C_y, C_z\}$ $\{i < j < k \cdots q < r \cdots y < z\}$, and the influence of C_i on C_r becomes $i_{i,r}$. Thus, $i_{i,r}$ is described as

$$i_{i,r} = \frac{|w_i \cap w_j \cap \cdots \cap w_r|}{|w_r|} \cdot i_{i,q}, \tag{7.3}$$

where $|w_r|$ denotes the count of terms in C_r and $|w_i \cap w_j \cap \cdots \cap w_r|$ denotes the count of propagated terms from C_i to C_q. Equation (3) means that $i_{i,q}$ affects $i_{i,r}$ in proportion to the count of propagated terms from C_i to C_r in the count of terms in C_r. Here, let I_{ξ_i} be the sum of influences diffused from C_i in ξ_i. Thus, I_{ξ_i} is described as

$$I_{\xi_i} = i_{i,j} + i_{i,k} + \cdots + i_{i,y} + i_{i,z}.$$

The influence of C_i is defined as the sum of I_{ξ_i} for all comment chains from C_i. Let Ξ_i be all comment chains that start from C_i, and the influence of C_i be D_{C_i}. Thus, D_{C_i} is described as

$$D_{C_i} = \sum_{\xi_i \in \Xi_i} I_{\xi_i}. \tag{7.4}$$

7.4.2 IDM for Participants

I define the influence of participant P to be the sum of the influence of P's comments. Let D_P be the influence of P, and κ_P be a collection of comments posed by P. Then, D_P is described as

$$D_P = \sum_{i \in \kappa_P} D_{C_i}. \tag{7.5}$$

7.4.3 IDM for Terms

The influence of comments is propagated by terms within those comments. Here, let us assume that all terms equally mediate the influence. In the case of C_i described above, $i_{i,r}$ is propagated by the medium of $\{w_i \cap w_j \cap \cdots \cap w_r\}$. Here, let the influence of term $t \in \{w_i \cap w_j \cap \cdots \cap w_r\}$ become $j_{i,r,t}$. Then, $j_{i,r,t}$ is described as

$$j_{i,r,t} = \frac{1}{|w_i \cap w_j \cap \cdots \cap w_r|} \cdot i_{i,r}, \tag{7.6}$$

where $j_{i,r,t}$ is the influence of t between adjacent comments C_q and C_r in ξ_i [1]. Here, let $J_{\xi_i,t}$ be the influence of terms in ξ_i. Then, $J_{\xi_i,t}$, which is measured by the sum of $j_{i,r,t}$ in ξ_i, is described as

$$J_{\xi_i,t} = j_{i,j,t} + j_{i,k,t} + \cdots + j_{i,y,t} + j_{i,z,t}. \tag{7.7}$$

The influence of t is defined as the sum of $J_{\xi_i,t}$ for all the comment chains including t. Let the influence of t be D_t, and all comment chains including t be Ξ_t. Then, D_t is defined as

$$D_t = \sum_{\xi_i \in \Xi_t} J_{\xi_i,t}. \tag{7.8}$$

7.5 Case Study

Here, the result of our preliminary experiment is reported. The IDM is applied to a small part of a comment chain in the Yahoo! JAPAN Message Boards [2]. The small comment chain was composed of 17 comments by 13 participants. The main topic was clothing made of fleece, especially relating to popular colors and Internet shopping. Below, I describe the results and considerations.

7.5.1 Discovery of Influential Comments

The flows of influence between comments are shown in Fig. 7.3, and the top five comments, in the order of diffusing influence (D_C) value, are shown in Table 7.1. The contents and the subsequent development are as follows.

Table 7.1. The top five comments in the order of D_C

Rank	Comment ID	D_C
1	#604	1.504
2	#615	0.574
3	#618	0.375
4	#614	0.337
5	#605	0.237

As you can see in Fig. 7.3, the comment chain started from the top-ranked comment #604. Actually, #604 offered a lot of cues for topics that triggered the other comments. The summary of #604 is as follows.

---

[1] A reader may consider a case having two paths (i.e. comment chains) between i and r, where $j_{i,r,t}$ can not be defined on a single path from i to r via q. However, we ignore such cases because people seldom respond to multiple messages with only one message.

[2] http://messages.yahoo.co.jp/index.html

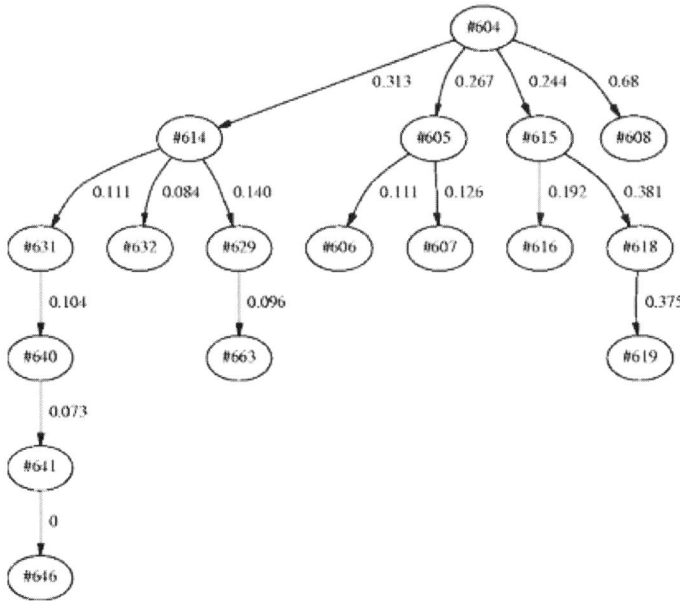

Fig. 7.3. The comment chain used in the case study. Nodes denote the comments, while directed links denote the flow of influence. The numbers beside the links show the values of the diffusing influence

[**Summary of #604**] In the weekend sale with limited stocks of fleece, pink fleeces were sold out in no time. I also bought a pale olive-colored one because I felt the color was unexpectedly beautiful. In my opinion, full-zippered ones were prettier than half-zippered ones.

The topics described in the second-rank comment #615 were also discussed in #618 and #619. This implies that #615 was an attractive and influential comment. The summary of #615's reply to #604 is as follows.

[**Summary of #615**] I am also interested in the pale olive-colored fleece. But I am worried that the color in the posters is pretty different from the color on UNIQLO's website. Which color is close to the real one?

The third-rank comment #618 also described the fleece. From the following summaries of #618 and #619, one can understand that #618 effected great influence on #619.

[**Summary of #618**] The pale olive-color of fleece in posters was different from the color on the UNIQLO website. I confirmed with my eyes that the color in the posters was the same color as the real item.

[**Summary of #619**] I had the same question, too. Thanks to your good information, I was able to decide to buy the pale olive-colored fleece. Thank you.

As shown in Fig. 2, where the fourth-ranked comment #614 was the beginning of the left part of the comment chain, #614's topic attracted participants' interest. The summary of #614 is as follows.

> **[Summary of #614]** Hi, I am a staff member at a UNIQLO shop. In my shop, the pink fleece was sold out in only a day. I think that Internet shopping may be good if you don't have enough time to go. You will not hesitate to purchase by the Internet shopping once you see the original.

The topics in the fifth-ranked comment #605 developed further in #606 and #607. The summary of #605 replying to #604 is as follows.

> **[Summary of #605]** Because of the ease to dress/undress, I think that the full-zippered fleece is more popular than the half-zippered one. I bought two favorite fleeces, which were colored peach and ultra marine. I heard from a UNIQLO staff member that pink fleeces were sold out in only half a day.

Intuitively, the importance of a comment seems to be measured by the number of replies. This simple measure of importance is sometimes true: #604's reply ranking as well as influence ranking were first, as shown in Table 7.1. However, the influence of #614 is less than #615 and #618, although the reply ranking of #614 is higher. This implies that the frequency a comment is replied to is not always important, and this phenomenon is well known in the field of social psychology. Kiesler et al. pointed out that the participants in online communities were apt to reply to slanderous comments that are almost meaningless [7.6].

From these considerations, we can understand that comments of high D_C values supply influential topics that can attract participants' interest and trigger participants' comments.

7.5.2 Discovery of Opinion Leaders

The flow of influence between participants, called the *human network*, is shown in Fig. 7.4, and the top five people in the order of values of diffusing influence (D_P) and their comments are listed in Table 7.2. The comments in Table 7.2 include all the influential comments discussed above. Therefore, the people extracted by D_P can be regarded as opinion leaders.

Interestingly, D_P was not in proportion to the frequency of posting/replying. The D_P of M086, the most frequently posting/replied to participant, was lower than M010 at the bottom of Table 7.2 regardless of the frequency (three times). To investigate the reason, let us look at comments #629, #631, and #641 posted by M086, below.

> **[Summary of #629, #631]** The pink fleece, which I bought last week, is so prominent that I am gazed at whenever I wear it while walking in the street. By the way, my boyfriend is longing to work at the UNIQLO shop.

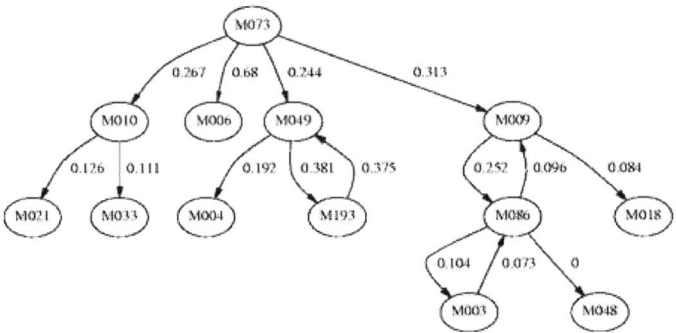

Fig. 7.4. Human networks in the comment chain in Fig. 2

Table 7.2. The top five participants in the order of D_P

Rank	Member ID	Comment ID	D_P
1	M073	#604	1.504
2	M049	#615, #619	0.574
3	M193	#618	0.375
4	M009	#614, #663	0.337
5	M010	#605	0.237

Could anyone give me any information about that? (*#629 and #631 were put together because their contents were almost the same.*)

[Summary of #641] Thank you for your response. Well, it may be right that my boyfriend should begin working part-time at first. What questions are asked at the interview? Is there any recipe that enables him to pass the interview?

The topic of pink fleece in #629 and #631 was replied to in one comment. However, the topic was already old-fashioned because it had been discussed enough in the previous comments #604, #605, and #614. In the IDM, the influence of a comment is reduced if the terms are not quoted. The low influence of #629 and #631 simply reflected the situation, i.e. the interest of participants had already moved on to other topics. The topic of working at UNIQLO in #629, #631, and #641 also did not develop very far. This means that most of the people were interested in clothing rather than working at a UNIQLO shop. From these investigations, we can consider that the comments of M086 did not arouse the interest of participants very much, and therefore M086 was not a suitable opinion leader. Talkative people did not necessarily become opinion leaders.

7.5.3 Discovery of Interesting Terms

The top 10 terms in the order of values of diffusing influence (D_t) are listed in Table 7.3, and the top 10 frequent terms are listed in Table 7.4. Summarizing the topics in the comment chain, the discussed topics are as follows.

– Pink and pale olive-colored fleece.
– Full-zippered and half-zippered fleeces.
– The difference between fleece colors of original garments and those appearing on posters and the Internet.

Comparing the terms in Table 7.3 with those in Table 7.4, it is apparent that the former terms were more specific in expressing the central topics of the comment chain, such as fleece, pink, and pale olive. On the other hand, the latter terms tended to be too general, such as buy, good, see, and sale. Summarizing the above discussions, I can conclude that the terms with high D_t were quite fascinating, and could be triggers for deep discussions. The terms with high frequency but low D_t were too common to understand the participants' potential sense of value.

Table 7.3. The top ten terms in the order of D_t

Rank	Term	D_t
1	Shop	0.172
2	Color	0.158
3	Fleece	0.145
4	The internet	0.145
5	Shopping	0.127
6	Original	0.127
7	Pink	0.124
8	Pale olive	0.099
9	Poster	0.061
10	UNIQLO	0.056

7.6 Experimental Evaluations

The case study was evaluated subjectively by the content-analysis approach. Here, the IDM is objectively evaluated by using *precision* (the ratio of correctly extracted targets to extracted targets) and *recall* (the ratio of correctly extracted targets to the targets that should be extracted) measurement [7.11], which are traditionally used in the field of information retrieval.

For the experiment, a message board was found on the Internet where participants were actively discussing the clothing of UNIQLO.com (http://www.uniqlo.com), which is one of the most popular brands of clothing in Japan. During the period from December 3, 1999 to July 27, 2001, the message board received 1161 comments

Table 7.4. The top ten frequent terms

Rank	Term	Freq.
1	Color	26
2	Buy	10
3	UNIQLO	9
4	Good	7
4	Work	7
4	Pink	7
4	Shop	7
4	See	7
9	Popular	6
10	Sale	5

by 430 participants. All the comments were downloaded and had the IDM applied to them. Note that the comments were written in Japanese, and there is no space between words in Japanese. Therefore, all the comments were converted into morphemes [7.8] in advance. In the following case studies, experimental results were translated from Japanese into English as the case may be.

Before experiments, all the downloaded comments were read thoroughly. Then, based on our intuition, 28 influential comments, nine opinion leaders, and 56 interesting terms that should be extracted were picked up. I first extracted 30 comments by the IDM, and, as a comparison, I also extracted 30 comments by reply-index (RI), which extracts comments that have many replies. RI is a widely used approach for ranking popular comments. The precision and recall values are shown in Table 7.5.

Table 7.5. Precision and recall for extracted comments

	IDM	RI
Precision	0.53	0.40
Recall	0.57	0.43

Next, 10 participants were extracted by the IDM. For comparison, 10 participants by RI, and 10 participants by post index (PI), which extracts frequently posting participants, were also extracted. PI is a conventional approach to extract talkative participants. The precision and recall values are shown in Table 7.6.

Table 7.6. Precision and recall for extracted participants

	IDM	PI	RI
Precision	0.70	0.60	0.50
Recall	0.78	0.67	0.56

Finally, 100 terms by IDM, TF (term frequency) [7.7], and TFIDF (term frequency inverse document frequency) [7.11] were extracted. TF and TFIDF are the most widely used approaches for information retrieval. The corpus used by TFIDF was produced from electronic articles in Mainichi newspapers in 1998 and 1999. A total of 236, 600 articles composed of 164, 790 word types were collected. The result of the precision and recall values is shown in Table 7.7.

Table 7.7. Precision and recall for extracted terms

	IDM	TF	TFIDF
Precision	0.45	0.27	0.34
Recall	0.8	0.48	0.61

In any case, it is clear in Tables 7.5, 7.6, and 7.7 that precision and recall values from the IDM were higher than the values of the other methods. These results mean the superiority of the IDM over other traditional approaches.

7.7 Conclusion

In this chapter, I proposed a new way of finding valuable information such as influential comments, opinion leaders, and interesting terms from threaded online discussions. The proposed approach (IDM) is one method of formalizing the diffusion process known as the 'word-of-mouth' phenomenon. Experiments showed that information on online communities' centers of attention was obtained more accurately by the IDM than by other traditional approaches.

Chance discovery begins from identifying rare or novel events with potentially significant consequences for decision making[7.9]. From this point of view, the IDM can be used as a tool for discovering chances by identifying and understanding influential comments, opinion leaders, and interesting terms.

References

7.1 Bordia P, Rosnow RL (1998) Rumor Rest Stops on the Information Superhighway: Transmission Patterns in a Computer-Mediated Rumor Chain, *Human Communication Research* 25:163–179
7.2 Gladwell M (2000) *THE TIPPING POINT: How Little Things Can Make a Big Difference*, Little Brown & Company, Boston, MA
7.3 Granovetter M (1973) Strength of Weak Ties, *American Journal of Sociology*, 8:1360–1380
7.4 Kaneko I (1996) The Great Hanshin-Awaji Earthquake and Network Organization Theory, *Proceedings of Innovative Urban Community Development and Disaster Management*, pp.233–241
7.5 Katz E, Lazarsfeld PF (1955) *Personal Influence*, The Free Press, New York, NY

7.6   Kiesler S, Siegel J, McGuire TW (1984) Social Psychological Aspects of Computer-Mediated Communication, *American Psychologist*, 39:1123–1134

7.7   Luhn HA (1957) Statistical Approach to the Mechanized Encoding and Searching of Literary Information, *IBM Journal of Research and Development* 1(40):309–317

7.8   Matsumoto Y, Kitauchi A, Yamashita T, Hirano Y (1999) Japanese Morphological Analysis System ChaSen version 20 Manual, *NAIST Technical Report, NAIST-IS-TR99009*

7.9   Ohsawa Y (2002) Chance Discoveries for Making Decisions in Complex Real World, *New Generation Computing*, 20(2):143–163

7.10 Rogers EM (1962) *Diffusion of Innovations*, The Free Press, New York, NY

7.11 Salton G, McGill M (1983) *Introduction to Modern Information Retrieval*, McGraw-Hill, New York, NY

7.12 Shibanai Y, and Ikeda K (1995) 'Buggy' Pentium Inside! How the News Diffused in the Networked World, *Proceedings of IEEE Workshop on Networked Relations*, pp.175–188

7.13 Siegel J, Dubrovski V, Kiesler S, McGuire TW (1986) Group Processes in Computer-Mediated Communication, *Organizational Behavior and Human Decision Processes 37*, pp.157–187

7.14 Wakefield J (2001) Catching a Buzz, *Scientific American*, November 2001, pp.22–23

8. Dimensional Representations of Knowledge in an Online Community

Robert McArthur and Peter Bruza

Distributed Systems Technology Centre, Brisbane, Australia
email: mcarthur@dstc.edu.au

Summary.

   Chance discovery in online communities is the serendipitous meeting of two people with a background or interest in common. It is a solution to some problem that the community has, but that solution must come from without. In this chapter, we separate the area into three facets, chance discovery of online communities, between communities, and within a community. This separation is adequate to capture most contemporary research. We examine and illuminate the technological case where computer systems have been designed to actively assist humans in the discovery process.

8.1 Introduction

Chance discovery in online communities has many facets. It is the serendipitous meeting of two people with a background or interest in common (the interest being subsidiary to the community's *raison d'être*). It is a solution to some problem that the community has, but that solution must come from without.

   In this chapter, we separate the area into three facets:

1. chance discovery *of* online communities;
2. *between* communities; and
3. *within* a community.

This separation is adequate to capture most contemporary research. To further narrow the field, we examine and illuminate the technological case where computer systems have been designed to *actively* assist humans in the discovery process.

   Before discussing the substantive topic, it behooves us to place a stake in the ground and outline our definition of online community. Our prior research [8.22] defines a conceptual model of online community identifying people and 'glue' as the primary components. The 'glue' is intergroup and interpersonal. Intergroup 'glue' is a combination of the purpose of the community, the commitment by the members, the context created in and by the community, and the infrastructure to support the community. Interpersonal 'glue' is, at least, the personal ties of the members. The more 'glue', the more the group can be identified as an online community. Preece [8.28] and Whittaker et al. [8.36] were helpful in refining the definition.

   The following discussion describes each of the facets and notes applicable recent work. A full bibliography is not possible in the available space. Therefore, only very recent work is highlighted, especially from the special issue of the Communications of the ACM of April, 2002 (*Supporting Community and Building Social*

Capital), desirous of providing a taste of contemporary research, and that linking each contribution to a facet will assist in understanding and situating the facets.

The first facet, discovery of online communities, can take the form of finding a community where none (seemingly) exists: that is, bringing together people into a community; or it may be in unearthing a community that does exist – an extension of the information-retrieval problem. Flake et al. [8.10] define a community on the Web as a set of Web pages that link to more pages in the community than to pages outside of the community. A graph-theory algorithm, maximum flow/minimum cut, is applied to determine Web communities that have not been formally identified. Their work uses only the linkages of the structure of pages, ignoring substantive content, to determine community.

Kumar et al. [8.17] also search for communities on the Web, reporting success in finding nascent communities with characterizable themes of which 65% were still in existence 18 months later. They use co-citation of URLs (actually URL subsets) to indicate that two pages are in the same community. The authors are very careful to eliminate existing communities, or even communities which are just starting – *"The underlying principle is that if many of the fans in a core come from the same Web-site, this may be an artificially established community serving the ends (very likely commercial) of a single entity, rather than a spontaneously emerging Web community."*. Interestingly, in contrast to Flake et al., they believe *'Web sites that should be part of the same community frequently do not reference one another. In well-known and recognized communities, this happens at times for competitive reasons and at others because the sites do not share a point of view... Even pages that would otherwise point to one another frequently do not – simply because their creators are not aware of each others' presence. '*.

To contrast these examples of our first facet where chance discovery clearly takes place, Hiltz and Turoff [8.15] summarize 'learning networks', a potential form of community, as *"... groups of people who use the Internet and Web to communicate and collaborate in order to build and share knowledge"*. However, just because people are finding something easier or better by being part of an online community or learning network, it may be that the information is *clearly* available inside the community though not outside. No chance discovery need take place.

An example of the second facet, discovery between communities, occurs when two communities are discussing a similar topic, unknown to each other. Matsumura et al. [8.19, 8.20] define community as people coming together with a shared interest, and that a Web page supported by, or linked to, the community satisfies the 'interest'. The Web is thus a mirror to humanity's path in the physical world, and they apply the *KeyGraph* algorithm to a search engine's result set to uncover trends in the communities, and thus the physical world. *KeyGraph*, one of the few techniques that are based on using the substantive content of the community, is described in more detail in Chap. 18.

Swanson and Smalheiser [8.32] present an interesting example of discovery between communities. Examining the medical literature by hand, they found a relationship between Raynaud's disease and a hitherto unknown cure, fish oil [8.31].

The two literatures (or communities) – that of the disease and that surrounding fish oil – were 'non-interactive': they seldom, if ever, cited each other. Other authors, e.g. Weeber et al. [8.34], have attempted to re-create this discovery automatically. Although the content of the community is used, and inference is done, it is debatable whether this constitutes *chance* discovery.

The third facet, discovery inside online communities, could be discovery of information that has been hidden (purposely or not), discovery of people, processes, etc. An example of this facet is described in Chap. 9, based on the work from this chapter.

Matsumura et al. [8.21] examine mailing list messages to discover influential comments that are stimulating people's interest. A model called *influence diffusion*, used on small data sets, measures the diffusion of words (or phrases) by summation of the diffusion of the word to all other messages in the list. However, the assumption that "*people's idea* [sic] *is expressed and propagated by the medium of terms* " is not supported by evidence – using 'concept' rather than 'term' may be more appropriate but make the model much harder to implement; also the simple summation function is not justified and is not compared with other, equally likely, functions.

Smith [8.30] notes " *Like many related conversational media, including email list and Web discussion boards, the problem is often not finding others who share your interests. Instead, the challenge is dealing with too much content of mixed value.*". That is, the problem is not the first facet, but the third. Smith, who also co-edited an important work about online communities [8.33] and is not likely to misjudge the difference between many Usenet groups (as used in [8.30]) with an online community, states that "*A key finding of collective action studies shows that mutual awareness of other participant's histories and relationships is critical to a cooperative outcome.*". While the study is only about applying metrics to Usenet groups, leaving the inferencing to the human user, it is an interesting first step in collecting information about the community. For example, "*The thread tracker report* [displays the 40 largest threads in a group] *often captures topics of broad interest to the community and rarely displays subjects that suggest spam or even off-topic subjects.*".

Briefly, two further examples of our third facet are Erickson et al. [8.9], who note that we can discover things from the subtle clues about the presence and activities of others. The social proxy section of their *Babble* system has some aspects of chance discovery: users report "*noticing a crowd 'gathering' or 'dispersing'. . .*". Although not presenting a solution, Nardi et al. [8.23] note two problems for which chance discovery can be of assistance: finding potential new members of social networks, and "*obtaining awareness information for distant contacts.*".

It is remarkable how much techniques that do not examine the substantial content of the community can accomplish. However, the limits imposed by ignoring what is being communicated and how is it being communicated, are both possible to, and necessary to, overcome.

The remainder of this chapter describes a theoretical underpinning of how to collect and process one form of the content of a community, *conversational impli-*

cature . Conversational implicature is a form of tacit knowledge that can be mined from the substantive content of the utterances between members of a community. It can be envisaged as a set of associations to relevant terms which, while being implied by the utterances in question, are not explicitly stated. In other words, conversational implicature is shared, though un-stated, context between the individuals involved around the utterances.

8.2 The Semantic Context Within Utterances

Consider the situation when Peter meets Rupert in the hall of a technology research institution[1]. Peter utters to Rupert, 'How is it going with John?' To an outsider, this utterance could mean many things, but Rupert can situate the utterance within the background knowledge shared with Peter. For example, Rupert infers that 'John' refers to 'John Smith', and is aware that John Smith works for Microsoft and is interested in licensing the product called GuideBeam. Such lifting of background context in relation to an utterance has been termed *conversational implicature* [8.13]. Conversational implicature is any meaning implied or understood from an utterance which goes beyond what is strictly expressed or entailed. Particularized implicatures are those holding only within a specific occasion or situation. Generalized implicatures hold, in principle, whenever an utterance is uttered. Our concern is how particularized implicatures, like the ones above, can be mined from a small, coherent set of utterances within a given situation.

Grice [8.13] posited that conversational implicature is fashioned out of the following:

- the linguistic meaning of the utterance;
- contextual information (i.e shared background or general knowledge);
- the utterance accords with the co-operative principle – that is, the utterer and receiver of the utterance are mutually engaged in an interaction of benefit to both.

We restrict our attention to utterances governed by the co-operative principle as it ensures a sufficient level of content and coherence in the underlying communication for the mining of conversational implicature.

Grice identified four categories in which various maxims underlie the co-operative principle:

1. Relation: the utterance should be relevant.
2. Quantity: the utterance should only be as informative as required.
3. Quality: the utterer only utters what (s)he believes to be true and supportable.
4. Manner: the utterer should be brief, orderly, and avoid obscurity and ambiguity.

What theories and techniques are applicable for the mining of conversational implicature from utterances? Perry [8.26, 8.27] introduces three levels of semantic context that we regard as being applicable to utterances from a theoretic perspective,

---

[1] We consider that communities of practice [8.35], as well as more informal communities, exist within most if not all institutions and companies.

while at the same time providing some pragmatic clues as to how utterances may be processed automatically. The first level is *presemantic context* which is needed to render a syntactic evaluation to an utterance. For example, 'John' falls within the syntactic category of being a 'first name of a person'. More generally, presemantic context involves determining what are nouns, verbs, etc. in an utterance. Syntactic ambiguities can arise: for example, 'Jill saw her duck under the table.'. Is 'her' an indexical or possessive pronoun? Such ambiguities are fairly rare in sets of utterances online, perhaps due to the 'manner' category underlying Grice's co-operative principle. Observe that even though an utterance may be ambiguous at the surface level, it may not be ambiguous in the light of shared background knowledge. Pragmatically, presemantic context can be computed using part-of-speech (POS) tagging technology.

The second level of context is referred to as *semantic context*. This involves attaching meaning to syntactic structures . Anaphora resolution is an example of this. Assume during presemantic context that 'her' has been identified as a indexical and not a possessive pronoun; contextual information can be used to assign a referent to the indexical to establish its meaning. For example, when 'I' is used in email, the author of the email can be assigned as the referent. It should be mentioned, however, that automatic anaphora resolution remains a thorny problem.

Semantic context deals with meaning. Our position is that the representations of information manipulated in the data-mining process should correlate with the representations used in human processing. That is, we are firmly in the semiotic-cognitive camp with respect to meaning. In particular, we adhere to the geometric metaphor of information representation advocated by Gärdenfors [8.14]. In our case, the geometric metaphor is realized by representing words as vectors, similar to models such as hyperspace analog to language and latent semantic analysis (each described later). Both of these models are numeric approaches to semantic representation, and originate from cognitive science.

The third level of Perry's context is termed *postsemantic context* . We agree with Penco [8.25] that postsemantic context is what is taken for granted in the linguistic interchange of a community. Conversational implicature is a particular manifestation of postsemantic context arising from utterances. Postsemantic context is closely related to the notion of tacit knowledge that is being investigated within the fields of online community and social information. These fields provide potential applications for the mining of utterances for conversational implicature.

Tacit knowledge is one representation of organizational knowledge and memory (e.g., [8.35, 8.16]). This knowledge is the residue of the processes and rationale of the organization at any particular point in time. It has been, with varying degrees of success, captured through formal knowledge-sharing systems such as distributed document repositories, but is now often seen as something that is developed and nurtured through the networking of people comprising different communities of practice [8.24, 8.35]. For example, when a pivotal person in an organization is suddenly not available, the ability to mine implicatures from their email utterances would be vital to determining to whom the person was talking, and about what. Secretaries may assist with the former, but only through examination of the physical artifacts of collaboration and discussion can the latter be attempted.

Section 8.3 deals with how words can be represented from a semiotic-cognitive perspective in the form of vectors representing aspects of both presemantic and semantic context. The next chapter details techniques for the mining of conversational implicature using these vector representations , and illustrations of the vector representations and a discussion of the performance of the mining techniques.

8.3 Representation of Utterance

Utterances must have a representation to facilitate the process of mining for conversational implicature. In light of the above discussion on presemantic and semantic context, and in accord with our semiotic-cognitive stance, we advocate representing words in utterances as dimensional structures . These are the basic carriers of the meaning of the word in question but, in addition, the dimensional structures will have presemantic (i.e. syntactic) information embedded. In this section, the automatic derivation of the dimensional structures from utterances will be detailed. The next section will show how they can be used for distilling conversational implicature.

In the example given in the introduction, Peter and Rupert share some background. Freyd [8.11] provides a psychological-theoretic frame of a viable communal (i.e. shared) knowledge representation:

"What seems common to most of the main approaches to semantic is an assumption that values of semantic components, or features, are critical to word meaning. What is relevant to shareability theory is that a smaller number of features seem to be used than the number of words.". (pp. 195-6)

"I am arguing that a dimensional structure for representing knowledge is efficient for communicating meaning between individuals. That is, a small-dimensional structure with a small number of values on each dimension is argued to be especially shareable, which might explain why such structures are observed.". (pp. 198-9)

Freyd's suppositions on the dimensional nature of shared knowledge are compatible with a recent, three-level model of cognition by Gärdenfors [8.14]. How information is represented in this model varies greatly across three different levels: connectionist, conceptual, and symbolic. It is the conceptual level that is of relevance to this chapter.

Within the conceptual level, properties and concepts are represented geometrically as points or regions in a space of dimensions. For example, the property 'color' can be represented as a region within the ternary space defined by the dimensions hue, chromaticity, and brightness. Gärdenfors takes the position that the meanings of words come from conceptual (i.e. dimensional) structures in people's heads. In addition, he adopts a socio-cognitive position that the meanings emerge from the conceptual structures harbored by individual cognition together with the linguistic power structure within the community. Of significant note is his adoption of Freyd's supposition: social interactions will constrain these conceptual structures. This has implications for computer-based representations because it may mean that relevant dimensions are not represented, or the value in a dimension may not be weighted sufficiently.

It is worth mentioning in passing that this constriction of the dimensional structure by the individual for social interaction relates closely to Grice's category and maxims of quantity mentioned in the introduction. We tend to economize our utterances, for example, by use of anaphora and liberal use of abbreviations made permissible by shared background. Why utter 'John Smith of Microsoft', when Peter already knows that Rupert knows about John and who he is. Therefore, Peter utters the more economical expression 'How's it going with John?' The constriction of conceptual structures, which are, after all, cognitive structures, may well be another facet of cognitive economy which has recently been posited as a major factor in human (abductive) reasoning [8.12].

To bridge the gap between cognitive dimensional structures and actual computational representations, we propose using a variant of hyperspace analog to language (HAL) [8.6]. HAL produces vectorial representations of words in a high-dimensional space that seem to correlate with the equivalent human representations. For example, word associations computed on the basis of HAL vectors seem to mimic human word-association judgments. HAL is "*A model that acquires representations of meaning by capitalizing on large-scale co-occurrence of information inherent in the input stream of language.*". It "*...correlated with lexical decision latencies from a word priming task*" and "*...simulations using HAL accounted for a variety of semantic and associative word priming effects that can be found in the literature...and shed light on the nature of the word relations found in human word-association norm data.*". In short, HAL vectors seem to be promising computational representations of word meanings from a semiotic-cognitive perspective.

This chapter develops upon the desire of Burgess et al. [8.6] to explore the nature of the dimensional spaces created in HAL, by extending the model with presemantic information in the form of syntactic structures (index expressions), part of speech, and basic anaphora resolution.

8.3.1 Vector Creation

The basic carriers of meaning are the vector representations of words in the utterance. These vectors are input into the mining process for conversational implicature. Figure 8.1 depicts how vector representations of words within utterances are derived from email messages.

Part of speech (POS)
POS (part of speech) tagging is a computationally efficient means of mapping arbitrary tokens into syntactic classes, determining basic linguistic information from a corpus. It is the means by which presemantic context can be automatically gleaned from utterances. The technology has matured to achieve high levels (95%+) of precision [8.8]. It is gathered by various methods (rule-based, probability-based, and memory-based being most common) all of which add part of speech tags–noun, verb, pronoun, etc.–to the original text.

Eric Brill's [8.3, 8.2] freely available rule-based POS tagger was used. Although it is slow, and perhaps not as accurate as some modern, commercially available taggers, it is well known and initially seemed adequate.

Index expressions

The second aspect of presemantic context is the syntactic structure of utterances. We employ index expressions to realize these syntactic structures. Index expressions are tree-based syntactic structures whereby the nodes correspond to terms and the edges correspond to connectors between terms. The language of index expressions was originally developed for the automatic parsing of document titles and abstracts for use in information retrieval [8.1, 8.4, 8.5].

The advantage of the index expression language is that its simple grammar allows for efficient parsing:

$$E \to T\{C\ E\}^*,$$

where T corresponds to terms like nouns and adjectives, and C corresponds to connectors between terms. The connectors are a predefined set of prepositions.

For example, the index expression corresponding to the text '*The effects of spreading pollution on the population of Atlantic salmon*' is depicted in Fig. 8.2.

During the parsing process, articles like 'the', 'a', etc. are ignored and the null connector '$\sim$' is identified as an unspecified relationship between terms. The key to creating the tree structure is the intuition that some connectors bind terms more strongly than others. For example, 'of' is a strongly binding connector, so when it is encountered during parsing, the parse tree is deepened. Conversely, connectors such as 'on' are less strongly binding, so such connectors lead to the broadening of the parse tree. For more details of index expressions, the reader is referred to [8.4].

As the index expression language was defined to parse document titles and abstracts, which normally feature sentences in the passive form, the language is not expressive enough to cover the range of expression appearing in arbitrary utterances. In particular, utterances are often in the active form which features a greater variety of connectors via various verb forms. In addition, the index expression language does not explicitly represent presemantic context. We therefore extended the index expression language to accommodate both issues.

In order to highlight these extensions, the sample sentence of Fig. 8.2 is depicted as an extended index expression in Fig. 8.3. Observe that the part-of-speech information is kept with each term. Nodes are important information-bearing terms (typically nouns), which may have qualifiers associated with them. Qualifiers are amounts (e.g. 'a', 'the', 'five', '12') or (e.g. 'Atlantic', 'blue', 'large'). Extended index expressions feature a much larger class of connectors which are basically drawn from propositions, verbs, or verb compounds. An example of a connector based on a verb compound is 'of spreading'. Verb compounds are identified using POS information according to the following rules:

– connector followed by imperative verb, e.g. 'to make' (join the verb to the connector forming a new connector). An imperative verb is a term tagged as VB,

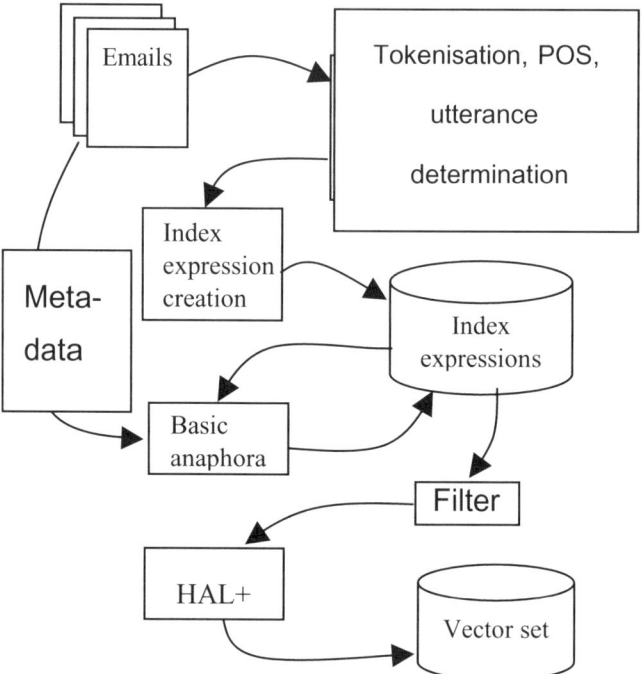

Fig. 8.1. Representing words in utterance as vectors

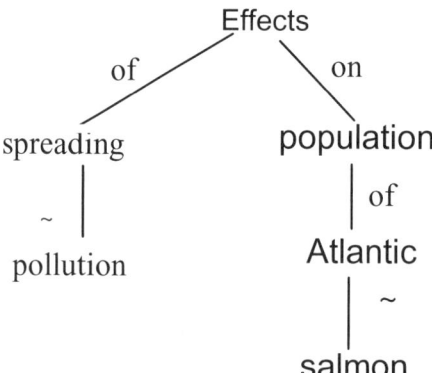

Fig. 8.2. Example of an index expression

VBD, VBN, VBP, VBZ, MD, HV, HVD, or HVZ, or is a term 'have', 'had', 'having', 'owns', or 'owned'.

- connector verb followed by connector, e.g. 'come from' (join both together to form a new connector). A connector verb is a term tagged as VB, VBD, VBN, VBP, or VBZ.
- present participle followed by imperative verb, e.g. 'having come' (join both together forming a new connector). A present participle is a term tagged as VBG. See above for the definition of imperative verb.
- two connectors together, e.g. 'to be' (join both connectors forming a single new connector).
- connector followed by a determiner then a connector, e.g. 'above these during' (remove the determiner). A determiner is a term tagged as DT.

These rules are applied recursively *up* the index expression tree.

Note that index expressions are not the only means for representing syntactic structures. Other syntactic representations could also be employed in their place, for example link grammars [8.29].

Extended index expressions are syntactic structures used to represent presemantic context in utterances. The next stage, which involves introducing aspects of semantic context, is the resolution of anaphora.

Basic anaphora resolution

Anaphora is the co-reference of one expression with an antecedent (*cataphora* is co-reference with a following expression) [8.37]. The antecedent provides the information necessary for interpreting the expression. An example is between the two sentences: 'A well-dressed man was speaking. He had a foreign accent.' The term 'He' in the second sentence is an anaphoric reference to the 'well-dressed man' in the first sentence.

Anaphora is common in utterances, and in email in particular. We do not attempt full anaphora resolution but implement an extremely basic algorithm: replace references to 'I' or 'me' with the sender of the email, and references to 'you' or 'your' with the receiver; these are part of the email meta-data and easily accessible.

No other anaphoric references are as easily determined as these, so terms such as 'he', 'we', 'they', 'it', etc. are left unchanged. We adopt this conservative approach as imprecise anaphora resolution would pollute the vector representations of some words with spurious dimensions.

HAL

Up to this point, the exposition of information representation has centered largely upon aspects of presemantic context. The second level of Perry's three levels will now be addressed, namely semantic context.

A human encountering a new concept derives the meaning via an accumulation of experience of the contexts in which the concept appears. This opens the door to 'learn' the meaning of a concept through how a concept appears within the context of other concepts. Following this idea, Burgess and Lund developed a representational model of semantic memory called hyperspace analogue to language (HAL), which automatically constructs a dimensional semantic space from a corpus

of text [8.7, 8.18, 8.6]. The space comprises high-dimensional vector representations for each term in the vocabulary. Given an n-word vocabulary, the HAL space
is a $n \times n$ matrix constructed by moving a window of length l over the corpus
by one word increment ignoring punctuation, sentence, and paragraph boundaries.
All words within the window are considered as co-occurring with each other with
strengths inversely proportional to the distance between them. After traversing the
corpus, an accumulated co-occurrence matrix for all the words in a target vocabulary is produced. Note that the word pair in HAL is direction sensitive, i.e. the
co-occurrence information for words preceding every word and co-occurrence information for words following it are recorded separately by its row and column
vectors. By way of illustration, the HAL matrix for the example text 'The effects of
spreading pollution on the population of Atlantic salmon' is depicted in Table 8.1
using a five-word moving window ($l = 5$):

Table 8.1. Example of a HAL matrix

	The	eff	of	spr	poll	on	pop	Atl	sal
The		1	2	3	4	5			
eff	5								
of	8	5		1	2	3	5		
spr	3	4	5						
poll	2	3	4	5					
on	1	2	3	4	5				
pop	5		1	2	3	4			
Atl	3		5		1	2	4		
sal	2		4			1	3	5	

This table shows how the row vectors encode preceding word order and the column vectors encode posterior word order. Our pilot studies revealed that it was not
useful to preserve order information, so, for our purposes, the HAL vector of a word
is represented by the addition of its row and column vectors. As an example of a
HAL vector derived from a large corpus, consider part of the normalized HAL vector for '*superconductors*' computed from a corpus of Associated Press news:

superconductors = $<$ U.S.:0.11 american:0.07 basic:0.11 bulk:0.13 called:0.15 capacity:0.08
carry:0.15 ceramic:0.11 commercial:0.15 consortium:0.18 cooled:0.06 current:0.10
develop:0.12 dover:0.06 electricity:0.18 energy:0.07 field:0.06 goal:0.06 high:0.34
higher:0.06 improved:0.06 japan:0.14 loss:0.13 low:0.06 make:0.07 materials:0.25 new:0.24
require:0.09 research:0.12 researching:0.13 resistance:0.13 retain:0.06 scientists:0.11
semiconductors:0.10 states:0.11 switzerland:0.06 technology:0.06 temperature:0.48
theory:0.06 united:0.10 university:0.06$>$
This example demonstrates how a word is represented as a weighted vector whose
dimensions comprise other words. The weights represent the strengths of association between 'superconductors' and other words seen in the context of the sliding
window: the higher the weight of a word, the more it has lexically co-occurred with
'superconductors' in the same context(s).

The quality of HAL vectors is influenced by the window size; the longer the window, the higher the chance of representing spurious associations between terms. Burgess et al. used a size of 10 in their studies [8.6]. In addition, it is sometimes useful to identify the so-called *quality properties* of a HAL vector. Quality properties are identified as those dimensions in the HAL vector which are above a certain threshold (e.g. above the average weight within that vector).

Burgess et al. were able to demonstrate the cognitive compatibility of HAL vectors with human processing via a series of word matching and word similarity experiments with notable results with respect to the semantic association between words [8.18]. Therefore, HAL vectors would seem to be a promising candidate for representing semantic context. However, there are a number of areas of concern regarding the employment of HAL vectors for deriving conversational implicature. Firstly, utterances provide a small amount of data from which to mine. Burgess et al. used a corpus of 160 million words for their experiments with HAL. They did note, however, that even though the dimensional space was very high, the salient dimensions were much fewer in number:

"At present we think that 100–200-dimensional space is an accurate estimate of what is required for simulating human memory. For some effects, as few as 10 vector elements can account for a significant amount of variance... In shortening the vectors, it is important to keep the vector elements that are most 'informative'."[8.7].

On the basis of this, it would seem that using HAL on small data sets would be justified provided that salient dimensions can be weighted accordingly.

The second area of concern is HAL's lack of explicit representation of presemantic context. Even though it has been shown how HAL can be used to derive some aspects of presemantic context, these results rely on large amounts of data. For this reason, we have enhanced the HAL representation with aspects of presemantic context explicitly via the use of POS and index expressions, as described earlier. More details of these enhancement will be provided shortly.

Finally, HAL was designed for finding global co-occurrence rather than a mixture of global and local co-occurrence, as we are performing. Burgess and Lund speculated: '*We suspect that global co-occurrence models, more so than local co-occurrence models, will better capture the richness of cognitive and language effects that are important to the comprehension process.*' [8.7]. Global co-occurrence refers to vector representations embodying word associations derived in the context of a whole corpus. For example, the vector representation of the word 'bank' may contain dimensions corresponding to both the financial sense of the word and the river-bank sense. In other words, global co-occurrence models like HAL produce an *accumulated* vector representation of a word summed across the various contexts (windows) in which the word appears. Small sets of utterances may not allow enough accumulation across windows to produce 'good' vector representations of words. Our assumption is that the use of presemantic context can reduce the reliance on larger amounts of data to produce 'good' vector representations.

We modified HAL because of the type of associations that we were interested in. In the example from the introduction, we are interested in the associations for

'John'. We would like to see associations like 'John Smith', 'works for Microsoft', and 'interested in Guidebeam'. HAL's global associations would merge all of John's associations accumulated across contexts into a single vector. There may be three John's mentioned, who work for different companies but who have some association with the topic in question and thus whose utterances should be represented in our sample set. HAL would not be able to distinguish locally, with our question of 'How's it going with John?', which John is being talked about. We believe that the local associations formed with our modified HAL have a greater chance of being closer to answering, though not fully, that question.

To distinguish Burgess et al.'s HAL, and our model, we describe our model as HAL+ from this point on.

We will use the sentence already described in Sect. 8.3 to visualize the differences between HAL and HAL+. The modified index expression of the example is (Fig. 8.3):

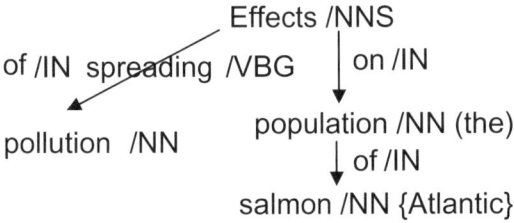

Fig. 8.3. Example of a structure used as input into HAL+

The differences between HAL and HAL+ are:

1. HAL slides a fixed window across the text. Syntactic structure is not taken into account. We slide a window over *contexts* in an index expression. A context is defined to be a path from the root node in an index expression to a leaf node. By way of illustration, the figure above features two contexts: [effects, pollution] and [effects, population, salmon]. As a consequence, lexical co-occurrence is determined by syntactic structure. In our example, the node 'effects' is only two 'window places' away from the node 'salmon', whereas HAL would treat these as having a distance of nine places.

2. It is important not to lose associations with words that are not nodes. For example, the term 'spreading' has a strong association with the nodal terms 'effects' and 'pollution'. In HAL+, each node in an index expression corresponds to a dimension. However, ancillary associations may be stored with each dimension. So, 'spreading' would be stored with both the 'pollution' and 'effects' dimensions, and 'Atlantic' would be stored with the 'salmon' dimension. Stop words and quantities are not considered information-bearing enough to store.

3. HAL's moving window ignores sentence boundaries. One of the reasons for this is to capture cross-sentence meanings and associations. We have already provided for some of these associations by our basic anaphora resolution. There-

fore, our implementation makes use of the previous unfiltered index expression only. In terms of our salmon example, if the previous sentence was 'Global ozone depletion', then the term 'effects' would have an association with 'ozone' and 'depletion'. We believe that within utterances relevant associations between terms tend to be more localized. For this reason, we did not deem it useful to represent cross-sentence associations beyond the scope of the previous sentence.

4. As there are fewer nodes in our modified index expressions compared to terms in the text, a window size of 10 seems too high. The higher the window size, the greater the chance of introducing spurious associations. We have chosen a window size of 6 in the interests of precision, but acknowledge that more work is required to fully justify a particular value.

5. The weighting function used by HAL is linearly decreasing. That is, the relationship between two terms is given a weight equal to the inverse of the number of places (terms) between them within the context of the window. Since we are trying to extract both global and local co-occurrence information, a non-linear weighting between terms was used dependent upon the POS information.

A multiplier for weights was applied in certain situations:

○ ×3 if both terms are proper nouns with a null connector between them. This is to intensify the association phrases involving proper nouns.

○ ×2 if both terms are nouns separated by a null connector. This is to intensify associations within noun phrases.

○ ×1.5 if both terms are nouns. This is to intensify the associations of noun phrases.

○ ×4 if the relationship is between a noun that precedes a PRP(preposition)-tagged term (such as 'you', 'he', 'we'). This is to intensify potential anaphoric references that we have not been able to deal with.

So 'population' and 'effects' would have their linear function weight of 6 (one node apart using window size = 6) multiplied by 1.5.

Table 8.2 depicts the HAL+ space of the example sentence. Compare this table with the HAL space for the same sentence given in Table 8.1. Observe how dimensions in the HAL+ space are restricted to significant, information-bearing terms as determined by the combination of presemantic context in the form of POS and syntactic structure. Associations such as between 'effects' and 'salmon', akin to viewing the sentence as being *about* the effects on salmon of something, are now weighted more strongly because the syntactic structure brings these terms closer together. Note that HAL (i.e. syntactic structure not being taken into account) would require a window size of at least 8 to find even the weakest of relationships between these two terms. Unfortunately, Table 8.2 does not reflect the desirable association between the terms 'pollution' and 'salmon'. This is because the window used to compute weighted associations slides over a context (a sequence of nodes from root to leaf). Alternative definitions of context are possible which would see these terms together; however we will not pursue them further here. Suffice to say, the definition of context used to compute weighted associations in HAL+ is conservative and errs on the side of pre-

Table 8.2. Example of a HAL+ matrix

	eff	poll	pop	sal
eff				
poll	9			
pop	9			
sal	7.5		9	

cision. In summary, HAL+ is an $n \times n$ matrix, where n corresponds to the number of unique nodes across all unfiltered index expressions. The value (i, j) represents the strength of association between word i and word j in the context of the utterance set. Each row i of HAL+ corresponds to the vector representation of word i. This vector is the carrier of presemantic and semantic context.

8.4 Closing Remarks

This chapter outlined a view of chance discovery in the online community using three facets: discovery of communities, between communities, and within communities. Describing other research in the area as being in one or more of these three facets showed that there was little use made of the content of the communication: most of the research is about the networks and linkages of people or Web pages. Therefore, since the content of the utterances between people (both inside and outside a community) *is* very important, we detailed a method for collecting and storing the utterances: a vector representation of words embodying aspects of both presemantic and semantic context. This representation is widely applicable to any communication of textual utterances.

The third aspect of semantic context in utterances is postsemantic context. Postsemantic context can be considered to be shared, though un-stated, context, or in other words, a form of tacit knowledge. Chapter 9 describes techniques for deriving postsemantic context based on the vector representation of words, illustrated by a set of email utterances, and pursues the vector representation to discover the ebbs and flows of socio-cognitive 'meaning' within the online community.

References

8.1 Berger F (1998) *Navigational query construction in a hypertext environment*, PhD thesis, Nijmegen University
8.2 Brill E Rule-based tagger. Available: http://www.cs.jhu.edu/~brill/ .
8.3 Brill E (1992) A simple rule-based part of speech tagger, in *Proceedings of the Third Conference on Applied Natural Language Processing, ACL*, pp. 152–155, Trento, Italy
8.4 Bruza PD (1993) *Stratified information disclosure: a synthesis between hypermedia and information retrieval*, PhD thesis, Nijmegen University
8.5 Bruza PD, Dennis S (1997) Query reformulation on the internet: empirical data and the hyperindex search engine, in *Proceedings of RIAO 97: Computer-Assisted Information Searching on Internet*, pp. 488–499 Montreal, Canada

8.6 Burgess C, Livesay K, Lund K (1998) Explorations in context space: words, sentences, discourse in *Discourse Processes*, 25: 211–257

8.7 Burgess C, Lund K (1997) Modelling parsing constraints with high-dimensional context space in *Language and Cognitive Processes*, 12:177–210

8.8 Charniak E (1993) *Statistical language learning*. MIT Press, Cambridge, MA, p.49

8.9 Erickson T, Halverson C, Kellogg WA, Laff M, Wolf T (2002) Social Translucence *Communications of the ACM*, 45(4):40–44

8.10 Flake GW, Lawrence S, Giles CL (2000) Efficient Identification of Web Communities. in *KDD*, (Boston, MA), ACM, pp,150–160

8.11 Freyd J (1983) Shareability: the social psychology of epistemology, *Cognitive Science*, 7:191 – 210

8.12 Gabbay D, Woods J (2000) Abduction, European Summer School on Logic, Language and Information. Published electronically:
http://www.cs.bham.ac.uk/~esslli/notes/gabbay.html

8.13 Grice P (1989) *Studies in the way of words*. Harvard University Press

8.14 Gärdenfors P (2000) *Conceptual Spaces: the Geometry of Thought*. MIT Press, London

8.15 Hiltz SR, Turoff M (2002) What makes Learning Networks Effective? *Communications of the ACM*, 45(4):56–59

8.16 Kellog WA (2001) Using socially translucent systems to support tacit knowledge in online communities, Keynote Talk in the *ECSCW2001 Workshop on Tacit Knowledge*, published electronically http://www.unite-project.org/ecscw01-tkm/

8.17 Kumar R, Raghavan P, Rajagopalan S, Tomkins A (1999) Trawling the Web for Emegring Cyber-communities, in *Proceedings of the 8th International World Wide Web Conference*, Elsevier Science, Amsterdam, The Netherlands, pp.403–415

8.18 Lund K, Burgess C (1996) Producing high-dimensional semantic spaces from lexical co-occurrence in *Behavior Research Methods, Instruments & Computers*, 28(2):203–208

8.19 Matsumura N, Ohsawa Y, Ishizuka M (2001) Discovery of Emerging Topics between Communities on WWW, *Lecture Notes in Computer Science* 2198, pp.473–482, Springer Verlag, Heidelberg, Germany

8.20 Matsumura N, Ohsawa Y, Ishizuka M (2001) Future Directions of Communities on the Web. in *New Frontiers in Artificial Intelligence* LNAI 2253, Springer Verlag, Heidelberg, Germany pp.435–443

8.21 Matsumura, N, Ohsawa, Y. and Mitsuru, I. Mining Mailing List Archives. Poster in 11th *Poster in International World Wide Web Conference*, Published only electronically http://www2002.org/posters.html

8.22 McArthur R and Bruza, PD (2001) The ABCs of Online Community, *Processsdings of Web Intelligence: Research and Development*, LNAI 2198, pp. 141–147, Springer Verlag, Heidelberg, Germany

8.23 Nardi BA, Whittaker S, Isaacs E, Creech M, Johnson J, Hainsworth J (2002) Integrating Communication and Information through ContactMap. *Communications of the ACM*, 45(4):89–95

8.24 O'Hara K, Brown B (2001) Designing CSCW technologies to support tacit knowledge sharing through conversation initiation:, In Workshop Note of *ECSCW2001 Workshop on Tacit Knowledge*, published electronically http://www.unite-project.org/ecscw01-tkm/

8.25 Penco C (1999) Objective and cognitive context, in *Modeling and using context: Proceedings of the Second International and Interdisciplinary Conference CONTEXT'99*, LNAI, v.1688, Springer Verlag, Heidelberg, Germany p.272

8.26 Perry J (1997) Indexicals and demonstratives, in *A companion to the philosophy of language,* Hales, B. and Wright, C. Eds. Oxford: Blackwell, pp.593–595

8.27 Perry J (1998) Indexicals, contexts, and unarticulated constituents, in *Proceedings of the 1995 CSLI-Amsterdam Logic, Language and Computation Conference*. Stanford: CSLI Publications, pp.1–16

8.28 Preece J (2000) *Online Communities: Designing Usability, Supporting Sociability*, Wiley, New York, NY

8.29 Sleator D, Temperley D (1993) Parsing English with Link Grammar, *Proceedings Third International Workshop on Parsing Technologies*, pp.277–292

8.30 Smith M (2002) Tools for Navigating Large Social Cyberspaces. *Communications of the ACM*, 45 (4):51–55

8.31 Swanson DR (1986) Fish oil, Raynaud's syndrome, and undiscovered public knowledge, in *Journal of the American Society for Information Science*, 30:7–18

8.32 Swanson DR and Smalheiser NR (1997) An interactive system for finding complementary literatures: a stimulus to scientific discovery, in *Artificial Intelligence*, 91:183–203

8.33 Smith M, Kollock P (eds) (1999) *Communities in Cyberspace*, Routledge, London

8.34 Weeber M, Klein H, Jong van den Berg L, Vos R (2001) Using concepts in literature-based discovery: Simulating Sawnson's Raynaud-Fish Oil and migraine-magnesium discoveries. *Journal of the American Society for Information Science and Technology*, 52(7):548–557

8.35 Wenger E (1998) *Communities of practice: learning, meaning and identity*, Cambridge University Press

8.36 Whittaker S, Isaacs E. and O'Day V (1997) Widening the net: workshop report on the theory and practice of physical and network communities, *SIGCHI Bulletin*, 29(3):27–30, ACM Press, New York, NY

8.37 http://www.sil.org/linguistics/GlossaryOfLinguisticTerms/WhatIsAnaphora.htm

9. Discovery of Tacit Knowledge and Topical Ebbs and Flows Within the Utterances of an Online Community

Robert McArthur and Peter Bruza

Distributed Systems Technology Centre, Brisbane, Australia
email: mcarthur@dstc.edu.au

Summary.

This chapter shows how to derive postsemantic context based on vector representations of words (described in Chap.8). The core problem is to discover relevant word associations in relation to seed words in the utterance. This may involve uncovering implicit associations or reweighting explicit associations more highly. The set of such associations forms a part of 'conversational implicature'. The chapter describes techniques for computing associations in a dimensional space that have shown promise in the literature. The goal is to provide some initial insights as to their usefulness for mining conversational implicature by applying them to a small set of email utterances.

9.1 Introduction

9.1.1 Minkowski Function

Gärdenfors states that the "*similarity of two stimuli can be determined for the distances between the representations of the stimuli in the underlying psychological space*" [9.4]. In practice, associations between words can be computed by calculating the similarity between their vector representations in a dimensional space. The distance between two words x and y in the HAL+ space can be calculated using the Minkowski distance measure:

$$d(x,y) = \sqrt[r]{\sum_{1}^{n} (|x_i - y_i|)^r},$$

where $d(x,y)$ denotes the distance between the HAL+ vectors for x and y. In the results reported later, we chose $r = 1.0$ as this value led to a sizeable correlation between vector similarity and cognitive effects [9.6]. Following Gärdenfors, the similarity $s(x,y)$ between HAL+ vectors for x and y was calculated as an exponentially decreasing function of distance:

$$s(x,y) = e^{-cd(x,y)}.$$

We employed a neutral value for the sensitivity parameter ($c = 1$).

The Minkowski function has shown some promise as a means of measuring associations between words that correlate with human performance on word-association tasks. For this reason, we chose it as one means of mining associations.

9.1.2 LSA

Latent semantic analysis (LSA) is a method of representing the meaning of words as vectors in a dimensional space reduced by singular value decomposition (SVD) [9.5]. The meaning can be considered "*as a kind of average of the meaning of all the passages in which it appears and the meaning of a passage as a kind of average of the meaning of all the words it contains*" [9.5].

The role of SVD is fundamental to LSA. The general claim is that similarities between words can be more reliably estimated in the reduced dimensional space than in the original one. The rationale is that contexts which share frequently co-occurring terms will have a similar representation in the reduced dimensional space, even if they have no terms in common.

For our purposes, the input to the LSA process is the $n \times n$ HAL+ matrix produced by the vector-creation process detailed in the previous section. We did not normalize the values in the matrix as advocated in [9.11] because pilot studies revealed a 6–9% improvement using un-normalized values. This is perhaps due to the small data set, but may also be due to the presemantic and semantic processing before HAL+.

After dimensional reduction, the weight (i, j) may be non-zero, whereas it was zero before dimensional reduction. Where positive, it suggests that word i is implied within context j. In other words, LSA can discover implicit associations, or strengthen/weaken existing associations. Such behavior is relevant for the mining of conversational implicature [9.3]: those associations that appear after dimensional reduction, or are strengthened by it, may be suggestive of postsemantic context [9.9, 9.10]. Due to space constraints, the dimensional reduction process will not be described further. The details can be found in [9.5].

After dimensional reduction, word associations are computed via the cosine function employed on vectors in the reduced dimensional space.

While LSA's ability to reflect word associations produced by human subjects has been consistently demonstrated in a variety of tasks, we note that LSA's creators warn:

"*Nonetheless, the relationship between some close neighbors in LSA space can occasionally be mysterious (e.g. 'verbally' and 'sadomasochism'. . .), and some pairs that should be close are not. It's impossible to say exactly why these oddities occur, but it is plausible that some words that have more than one contextual meaning receive a sort of average high-dimensional placement that out of context signifies nothing, and that many words are sampled too thinly to get well placed.*" [9.5].

9.1.3 Information Flow

The Minkowski function is a means of computing the similarity between words. The information flow between words, on the other hand, has shown promise in deriving implications between words [9.8]. By way of illustration, consider the word NEC (denoting the Japanese technology corporation). Similarity computations based on this word on the Reuters 21 758 collection tended to produce the names of other technology companies: 'Intel', 'Unisys', etc. Information flow, on the other hand, produced terms such as 'computer', 'corp', 'electronics', etc., which show a markedly different character. They have been termed 'information containment' relationships because 'NEC' is considered to contain/carry the information 'computer'. Information flow is a realization of Barwise and Seligman's theory of information inference using vector representations of words [9.1]. For more details see [9.8].

Information containment relationships are computed by ascertaining the degree of inclusion between respective vector representations of words. The degree of inclusion of word x (termed the source) in the word y (termed the target) is computed in terms of the ratio of intersecting quality properties of the vector representations for words x and y to the number of quality properties in the source:

$$\text{degree}(x, y) = \frac{\displaystyle\sum_{i \in (QP_\mu(x) \wedge QP(y))} x_i}{\displaystyle\sum_{i \in QP_\mu(x)} x_i}.$$

In terms of the experiments reported below, the set of quality properties $QP_\mu(x)$ in the source HAL+ vector is defined to be all dimensions with above average weight. The set of quality properties $QP(y)$ in the target HAL+ vector is defined to be all dimensions with positive weights. These definitions have been determined empirically in information flow experiments to drive automatic query expansion in connection with reasonably large text corpora [9.12]. It is an open question whether these definitions are applicable to small utterance sets.

Our reason for including the information flow model is that it has demonstrated inferential character in uncovering implicit term associations for use in automatic query expansion [9.12]. We hypothesized that this model may be able to discover implicit word associations relevant to conversational implicature.

9.2 Illustration I: Discovering Tacit Knowledge

9.2.1 Data Set

We used a small data set of four emails to illustrate the representation and methods outlined. This small example is drawn from the real-life of our research organization. We believe that the type and contents of the emails are representative of their ilk, namely, those from one member of an IT company to another about a particular work topic.

E1 : from Peter to Rupert

Rupert,

I just met this guy, John Smith. He works for Microsoft and is interested in Guidebeam. Can we get a contract for an evaluation licence to him soon?

Cheers
Peter
E2 : from Rupert to Peter
*Sounds good. **I'll send him something ASAP.***
Rupert
E3 : from Rupert to John Smith
Hi John,
Peter Bruza said you are interested in evaluating Guidebeam. Can you please sign and return the attached licence agreement and we can send you the software.
Rupert
Account Manager
E4 : from John Smith to Rupert
Rupert,
I've faxed the signed agreement to you.
John

The final utterance is a question from Peter to Rupert:
'*Hi Rupert. How's it going with John?*' Utterances in bold are deemed, via a simple discourse-level heuristic, to be worth using with HAL+ to generate vectors. The above terms are converted into their anaphoric reference based on the sender and receiver of the message.

9.2.2 Representation of Data

The following vectors are representative of those gathered by applying HAL+ to the index expressions created from the utterances. HAL+ had a dimensionality of 20. We were hampered by errors in the POS tagging and, as these severely compromised the results, we manually modified the POS for four terms (e.g. 'interested' was tagged as an adjective instead of a verb)[1].

John : Smith:18, guy:12, He:12, Peter*:10, Microsoft:7.5, Guidebeam:6 ($\mu = 10.9$, $\sigma = 4.2$)
John* : agreement:38, licence:28, Bruza:22, Guidebeam:22, Peter:20, software:16, John*:16, Can:14, we:10, Rupert*:5 ($\mu = 19.1$, $\sigma = 9.3$)
licence : John*:28, we:18, Can:16, him:12, agreement:12, evaluation:12, Guidebeam:9, contract:7.5, Bruza:6, software:4.5 ($\mu = 12.5$, $\sigma = 6.9$)

* - terms derived through basic anaphora resolution.

The vectors already exhibit many relevant associations for conversational implicature, e.g. Microsoft and Guidebeam in the 'John' vector. However, some of these are not weighted highly. It is one of the tasks of the mining techniques to remedy this.

---

[1] Note that the online demonstration of a modern commercial tagger correctly identified these tags.

Also, while it may seem as though we have lost information, in reality there is presemantic context hidden in the vector in the form of connector verbs and adjectives associated with each noun. We have not shown these for clarity, but an example is that the dimension 'Microsoft' in the 'John' vector has an association of 'works for'. These are not used in the following mining techniques but assist viewing output and interpreting a particular vector's dimensions.

The 'John*' vector, derived by anaphora resolution, is guaranteed to be that for the person John Smith (e.g. 'I' in email from John Smith is mapped to the term 'John*'). It is likely that the vector representation of 'John' is also that of John Smith due to the highly weighted 'Smith' dimension[2]. We have chosen to use vector addition of the normalized vectors to join the two vectors together, resulting in a single vector representing the word 'John':

John : Smith:25.4, agreement 22.8, Guidebeam:21.7, guy:16.9, He:16.9,
  licence:16.8, Peter*:14.1, Bruza:13.2, Peter:12.2, Microsoft:10.6, John:9.6,
  software:9.6, Can:8.4, we:6.0, Rupert*:3.0
  $(\mu = 13.8, \sigma = 6.3)$

Even though these HAL+ vectors are promising, possessing most if not all of the relevant associations for conversational implicature, some issues are still outstanding. The major problem is how to extract the quality dimensions from the vector: for example, in the 'John' vector, the Microsoft dimension is weighted quite low less than the mean. Also, some desirable associations are not represented at all, such as 'contract'. It is the task of the mining techniques to discover and weight highly those associations relevant to conversational implicature.

9.2.3 Methods Applied

We applied all four methods to this data: Minkowski similarity, LSA, LSA with cosine, and information flow. Although each method was applied to the entire dataset, for brevity we have only shown results for two vectors: the combined 'John' vector, which is of most interest in looking at the utterance 'How's it going with John?', and the 'licence' vector as an example of another term with some interesting associations. Only the six dimensions with the highest weights are shown.

It is clear that the dimensions' weights are wholly dependent upon the HAL+ weighting function. Since the function is so important, and further tuning is required to determine generally applicable results, we have applied the Spearman rank coefficient to measure the difference in rank of the dimensions rather than comparing their particular weights.

After applying each method, the resultant list of vectors that are, in turn, 'similar to' each of these two example vectors is compared to the 'perfect' resultant vector (see below) using the Spearman rank coefficient (r). Values in parentheses after the Spearman rank are also Spearman rank coefficients, but have been calculated on part

---

[2] In general, we cannot be certain that a 'John' vector represents a single individual (in this case, John Smith); it may relate to other John's as well.

of the vector only either only those dimensions whose values were above the mean, or at least the highest six dimensions (as our perfect vectors have at most six ranks). Mean and standard deviation are given for the HAL, HAL+, and LSA results.

Table 9.1 depicts, in rank order, the associations that form the basis of conversational implicature around 'John' and 'licence' within our example. Associations of the same rank are on the same line. The ranks were determined by our expectation of highly weighted associations of the candidate terms in the utterance set. The table constitutes the solution against which the mining techniques can be measured.

Table 9.1. 'Perfect' vectors

	Perfect vector (rank only)		**Perfect vector** (rank only)
John	Smith He, guy Microsoft, Guidebeam licence software agreement	licence	evaluation, Guidebeam contract, agreement, software

It is interesting to note, in passing, that the highest associations for 'John' are relatively independent of context and more permanent: last name is 'Smith', gender is male ('he' and 'guy'), and that he works for 'Microsoft'. This could be seen as a global concept or generalized implicature. The lower associations are more dependent on context and impermanent (local context).

We applied LSA to the HAL+ vector set. In light of the results, we report both with and without applying the cosine similarity. We reduced the dimensionality of the matrix from 20 to various small values. The best results, for a reduction to three dimensions (denoted by $t = 3$), are shown. A tentative algorithm to determine a cutoff would be to take all values larger than one-third of the maximum value.

9.2.4 General Remarks

We begin with an examination of Table 9.4. The superior performance of HAL+ over HAL is an indication that the incorporation of presemantic context does lead to improved representations.

The most interesting result is that LSA is the only method to improve upon the Spearman coefficient of HAL+ – about +16%. In comparison to all methods, the HAL+ coefficient is quite good at $r = 0.7$, and the LSA better at $r = 0.8$. HAL+ is markedly better than HAL (0.68 vs. 0.19). Other than LSA, the other methods simply do not perform because they do not highlight associations that we would expect to see.

The association between John and Microsoft is lifted (from HAL+) by all methods from 10th rank to either 2nd or 3rd. By itself this augurs well, but is not repeated generally for all quality dimensions. We do not know why Microsoft alone is treated in this way.

As Table 9.2 shows, LSA introduces five new dimensions: evaluation (3.1), contract (6.7), him (−0.38), something (−0.77), and ASAP (−1.0). The only dimension that was 'lost' (changed to a negative association) was 'Rupert*', which was already the lowest dimension of the HAL+ vector. The major rises in dimensions' weights by LSA were: Microsoft (+104%), John (+29%), and Guidebeam (+25%). The worst falls were Rupert* (−125%), Bruza (−47%), and agreement (−42%).

Table 9.2. 'John' vector before and after LSA

Dimension	Weight		Δ	Δ%
	Before	After		
Microsoft	10.6	21.6	+11.0	104%
John	9.6	12.4	+2.8	29%
Guidebeam	21.7	27.2	+5.5	25%
He	16.9	15.0	−1.9	−11%
guy	16.9	14.2	−2.7	−16%
Smith	25.4	21.4	−4.0	−16%
Peter*	14.1	11.7	−2.4	−17%
licence	16.8	12.9	−3.9	−24%
Can	8.4	6.2	−2.2	−26%
we	6.0	3.8	−2.2	−36%
software	9.6	6.0	−3.6	−38%
Peter	12.0	7.1	−4.9	−41%
agreement	22.8	13.2	−9.6	−42%
Bruza	13.2	6.9	−6.3	−47%
Rupert*	3.0	−0.8	−3.8	−125%
contract	−	6.7		
evaluation	−	3.1		
him	−	−0.4		
something	−	−0.8		
ASAP	−	−1.0		

The dimensions of the John vector whose values are greater than the mean are given in Table 9.3. Only the LSA results are shown. We have included the associated information stored with each dimension to show how this aids the understanding of the vector's data.

The results for the 'licence' vector, provided as another interesting term, are quite different (Table 9.5). HAL+ only marginally improved on HAL ($r = 0.22$ vs. $r = 0.17$), while the best method's Spearman coefficient was the Minkowski similarity ($r = 0.56$). LSA was very poor: $r = 0.01$.

These results may be so radically different to 'John' as the two terms, both nouns, are very different *things*. It seems plausible that different types of concepts people (John) vs. a physical object (a licence) may have associations that are best determined by a variety of methods. That these methods have succeeded is not in

Table 9.3. Results for the 'John' vector

	LSA $t = 3$
John	Guidebeam (interested in)
	Microsoft
	Smith
	He (works for)
	guy
	agreement (have faxed)
	licence (sign, return)
	John
	Peter* (said are interested evaluating)

doubt: the increase in the r value from 0.19 (HAL) to 0.79 (LSA) in the 'John' vector, and from 0.17 (HAL) to 0.56 (Minkowski) in the 'licence' vector is marked.

The relative poor performance of the information flow model is somewhat surprising given that it has had notable success in uncovering implicit word associations for use in automatic query expansion [9.12].

9.3 Illustration II: The Ebb and Flow of 'Meanings'

The third facet of chance discovery within the online community (Chap. 8) is discovery within a community. One aspect, illustrated in Sect. 9.2, is of the un-earthing of tacit knowledge in the form of postsemantic context. Another aspect surrounds the evolution of socio-cognitive 'meaning' in the community. The 'meaning' is transmitted within the community through the utterance, sometimes in the tacit knowledge but often explicitly, and can be represented by dimensional structures like the HAL+ vectors. The change in these dimensional structures, due to the continual stream of utterances within the community, can be considered to be changes in the 'meanings' of words shared by the community.

It has been argued by Gärdenfors [9.4] that sharing of knowledge imposes constraints on individual cognitive representations of information within a co-evolutionary process between social knowledge structures and individual ones. In this regard, Freyd [9.2] hypothesizes that over time the co-evolution process will lead to dimensional structures that stabilize with a small number of values represented on each dimension. Examination of the dimensional structures over time can provide a rich tapestry of 'meanings' before the stabilization occurs. It may allow the chance discovery of the stabilized meaning to be detected before the community is aware of the fact, assisting the community to decide on the 'meaning' or begin the discussion anew. It can also provide an explanatory mechanism for new community members or those wishing to know how a 'meaning' evolved.

Changes in the socio-cognitive 'meanings' inherent in the vector representations emerge in three ways:

Table 9.4. Results for the 'John' vector

	HAL [0, ∞)	HAL+ [0, ∞)	Minkowski [0.074, 1]	LSA $t = 3$ [0, ∞)	LSA $t=3$+cosine [0, 1]	Information flow [0, 1]
John	you (17)	Smith (25.4)	Can (0.45)	Guidebeam (27.2)	He (0.93)	Guidebeam (0.77)
	peter (10)	agreement (22.8)	Microsoft (0.44)	Microsoft (21.6)	Microsoft (0.91)	Can (0.67)
	guy (10)	Guidebeam (21.7)	He (0.44)	Smith (21.4)	guy (0.88)	Microsoft ((0.57)
	Smith (10)	guy (16.9)	we (0.43)	He (15.0)	we (0.86)	we (0.56)
	ii (10)	He (16.9)	guy (0.42)	guy (14.2)	Can (0.85)	He (0.54)
	He (9)	licence (16.8)	Peter* (0.40)	agreement (13.2)	Peter* (0.81)	guy (0.47)
	$r = 0.19$ $r = (0.10)$	$r = 0.68 (0.70)$	$r = 0.64 (0.64)$	$r = 0.79 (0.81)$	$r = 0.52 (0.55)$	$r = 0.54 (0.58)$
	$\mu = 7.5$	$\mu = 13.8$		$\mu = 9.3$		
	$\sigma = 2.8$	$\sigma = 6.3$		$\sigma = 8.0$		

Table 9.5. Results for the 'licence' vector

	HAL [0, ∞)	HAL+ [0, ∞)	Minkowski [0.074, 1]	LSA $t = 3$ [0, ∞)	LSA $t=3$+cosine [0, 1]	Information flow [0, 1]
licence	and (16)	John (28)	software (0.15)	John (29.4)	software (0.97)	can (0.87)
	Can (13)	we (18)	Peter (0.13)	Can (15.5)	Bruza (0.96)	we (0.81)
	the (13)	Can (16)	Guidebeam (0.10)	we (13.8)	agreement (0.96)	Guidebeam (0.80)
	we (12)	him (12)	agreement (0.10)	licence (9.9)	Guidebeam (0.95)	John (0.75)
	agreement (10)	agreement (12)	Bruza (0.09)	Guidebeam (7.7)	Peter (0.93)	agreement (0.60)
	attached (10)	evaluation (12)	evaluation (0.08)	evaluation (7.1)	Rupert* (0.82)	software (0.59)
	$r = 0.17 (0.01)$	$r = 0.22 (0.02)$	$r = 0.56 (0.62)$	$r = 0.01 (0.02)$	$r = 0.45 (0.54)$	$r = 0.30 (0.24)$
	$\mu = 8.4$	$\mu = 12.5$		$\mu = 6.1$		
	$\sigma = 3.1$	$\sigma = 6.9$		$\sigma = 7.1$		

1. change in the number of non-zero dimensions of the vector; the number of contexts the word is used in increases/decreases, thus we say it has acquired more or less socio-cognitive 'meaning'. With respect to Freyd's hypothesis, stabilization of a representation can be detected when the dimensionality of a word's representation does not grow, or grows little, over time;

2. change in the values of an individual dimension; particular contexts that the word is used in occur more or less often;

3. increase in dimensionality of the vector space; the vocabulary used by the community increases with the use of new words.

Emergent topics of interest will show a rapid increase in dimensionality (1. above) over a time period. This also holds for the HAL+ representations of individuals that are involved in an emergent topic. This phenomenon could be exploited for the purposes of chance discovery *within* a community. For example, the discovery of emergent topics together with the individuals involved will allow the timely fostering of social networks that contribute to community 'glue' [9.7]. Paired with this is the possibility to explicitly extract information of the sort individual A knows about (or is passionate about) topic X. Such information may be invaluable to new members of the community. On the practical side, such a social network may take their discussion into a side 'forum' thereby relieving the main forum of the community of unnecessary detailed discussion.

We envisage that the HAL+ vectors of words could allow the possibility of chance discovery *between* communities. If the semantic Web comes to fruition, it is likely that online communities will maintain their own ontologies that represent their community knowledge [9.8]. HAL+ vectors can be associated with an ontology term, thereby attributing to it the current 'social-semantic' representation of the term in the community. Such representations can lead to a more precise determination of whether two different online communities share topics of interest.

To illustrate the potential, we ran HAL+ over a topic-based mailing list and examined the vectors over a limited time to attempt to discern the ebbs and flows of 'meanings'.

9.3.1 Data Set

The multi-hull mailing list[3] discusses issues around the topic of boats with more than one hull – predominantly catamarans and trimarans. It is typical of thousands of topic-based mailing lists. In May, 2001, a discussion broke out on the nature of a different type of multi-hull – a proa – and one proa in particular by the name of *Harry*. To visualize the ebb and flow of the change of topics over time, the model was applied to messages for April, May, and June, 2001 (*Harry* was not discussed again for the rest of 2001). General information about the mailing-list data is presented in Table 9.6.

Some samples to provide a flavor of the discussions about *Harry* are:

---

[3] http://www.steamradio.com/mailman/listinfo/multihulls

Table 9.6. Multi-hull mailing list statistics

2001	Number of msgs	Size of pre processed file (bytes)	Number of unique senders	Number of msgs from the top three senders	Number of senders sending		Number of msgs in which 'harry' is mentioned (excl. quotes)
					50%+ of msgs	>1 msg	
April	561	551 062	163	27, 21, 19	23 (14%)	78 (48%)	0
May	348	391 415	117	24, 14, 12	19 (16%)	54 (46%)	5
June	611	707 537	174	39, 30, 22	21 (12%)	72 (41%)	14

- *β was more than happy to feature **harry** on his Web page as a Pacific proa, until he decided that I (not **Harry**) was belligerent, at which time he removed it. His despair at my 'misuse' of the name has only sprung up since **harry** has been proven to do exactly what it was designed to do and I have successfully started selling plans.*
- *Is it safe to presume that if α starts referring to **Harry** as an ndrua, and quotes his perceived advantages as those resulting from the ndrua layout and not from a Pacific proa basis, you will no longer have those concerns which you have so clearly laid out?*
- ***Harry** was designed around some very specific parameters. What type of boat this led to was never a concern. Pacific proa is the nearest terminology in current use, hence I (and every other proa-interested person I have spoken to) call it a Pacific proa.*
- ***Harry** is far more a Pacific (European title) proa (adapted European term) than it is a traditional Pacific Island craft.*
- ***Harry's** function is to provide comfortable, fast, easy, low-cost weekend cruising for my wife and I.*
- *α continues to muddy the water by referring to **Harry** as a Pacific proa and making claims that really only apply when you have a small hull to weather, not a heavy one.*

Note that the names of some protagonists have been depicted as Greek letters (α, β, etc.) as their exact names are not relevant.

9.3.2 Method

The mailing-list messages were preprocessed before running HAL+:

- All quotes were removed (defined as message body lines starting with '>') as were the quote headers (text like 'On Sat, XXXX wrote:')
- Trailing garbage was removed (excluding signatures)
- Simple possessives (e.g. Robert's) were transformed into their singular (e.g. Robert)
- all words were converted to lower-case

Index expressions were generated and basic anaphora resolution performed. Only anaphoric references based on the sender could be processed, as the 'To:' field of the messages was the mailing list rather than individuals. This added to the noise in the final vectors as 'you' and 'your' occurred regularly. For clarity, these terms are not shown in the vectors.

The three methods by which 'meanings' emerge (Sect. 9.3) are studied for two particular terms, *harry* and 'proa', and the general case over all vectors over the three months. Two terms are used that need explanation: *substantive dimensions* are those dimensions of a vector with the highest values, while *substantive vectors* have the largest *number* of non-zero dimensions of all vectors in a month.

9.3.3 Representation

To give a flavor of the HAL+ vectors, those for *harry* and 'proa', sorted by largest dimension, are shown:

April harry : <no mention>
May harry : α:73, proa:39, hull:37.5, windward:31.5, type:31.5, happy:24, name:18, significant:18, drag:15, weight:15, plans:13.5, european:13.5, naïve:12, belligerent:12, pacific:12, catamaran:12...
   (147 dimensions, $\mu = 11.2$, $\sigma = 10.2$)
June harry : α:286.5, type:148.5, proa:139.5, γ:63, β:62, claims:61.5, boat:57, u:56, cost:52.5, photo:43.5, hull:37.5, weather:36, summer:34.5...
   (356 dimensions, $\mu = 17.2$, $\sigma = 28.9$)

April proa : limited:12, rowing:12, fixed:12, tri:12, skiffs:10.5, similar:10, area:9, lake:9...(22 dimensions, $\mu = 9.9$, $\sigma = 4.5$)
May proa : hull:214.5, proa:189, α:188.5, β:86.5, boat:84, catamaran:82.5, leeward:73.5, windward:27 ...(527 dimensions, $\mu = 20.3$, $\sigma = 36.4$)
June proa : cat:765, cruiser:682.5, racer:537, proa:432, α:406, tri:366, hull:286.5, β:214.5, atlantic:177, windward:165, sailing:144...
   (1016 dimensions, $\mu = 26.5$, $\sigma = 58.6$)

9.3.4 Discussion

Change in 'meaning' for the term *harry*
The change in the number of non-zero dimensions of the HAL+ *harry* vector is 0 (April) → 147 (May) → 356 (June). Clearly one could term *harry* as an emergent

concept – at least over the April–June period: it increasingly acquires new socio-cognitive 'meaning' and is far from stabilizing. Similarly, the 'proa' vector increases 22 (April) → 527 (May) → 1016 (June) performing in the same way. This indicates that it is not simply *harry* that is being discussed, but also things about proas in general.

Interestingly, the two most substantive emergent dimensions for *harry* (Table 9.7, column 1) are references to people. The discussion surrounding *harry* becomes increasingly personal in June leading to more use of 'I', 'me', and 'my'. Our simple anaphora resolution uncovers this, which would normally be lost as stop-words. The dying dimensions reflect that only five messages in May mentioned *harry* and so particular language in one utterance of May, such as the references to 'European', is not continued in June.

Table 9.7. Dimensions' differences from HAL+ vectors between May and June (top substantive only)

Harry's dimensions (May–June)			
Substantive emergent	Substantive dying	Highest Δ	
		+ve	−ve
δ	significant	boat	**windward**
β	**European**	**cost**	plans
claims	1430	one	name
u	fuller	proa	lee
pac	lbs	α	case
photo	naïve	sailing	
option		type	
way		up	

The change in the values of *harry*'s dimensions (Table 9.7, columns 3 and 4) reflect the change of contexts. For example, the 'cost' dimension had a value of 6 in May but rose sharply to 52.5 in June as discussion of both the price and other costs of proas in general and *harry* in particular started. It is interesting to note that the discussion of whether *harry* can 'go to windward' shows the largest decrease. Instead, the conversations moved to proas in general with the 'windward' dimension of the 'proa' vector increasing from 27 (May) to 165 (June).

Change in 'meanings' within the community

The changes in 'meanings' within the community are condensed in Table 9.8. The change in the number of non-zero dimensions of the most substantive vectors is shown in columns 3 and 4. The term 'multi-hull' is clearly one for which the meaning has probably crystallized within the community – this is the *raison d'être* of the community! It is interesting, though, that this is visible in such a small sample (over only two months). Other (potentially) crystallizing terms are generic like 'systems' or 'members'. Some terms are simply artifacts of the short time period, like 'david'.

The most change of substantive vectors – terms whose 'meaning' is far from crystallizing – nicely shows 'proa' ranked highly. The term *harry* is not similarly

Table 9.8. Change in the number of dimensions in vectors between May and June (top 18 substantive)

Substantive emergent vectors	Substantive dying vectors	Least change of substantive vectors (stable)	Most change of substantive vectors (nouns only)
Roy Mills	wingover	for	boat
ollier	meridian	engine	**proa**
75ft	fossett	**multihull**	cat
macalpine-downie	masthead	carbon	sailing
D Goodgame	Arjan Bok	thanks	time
stiletto	watt	**systems**	use
magic	stuart	drag	tri
antrim	set-up	wind	boats
kite	screecher	the	need
woods	lights	**david**	**hull**
wharram	propulsion	as	**sails**
marples	darbyshire	barker	bill
Donald McHardy	depth	**members**	way
seaclipper	autohelm	skipper	years
Koehler	ray	feet	**race**
zeeman	valves	brian	**design**
verne	stripes	boat	class
jules	storage	we	speed

ranked as only 14 out of 600+ messages mentioned it. However, most of the remaining substantive vectors seem to be generic terms ('hull', 'sails', 'race', 'design') whose meaning, like that of 'systems', we would expect to be crystallizing. This could again be an artifact of the short time period, but may also be more complex: the whole reason for the community is to discuss around the topic of multi-hull boats. Obviously some terms or concepts return again and again in any community, and many never fully crystallize; they perennially evolve as new members question the endoxa.

The increase in dimensionality of the vector space is presented as column 1. Most of the substantive vectors are highly ranked because our analysis is only over two months of the community's utterances: 'woods', 'wharram', and 'marples' are boat designers who feature regularly, but not each month; 'stiletto', which is the name of a particular design of boat, recurs on occasion ('stiletto' occurs in all months of 2001 except May, July, and November). However, the term 'zeeman' is the name of a particular boat and is a truly new term for the community (2000 2001): it did not occur except for one reference in June 2001.

General comments on applying HAL+

The POS tagging was unfortunately responsible for many errors that were apparent when viewing the HAL+ representations. This led to incorrect index expressions and further 'noise' in the vectors. Better POS taggers are available and should be considered necessary.

However, the index expressions themselves could not always represent the grammar and language used in the mailing list (see Sect. 9.3.1 for examples of the difficulty in parsing the utterances). At times the index expression was an incorrect path expression, which lost the advantage of the richer structure imparted by correctly formed, branching expressions. Other times the index expressions branched at inappropriate times. It may be that an intermediate style, including POS information and some further structural knowledge of sentences though without the expense of index expressions, may be better in studying mailing-list utterances.

A number of issues related to anaphora resolution became clear. Firstly, authors often used the ability to quote a passage and then make a remark, as a way of allowing the human reader to resolve anaphora. Often this wasn't as simple as using a pronoun like 'it' in the first sentence of the remark. Usually the entire remark had specific (and sometimes vague) references back to the quotation. These would also occur at some distance from the quotation. It is not clear how such references could be accurately and automatically disambiguated.

The second anaphora issue concerned the use of 'you', 'yours', 'them', 'we', etc. Unlike email messages in which the sender is (usually) clearly defined, these terms could not be so easily understood. To see the scale of the problem, the full *harry* vector for May was:

α : 73, **it:44**, **his:42**, proa:39, **what:38, you:38, than:38**, hull:37.5, **much:36, that:34**, windward:31.5, type:31.5, **but:26**...

A clear result of this is that the information carried in the anaphoric references is important. Without it, the people represented by 'I' would not be resolved and there would not be an association to them in the vector. Similarly, as the above references could not be resolved, vital information is lacking in the vectors. Therefore, we believe that the common practice of elimination of stop-words is probably unacceptable in creating a good vector representation from small sets of utterances. However, leaving the terms as above merely increases the noise in the representation. A potentially easy solution such as increasing the window size of HAL to try and capture the references would, in all likelihood, simply increase the noise without necessarily capturing the reference. Semantic processing of small utterance sets needs improvement by very good anaphora resolution. As much as possible should be attempted.

9.4 Conclusions and Further Research

Chance discovery in the online community can be separated into three facets: discovery of the community, between communities, and within a community. There is an increasing amount of research that can be situated in these facets, but most are interesting uses of statistical information between people, communities, Web pages, etc. There is little or no examination of how the substantive content of the utterances within a community, or between communities, could be used to facilitate the discovery of chances. In this chapter, and the previous one, we take the position that chance discovery in a community setting is intimately connected to creating awareness of relevant (e.g. emergent) topics, or people, and their associations. The topics, references to people, and the associations between these (people to topic(s), and people-to-people) are present, either implicitly, or explicitly, within the utterances flowing within a community. By way of illustration: Peter says to Rupert, 'How is it going with John?' Implicit to this utterance are associations to 'John Smith' from Microsoft who has been sent a 'licence agreement' for a product called 'Guidebeam'. If such associations can be brought to light, this opens the possibility for chance to manifest. For example, somewhere else there is Naomi, who needs a licence agreement for Guidebeam. She can now potentially be brought into connection with Rupert.

The implicit associations referred to in the example are forms of tacit knowledge termed conversational implicature. Conversational implicature is any meaning implied or understood from an utterance that goes beyond what is strictly expressed. This chapter deals with the mining of conversational implicature from small sets of coherent utterances. Conversational implicature is viewed as postsemantic context. The core problem is to discover relevant word associations in relation to seed words from the utterances. This may involve uncovering implicit associations, or reweighting explicit associations more highly. Four techniques are described for computing postsemantic context: latent semantic analysis (LSA), LSA+cosine, minkowski, and information flow.

The four techniques were applied to a typical set of email utterances. While this set is only an illustration, and not the basis of a systematic study, it may be concluded that conversational implicature can be mined successfully. More specific conclusions are:

- Vector representations embodying presemantic context are a better basis for mining conversational implicature than vector representations without presemantic context. This is somewhat surprising given the success of the latter within the cognitive science literature.
- Basic anaphora resolution permits more expressive representations of named entities which allow a better representation for computing word associations
- LSA can bring to light interesting implicit associations, though it can not be relied on to be the sole technique for mining conversational implicature.
- Information flow does not seem to produce encouraging results. This could be due to the low dimensionality of the data.

Even though this article has focused on small sets of utterances, we feel that the representation and associated techniques are applicable to larger sets of utterances, for example the email archive of an organization. In both cases the data can be viewed as a huge set of utterances which can be partitioned into smaller sets based on the utterer, the receiver of the utterance, and the topic(s) being uttered. Our future work is directed at scaling the representation and techniques presented in this article towards the mining of tacit knowledge and the capture of organizational memory.

Departing from a socio-cognitive stance, a major contribution of this chapter (and the preceding one) is a pragmatic representation of words that embodies aspects of presemantic and semantic context. In a nutshell, we feel that the dimensional representations presented are approximations of socio-cognitive 'meanings' of words within the context of the community. It is crucial to recognize that such 'meanings' are not static – they evolve. By analyzing how dimensional representations of words evolve, it is possible to identify emergent topics, and associated people. Conversely, it is possible to detect 'meanings' that are stabilizing, which is suggestive of community knowledge that is crystallizing.

This chapter has presented techniques for tracking the meanings of words via automatic processing of community utterances. Even though noise in the underlying representations is currently a problem, the analysis of the ebb and flow of meanings within a typical online community is a promising and interesting area of investigation for further pursuit.

The notion of chance discovery in relation to online communities is an emergent and intriguing topic. As our information environment becomes ever more complex, it seems we are losing awareness. For example, disciplines are becoming more and more specialized; the individual is becoming more and more insular. We will need communities to help enhance our awareness. The discovery of chances will be crucial in this regard.

References

9.1 Barwise J, Seligman J (1997) *Information flow: the logic of distributed systems*, Cambridge University Press, Cambridge

9.2 Freyd J (1983) "Shareability: the social psychology of epistemology", *Cognitive Science*, 7: 191210

9.3 Grice P (1989) *Studies in the way of words*. Harvard University Press, Cambridge, MA

9.4 Gärdenfors P (2000) *Conceptual Spaces: the Geometry of Thought* MIT Press. London, UK

9.5 Landauer TK, Foltz PW, Laham D (1998) An introduction to latent semantic analysis, in *Discourse Processes*, 25:259-284

9.6 Lund K, Burgess C (1996) Producing high-dimensional semantic spaces from lexical co-occurrence in *Behavior Research Methods, Instruments & Computers*, 28(2):203–208

9.7 McArthur R, Bruza PD (2001) The ABCs of Online Community, in *Web Intelligence: Research and Development*, LNAI 2198, Springer Verlag, Heidelberg, Germany, pp.141-147

9.8 McGuiness DL (2002) Ontologies Come of Age. In: Fensel D. et al (eds) *Spinning the Semantic Web: Bringing the World Wide Web to Its Full Potential*, MIT Press, Cambridge, MA

9.9 Perry J (1997) Indexicals and demonstratives, In: Hales B, Wright C (eds) *A companion to the philosophy of language*, Blackwell, Oxford, UK, pp.593-595

9.10 Perry J (1998) Indexicals, contexts, and unarticulated constituents, In *Computing Natural Language*, CSLI Lecture Note No.81, pp.1-11, CSLI Publications (Stanford, CA)

9.11 Song D and Bruza PD (2001) Discovering information flow using a high dimensional conceptual space, in *Proceedings of the 24th Annual International ACM SIGIR Conference*, pp. 327-333

9.12 Song D, Bruza PD (2002) "Towards a theory of context sensitive information inference", accepted for publication in *Journal of the American Society for Information Science and Technology*, pp.326–339

10. Agent Communications for Chance Discovery

Peter McBurney[1] and Simon Parsons[1,2]

[1] Department of Computer Science,
University of Liverpool,
Chadwick Building, Peach Street,
Liverpool L69 7ZF, UK
email:p.j.mcburney@csc.liv.ac.uk

[2] Department of Computer and Information Science,
Brooklyn College, City University of New York,
2900 Bedford Avenue,
Brooklyn, NY 11 210, USA
email:parsons@sci.brooklyn.cuny.edu

Summary.

This chapter considers chance discovery and management in a community of intelligent, autonomous, software agents, where agents may have differing beliefs and intentions. For such a community of agents, we derive a set of five requirements for the design of languages and protocols for communications between the agents when discussing chance discovery and management. We then use these requirements to assess two proposals in the multi-agent systems community for agent communications: generic languages, such as the FIPA ACL, and dialogue game protocols. The latter are found to have greater potential capability to support dialogues over chance discovery and management between autonomous agents.

10.1 Introduction

Chance discovery and management in many real-world situations involves more than one entity. Identifying the causes of failure in a telecommunications network, for example, may involve co-ordination between a number of companies and organizations, each owning physical facilities or providing services over part of the network concerned. Each company in the network may have only a partial view of the whole network, with no one having an entire view. Moreover, each may have information about its portion of the network which is commercially sensitive, and so therefore cannot be shared with the other participants. In such circumstances, relevant knowledge may be distributed and decision making may require collaboration between the various entities involved. To enable effective chance discovery and management here the entities involved need to be able to communicate with one another. If processes of chance discovery and management are to be automated, then communications languages and protocols are necessary for the computational entities involved. In this chapter, we present the current state of research on communications languages in the multi-agent systems community, and discuss its relevance for chance discovery and management.

The chapter begins in Sect. 10.2 with a discussion of the requirements which chance discovery and management (CDM) place on languages and protocols for agent communications. We then consider, in Sect. 10.3, one of the main proposals for a generic agent communications language, FIPA ACL, against these requirements. This is followed, in Sect. 10.4, with a similar discussion of dialogue game protocols, which researchers in artificial intelligence(AI) have adopted recently from philosophy as the basis for agent communications. Both Sects 10.3 and 10.4 also serve as introductions to the specific proposals, for those unfamiliar with these areas of AI research. Section 10.5 concludes the chapter with a discussion of future work.

10.2 CDM Requirements

We begin by assuming we have a community of autonomous computational agents, each with a partial view of some domain. Our notion of agent is that presented in [10.47]: an agent is situated in a specific environment and there exhibits four characteristics: autonomy of decision making; social awareness; reactivity to events in its environment; and proactivity, as it seeks to actively achieve some goals within its environment. A community of such agents is a multi-agent system where two or more autonomous computational agents interact together. An example could be a network of water data monitors, each responsible for water and flood control in part of some geographic region. In this example, each monitor may be able to collect local information, send and receive information to and from other monitors, and take local action, such as opening a dam sluice to release excess water. Actions may be taken independently or in collaboration with other monitors.

Given such an agent community, we will assume throughout that the relevant domain of knowledge is partitioned, with each agent in the community having only knowledge of, or power in, some proper subset of the domain. For example, the partition of the domain could be based on geography, as in the case of distributed water control monitors, or based on time, as when Earth-based radio transceiver stations are only able to observe orbiting satellites for several hours each day, or based on a conceptual hierarchy, as when telecommunications service providers do not have access to lower or higher layers in the physical network on which they operate. In addition, the participating agents may have different goals and intentions, which may arise as a result of different perceptions of their economic interests. These goals may conflict. Even if all the agents in the community are rational, in the economic-theoretic sense that each seeks to maximize its expected utility [10.6], their goals may be still be in conflict. A provider of value-added telecommunications services such as voicemail, for example, may only be able to increase its profit margin at the expense of the provider of the physical infrastructure underlying the value-added service.

Imagine such an agent community is engaged in the discovery and management of chance events, such as risks or opportunities. What requirements does this mission place on the communications languages and protocols between the participating agents? Firstly, the language would need to be able to transmit domain-specific information in an appropriate form between participants. But simply being able to transmit information is not sufficient. Because the members of the community each have only partial views of the domain, what appears to one agent as a chance event, a risk of flood say, may not appear that way to other agents in the community. Therefore agents also need the ability to question each other about the information transmitted, to give reasons for their respective beliefs and intentions, and to challenge and contest these. In other words, the participants need the ability to argue with each other concerning the messages they transmit, and this is the second requirement of an agent communications language for CDM.

Thirdly, the ability to present arguments and counter-arguments means that any communications protocol used by the agents must encode some theory of debate or argument, what is called a logic of argumentation [10.13]. Under what circumstances, for example, must an agent who asserts that some belief is true be then forced to defend this belief against questions or attacks by other agents? How many such questions may be asked, if the discussion is not to run forever? When must a debate end, and a resolution be sought? It is issues such as these which a theory of argument addresses. As the philosopher Jürgen Habermas wrote [10.19, p. 22]: *"The logic of argumentation does not refer to deductive connections between the semantic units (sentences) as does formal logic, but to nondeductive relations between the pragmatic units (speech acts) of which arguments are composed."*.In a later chapter in this volume, we discuss logics of argumentation in more detail (Chap. 11 by Parsons and McBurney).

Being able to persuade another agent to adopt a new intention is of little value if the communications language does not permit that agent to express any changes to its beliefs or intentions. Thus, a fourth requirement is that the communications language should enable the participating agents to articulate relevant changes in their internal states, a process we have called the expression of *self-tranformation* [10.33], following [10.18]. This might seem a trivial requirement, but, as will be seen in the next section, not all agent communications languages facilitate this.

Awareness of the possibility of rare events should lead rational participants to efforts to prevent or encourage their occurrence, and/or to mitigate or enhance their consequences [10.36]. Thus any discussion between agents about chance discovery leads naturally to discussions about chance management. But chance management may require action by more than one agent in a community, as when several flood-control devices need to co-ordinate their actions to prevent a flood. The assumption we have made of of agent autonomy means that, in general, agents cannot be ordered to adopt a specific intention, but must be persuaded.

In earlier work [10.30], we considered the nature of dialogues about chance discovery and management relative to a standard typology of dialogues in the philosophy of argumentation. This typology, due to Walton and Krabbe [10.44], identifies six primary types of dialogue, based upon the information the participants have at the commencement of a dialogue (of relevance to the topic of discussion), their individual goals for the dialogue, and the goals they share. **Information-seeking dialogues** are those where one participant seeks the answer to some question(s) from another participant, who is believed by the first to know the answer(s). In **inquiry**

dialogues the participants collaborate to answer some question or questions whose answers are not known to any one participant. **Persuasion dialogues** involve one participant seeking to persuade another to accept a proposition he or she does not currently endorse. In **negotiation dialogues**, the participants bargain over the division of some scarce resource. Here, the goal of the dialogue – a division of the resource acceptable to all – may be in conflict with the individual goals of the participants. Participants of **deliberation dialogues** collaborate to decide what action or course of action should be adopted in some situation. Here, participants share a responsibility to decide the course of action, or, at least, they share a willingness to discuss whether they have such a shared responsibility. Note that the best course of action for a group may conflict with the preferences or intentions of each individual member of the group; moreover, no one participant may have all the information required to decide what is best for the group. In **eristic dialogues**, participants quarrel verbally as a substitute for physical fighting, aiming to vent perceived grievances.

Most actual dialogue occurrences – both human and agent – involve mixtures of these dialogue types. In our earlier work [10.30] we observed that dialogues about chance events – chance discoveries – are forms of inquiries. However, unlike the inquiries in the Walton and Krabbe classification, they are not disinterested searches for truth; instead, they involve a search overlaid with a value system. Thus, a discussion about possible risks of some system involves a search not for all possible outcomes, but only for those with negative consequences. Thus, chance discovery dialogues differ from inquiries. Chance management dialogues, in contrast, are dialogues about what to do to prevent, encourage, mitigate, or enhance a chance event, and so may be viewed as deliberation dialogues. However, as with most deliberations, they may have information-seeking, inquiry, persuasion, or negotiation subdialogues embedded within them.

Despite a full discussion and exchange of arguments and counter-arguments between participants, agents may still disagree about what to believe and about what actions to take, if any. If a decision must be made by the community about action, then the community needs some means to resolve differences between its members. Such a mechanism could be as simple as a voting process, as proposed in [10.23], or an argument-classification system, as in [10.31][1]. Alternatively, a resolution mechanism may rely on a more complex process of collaborative deliberation, as in [10.21]. The resolution process adopted will embed or reflect a particular structure of social and political relationships between the participants, such as democracy [10.10] or an organizational hierarchy [10.38]. Whatever political structure and resolution mechanism adopted by the agent community, the communications language needs to support this; if, for example, a voting mechanism is used, then agents need to be able to express their votes subject to whatever other conditions, such as confidentiality and verifiability, are desired. This is the fifth requirement of an agent communications language for CDM: that it enables an appropriate mechanism for resolution of differences of beliefs or intentions of the participants.

---

[1] The argumentation-classification approach is explained in Chap. 11 by Parsons and McBurney.

Finally, as with the design of any artificial language, there may be requirements regarding computational simplicity, non-redundancy, transparency, etc.; we are not concerned with these issues in this chapter. It is also worth noting that in classical mathematical communications theory the semantics of any information transmitted is ignored in the design and analysis of the communications mechanism. For example, Shannon states [10.43, p. 31]: *"Frequently the messages have meaning; that is they refer to or are correlated according to some system with certain physical or conceptual entities. These semantic aspects of communication are irrelevant to the engineering problem."*. By contrast, our discussion above reveals this domain as one where the design of the communication mechanism is necessarily influenced by the meaning and purposes – the semantics – of the messages to be transmitted.

10.3 Generic Agent Communications Languages

In this section we consider a generic agent communications language from the perspective of the requirements listed above. Over the last decade two major proposals have been advanced for agent communications languages, the knowledge query and manipulation language (KQML) [10.28], which arose from work sponsored by the US Government's Defense Advanced Research Projects Agency (DARPA), and the Foundation for Intelligent Physical Agents' agent communications language [10.15], the FIPA ACL, which arose from attempts to develop an industry standard for agent communications. Initially, KQML existed without a defined semantics, criticism of which led to the FIPA ACL being defined from 1995. Since the two languages are similar [10.24], we focus attention here only on the FIPA ACL. We first describe the language, and then assess it against the CDM requirements of the previous section[2].

10.3.1 FIPA ACL

The FIPA ACL has a two-level hierarchy for communications messages. At the lower, or 'inner' level, the content of messages may be expressed in any logical language agreeable to the participants. Such content is wrapped in one of 22 possible locutions , the FIPA communicative act library (CAL), which together comprise an 'outer' level. For instance, the *inform* locution has the following syntax [10.15, p. 11]:

```
(inform
    :sender (agent-identifier :name j)
    :receiver (set (agent-identifier :name i))
    :content
        'weather (today, raining)'
    :language Prolog)
```

---

[2] More details regarding FIPA ACL and KQML may be found in [10.24, 10.28, 10.46].

In this example the content of the message is *'weather (today, raining)'*, while the locution in which this content is wrapped is *inform*. The syntax of the locution indicates that the language in which the message content is expressed is Prolog. The syntax also indicates that the sender of the message is identified by agent-identifier symbol j, while the intended recipient is identified by symbol i.

We now list the names of the 22 locutions of the FIPA CAL, together with the summary descriptions of their intended purposes taken from the current FIPA specification [10.15]. Where the syntax of the locution differs from its name, the syntax is included in parentheses after the name.

Accept Proposal (accept-proposal) The action of accepting a previously submitted proposal to perform an action.

Agree The action of agreeing to perform some action, possibly in the future.

Cancel The action of one agent informing another agent that the first agent no longer has the intention that the second agent perform some action.

Call for Proposal (cfp) The action of calling for proposals to perform a given action[3].

Confirm The sender informs the receiver that a given proposition is true, where the receiver is known to be uncertain about the proposition.

Disconfirm The sender informs the receiver that a given proposition is false, where the receiver is known to believe, or believe it is likely that, the proposition is true.

Failure The action of telling another agent that an action was attempted but the attempt failed.

Inform The sender informs the receiver that a given proposition is true.

Inform If (inform-if) A macro action for the agent of the action to inform the recipient whether or not a proposition is true.

Inform Ref (inform-ref) A macro action for sender to inform the receiver the object which corresponds to a descriptor, for example, a name.

Not Understood (not-understood) The sender of the act (for example, i) informs the receiver (for example, j) that it perceived that j performed some action, but that i did not understand what j just did. A particular common case is that i tells j that i did not understand the message that j has just sent to i.

Propagate The sender intends that the receiver treat the embedded message as sent directly to the receiver, and wants the receiver to identify the agents denoted by the given descriptor and send the received *propagate* message to them.

Propose The action of submitting a proposal to perform a certain action, given certain preconditions.

---

[3] Note that, as throughout this list, the wording here is that of the FIPA CAL specifications [10.15]. The two instances of the word *'action'* in this statement have two different meanings, the first referring to communicative actions in the FIPA Communicative Act Library executed in the course of dialogues by agents adhering to the FIPA ACL, while the second refers to actions undertaken by the agents outside a dialogue. Strangely, these different meanings are not made explicit in the specifications.

Proxy The sender wants the receiver to select target agents denoted by a given description and to send an embedded message to them.

Query If (query-if) The action of asking another agent whether or not a given proposition is true.

Query Ref (query-ref) The action of asking another agent for the object referred to by an [*sic*] referential expression.

Refuse The action of refusing to perform a given action, and explaining the reasons for the refusal.

Reject Proposal (reject-proposal) The action of rejecting a proposal to perform some action during a negotiation.

Request The sender requests the receiver to perform some action. One important class of uses of the request act is to request the receiver to perform another communicative act.

Request When (request-when) The sender wants the receiver to perform some action when some given proposition becomes true.

Request Whenever (request-whenever) The sender wants the receiver to perform some action as soon as some proposition becomes true and thereafter each time the proposition becomes true again.

Subscribe The act of requesting a persistent intention to notify the sender of the value of a reference, and to notify again whenever the object identified by the reference changes.

The use in these summaries of the word '*act*' to describe locutions is not accidental. The FIPA ACL has been given an axiomatic semantics, called *SL* (for *semantic language*), based on speech act theory [10.15, Informative Annex A][4]. This theory, due originally to a philosopher of language, John Austin [10.7], and extended by his student, John Searle [10.42], considers spoken human utterances as actions, in so far as they may change the state of the world. Some utterances do this demonstrably as when countries issue *declarations of war* on other countries. Other utterances change the world un-observably, as when one person informs another of something he or she does not already know, and thereby causes a revision to the listener's beliefs. Austin and Searle classified spoken utterances by their intended and actual effects on the world (including the internal mental states of those hearing the utterances), and developed preconditions for those effects to be realized. Speech act theory was then used to provide a semantics for agent utterances, particularly in work by Philip Cohen, Hector Levesque, and Raymond Perrault [10.11, 10.12]. Drawing on this work, the FIPA ACL semantic language *SL* is a modal logic formalism which defines pre and postconditions for the 22 locutions of the FIPA Communicative Act Library in terms of the beliefs, uncertain beliefs, and desires of the agents participating in the dialogue. For example, the *inform* locution, in which one agent tells another some proposition, may only be uttered if the first agent believes the proposition to be true. This is a semantic condition for the locution[5].

---

[4] For a description of different types of program semantics, see e.g. [10.34].

[5] Following Searle [10.42], this is termed a *sincerity* condition. Because beliefs and desires are internal states of the agents concerned, expressions by an agent of its beliefs, desires,

10.3.2 Assessment of FIPA ACL

What is one to make of the FIPA ACL as a protocol for chance discovery and management? We answer this question by considering FIPA ACL against each of the five CDM requirements presented in Sect. 10.2.

The first requirement is that the language support the transmission of domain-specific information. This FIPA ACL does, and it does so with no limitations on the content of messages. However, it is worth noting that the 22 locutions of the FIPA Communicative Act Library are not situated in any conversational context. It is as if the agents using them have always been connected to each other, are so now, and always will be, and every so often one agent sends a message to one or more of the others, an act which may initiate an exchange of further messages. The image that comes to my mind is that of a group of prisoners chained together for a lengthy period in a remote location, with only each other for company; consequently, they know each other's topics of conversation in depth[6]. In such a context there is no need for locutions which express a desire to enter or leave a dialogue, or for introductions, or for statements of what the conversation will be about. Only one locution, *cfp*, appears to initiate a conversation on a specific topic, in this case a call for proposals to undertake some action. This absence of entry and withdrawal locutions may or may not be appropriate for specific CDM domains.

The second CDM requirement is that the language support the questioning and defense of claims, and the giving and receiving of reasons for beliefs and intentions. Against this requirement, FIPA ACL is inadequate. Both this language and KQML grew initially from efforts by DARPA to develop technologies for *knowledge sharing* [10.24]. Such a conceptual paradigm explains why fully 11 of the 22 locutions seek to request or send information (*cfp, confirm, disconfirm, inform, inform-if, inform-ref, propagate, proxy, query-if, query-ref, subscribe*). Despite this, the language has not been designed with the possibility that such information may be rationally questioned or challenged. An agent receiving an *inform(ϕ)* message who is unsure about the truth of its contents ϕ, or who does not hold the belief that ϕ is true, has few options to express these views. That agent may utter *disconfirm(ϕ)*, which indicates disagreement, or perhaps *not-understood(inform(ϕ))*, which indicates that the second agent did not understand the *inform(ϕ)* message. For the second agent to question the first agent, the *disconfirm(ϕ)* response may be too strong, while the *not-understood(ϕ)* response may be too weak; how may the second agent simply ask the reasons for believing ϕ to be true? And, how may the first agent respond with an argument for ϕ? In any case, there is no obligation on the first agent to explain or justify its endorsement of or belief in ϕ upon such requests. The design of the language appears to have been undertaken without thought that an agent may seek to persuade another to change its beliefs.

---

or intentions can not be verified by other agents without access to the code of the speaker. This is not possible in most open agent systems, and so the issue of verifiability of agent communications languages – determining that participants actually conform to the semantic requirements of the language – is problematic [10.45].

[6] Jean-Paul Sartre's play *'No Exit'* describes just such a situation [10.41].

Thus, although FIPA ACL permits domain-specific information to be transmitted, thus satisfying the first CDM requirement, the language does not facilitate the questioning and contestation of such information, or the presentation of arguments for or against specific statements. Thus, the second CDM requirement is not satisfied. Likewise, the absence of rules regarding connecting locutions, such as the obligations of agents who assert some statement to defend it, mean that the third CDM requirement is also not satisfied, the requirement for a theory of argument. Users of FIPA ACL may, of course, agree themselves such a theory and use this over and above the language, in the same way that FIPA protocols for auctions have been defined, for example [10.16, 10.17], but the language has not been designed with such higher-level protocols in mind. The absence of a logic of argumentation for the FIPA ACL locutions means that any locution may follow any other. In addition to allowing belligerent or malicious participants to disrupt conversations, this also complicates the task of analysis of utterances and conversations. After all, any given sequence of utterances in the FIPA ACL may be followed by any one of the 22 locutions, and each of them in turn followed by one of 22 locutions, and so on. This creates a state-explosion problem for participants analyzing a sequence of utterances to decide what locution to utter next, or seeking to infer the future utterances of other participants.

The fourth CDM requirement is that the language permits agents to express *self-transformation*. The sincerity condition on the *inform* locution in FIPA ACL means that agents may only utter propositions they believe to be true. Changes of belief in a statement ϕ could only be expressed by successive utterances, as in:

inform(ϕ)

inform($\neg\phi$)

But there is no requirement or even possibility for an explicit retraction of the first statement before utterance of the second, since there are no explicit retraction locutions. If other agents receive these two *inform* statements, are they to conclude that the speaker has changed its belief about ϕ, or that the speaker is merely expressing inconsistent beliefs? Indeed, hearing such successive utterances may be indicative of other states in the speaker's mind: madness, or malicious behavior, or simply buggy code. The absence of any argumentation-theoretic rules in FIPA ACL mean that all of these conclusions are possible, and so further conversational overhead is required for listeners to determine that the speaker has indeed undergone a self-transformation. Moreover, FIPA ACL assumes that beliefs are two-valued (true or false), and so agents cannot express degrees of belief. In addition, the making of tentative suggestions or uncertain proposals does not appear possible. In short, FIPA ACL does not readily provide the means by which participants may express any self-transformation they may undergo as a result of discussion with other agents.

In Sect. 10.2, we cited Habermas's characterization of a logic of argumentation as a set of relations between speech acts. Looking at the FIPA ACL locution definitions, we see (as mentioned above) that 11 of the 22 locutions seek to request or transmit information; and that four involve negotiation (*accept-proposal, cfp, propose, reject-proposal*); six involve the performance of actions (*agree, cancel, refuse, request, request-when, request-whenever*) and two error handling (*failure, not-understood*)[7]. Our conclusion, looking at these locutions, is that this language has been designed primarily to enable information transfers and purchase transactions, not the rational discussions about beliefs and intentions required for dialogues over chance discovery and management. Purchase transactions may be initiated by a *cfp(.)* locution, which may elicit a *propose(.)* response and then a series of counter-proposals, and/or requests for information, etc. Following successful conclusion of these exchanges, the participating agents may report on the action they agreed to in the purchase transaction by means of the various action locutions. However, the absence of a theory of argument means that rational discussion about these transactions is not possible.

The fifth CDM requirement is that the language support appropriate resolution mechanisms between agents. FIPA ACL does not include any such mechanisms in its definition, but these could readily be defined to overlay the language, in the same way that certain auction protocols have been defined, e.g. [10.16, 10.17]. There would appear to be no obstacles, other than the weaknesses identified above, which would preclude such overlaid protocols.

In conclusion, we believe the FIPA ACL as it currently stands is unsuitable as a protocol for rational discussion between autonomous agents for chance discovery and management. It provides limited support for the rational questioning of others' beliefs or intentions, or the means to persuade them to change their beliefs and intentions; it provides limited means for agents whose beliefs change to express those changes; it provides no means to express tentative suggestions, group intentions, or arguments for positions. In other words, the protocol does not readily support inquiry, persuasion, or deliberation dialogues or negotiation dialogues other than simple purchase transactions. Since, as we noted in the previous section, chance discovery and chance management dialogues are types of inquiry and deliberation dialogues respectively, FIPA ACL, as it stands, does not support CDM.

10.4 Dialogue Game Protocols

In this section, we consider another approach to the design of agent communications languages which has recently received attention from researchers in AI. This involves the use of formal dialogue games, which are interactions between two or more players where each player 'moves' by making utterances, according to a defined set of rules. Although their study dates at least from Aristotle [10.5], they have found recent application in philosophy, computational linguistics, and artificial intelligence. In philosophy, dialogue games have been used to study fallacious reasoning [10.20, 10.27] and to develop a game-theoretic semantics for intuitionistic and classical logic [10.26] and quantum logic [10.35]. In linguistics, they have

---

[7] We count *cfp* in both the information and the negotiation categories. This categorization of locutions is based on that of [10.46].

been used to explain sequences of human utterances [10.25], with subsequent application to machine-based natural language processing and generation [10.22], and to human–computer interaction [10.8]. Within computer science and AI, they have been applied to modeling complex human reasoning, for example in legal domains [10.9], and to requirements specification for complex software projects [10.14]. Dialogue games differ from the games of economic game theory [10.37] in that payoffs for winning or losing a game are not considered, and because there is no use of uncertainty measures, such as probabilities, to model the possible moves of opponents. They also differ from the abstract games recently used as a semantics for interactive computation [10.1], since these latter games do not share the rich rule structure of dialogue games, nor are they intended to have themselves a semantic interpretation involving the sharing of beliefs and intentions among a group of agents.

10.4.1 Formal Dialogue Games

We now present a model of a generic formal dialogue game in terms of the components of its specification, taken from [10.32]. We first assume that the topics of discussion between the agents are represented in some logical language, whose well-formed formulae are denoted by the lower-case Roman letters, p, q, r, etc. A dialogue game specification then consists of the following elements:

Commencement rules: rules which define the circumstances under which the dialogue commences.

Locutions: rules which indicate what utterances are permitted. Typically, legal locutions permit participants to assert propositions, permit others to question or contest prior assertions, and permit those asserting propositions which are subsequently questioned or contested to justify their assertions. Justifications may involve the presentation of a proof of the proposition or an argument for it. The dialogue game rules may also permit participants to utter propositions to which they assign differing degrees of commitment, for example: one may merely *propose* a proposition, a speech act which entails less commitment than would an *assertion* of the same proposition.

Combination rules: rules which define the dialogical contexts under which particular locutions are permitted or not, or obligatory or not. For instance, it may not be permitted for a participant to assert a proposition p and subsequently the proposition $\neg p$ in the same dialogue, without in the interim having retracted the former assertion.

Commitments: rules which define the circumstances under which participants express commitment to a proposition. Typically, the assertion of a claim p in the debate is defined as indicating to the other participants some level of commitment to, or support for, the claim. Since the work of the philosopher Charles Hamblin [10.20], formal dialogue systems typically establish and maintain public sets of commitments, called *commitment stores*, for each participant; these stores are usually non-monotonic, in the sense that participants can also retract committed claims, although possibly only under defined circumstances[8].

Termination rules: rules that define the circumstances under which the dialogue ends.

Dialogue game protocols have been articulated for each of the rule-governed primary types of dialogues in the typology of Walton and Krabbe: e.g. information-seeking dialogues [10.22, 10.44]; inquiries [10.31]; persuasion dialogues [10.2]; negotiation dialogues [10.22, 10.29]; and deliberations [10.21]. However, as mentioned earlier, most real-world dialogues (whether human or agent) involve aspects of more than one of these primary types. Two formalisms have been suggested for computational representation of combinations of dialogue: the *dialogue frames* of Chris Reed [10.39], which enable iterated, sequential, and embedded dialogues to be represented; and our own *agent dialogue frameworks* [10.32], which permit iterated, sequential, parallel, and embedded dialogues to be represented. Both these formalisms are neutral with regard to the modeling of the primary dialogue types themselves, allowing the primary types to be represented in any convenient form, and allowing for types other than the six of the Walton and Krabbe typology to be included.

As an example of a dialogue game protocol, we present in outline a protocol developed by one of us with Leila Amgoud and Nicolas Maudet [10.2]. This protocol is based on James MacKenzie's philosophical dialogue game *DC* [10.27], a game for two players, both subject to the same rules. *DC* enables the participants to argue about the truth of a proposition and was designed to study the fallacy of begging the question (*petitio principii*, or circular reasoning). The agent interaction protocol of [10.2] based on *DC* has four distinct locutions: *assert, accept, question*, and *challenge*; these can be instantiated with a single proposition, and also, for the locutions *assert* and *accept*, with a set of propositions which together constitute an argument for a proposition. Thus the participants may communicate both propositional statements and arguments about these statements, where arguments may be considered as tentative proofs (i.e. logical inferences from assumptions which may not all be confirmed). The locutions of this protocol are similar to those of *DC* except that they do not include a locution for retraction of assertions (called *withdrawal* in *DC*). As with MacKenzie's game, the protocol of [10.2] establishes commitment stores which record, in full public view, the statements each participant has asserted. The syntax for this protocol has only been provided for dialogues between two participants, but could be readily extended to more agents, as the same authors did subsequently in [10.3].

Amgoud et al. demonstrate that their system enables persuasion dialogues, inquiry dialogues, and information-seeking dialogues [10.2]. However, as the authors note, to permit negotiation dialogues, the protocol requires additional locutions[9].

---

[8] It is worth noting here that more than one notion of *commitment* is present in the literature on dialogue games.

[9] It may also require additional locutions for deliberation dialogues, although the authors suggest otherwise.

These are presented in a subsequent paper by the same authors [10.4], in which three additional locutions are proposed, *request, promise*, and *refuse*, making seven in all. In addition to instantiation with propositions and with arguments for propositions, several of these locutions can also be instantiated with a two-valued function expressing a relationship between two resources. For example, the locution *promise*$(p \Rightarrow q)$ indicates a promise by the speaker to provide resource q in return for receiving resource p.

10.4.2 Assessment of Dialogue Games

Given the number and variety of dialogue game protocols which have been proposed, it is not possible to assess each of them against the CDM requirements for agent communications languages of Sect. 10.2. However, we may attempt an assessment of generic game protocols against each of the five requirements.

The first CDM requirement is that the language support the transmission of domain-specific information. Because all formal dialogue games proposed so far in AI have adopted the same two-layer structure of *message contents* wrapped inside *locutions*, then this requirement is generally satisfied. As with FIPA ACL and KQML, the message contents may be expressed in any formalism agreed between the participants, and so may support the transmission of domain-specific information. The second CDM requirement is that the language support the questioning and defense of claims, and the giving and receiving of reasons for beliefs and intentions. Dialogue game protocols have their origin in the philosophy of argumentation, and so this requirement has typically been satisfied[10]. The same conclusion applies to the third CDM requirement, which is that the agent communications language encode some theory of argument.

The fourth requirement is that the language permits agents to express self-transformation. Here, the assessment depends very much on the locutions permitted in the game, and their syntactic, semantic, and pragmatic rules of use. For example, in [10.33], we showed that the extent to which certain dialogue game protocols in AI permit self-transformation differs from one protocol to another. The fifth CDM requirement is that the language support appropriate resolution mechanisms between agents. In general, as with FIPA ACL, dialogue game protocols are sufficiently flexible to permit any agreed resolution method to be overlaid.

We conclude from this brief analysis that well-designed dialogue game protocols are better able to support agent communications for chance discovery and management dialogues than are generic agent communications languages such as FIPA ACL.

---

[10] It is interesting to observe that the main application of formal dialogue games before modern times was their use for oral disputation between philosophers [10.20].

10.5 Conclusions

In this chapter, our focus has been on the design of languages and protocols for communications between autonomous software agents engaged in joint chance discovery and management over some domain. We have assumed that the agents may have only partial views of the domain and may have differing interests and objectives. Thus, their beliefs and intentions may differ, and may even be in conflict. Consequently, any discussion between such a community of agents concerning possible chance events and any chance management actions may require considerable debate and argument. Arising from this conclusion, we proposed five requirements that any agent communications language would need to satisfy in order to fully enable agent dialogues about chance discovery and chance management.

We then considered two broad categories of proposals for agent communications languages from the multi-agent system community with respect to these five requirements: generic agent communications languages and formal dialogue game protocols. For the former category, we took the agent communications language of the Foundation for Intelligent Physical Agents, FIPA ACL, as a typical example. Although it met two of the requirements, we concluded that FIPA ACL was inadequate to the task of supporting dialogues over chance discovery and management. In contrast, dialogue game protocols have much greater potential to support CDM discussions, although this will depend on the specific locutions and rules of the dialogue game in any particular case.

Chance discovery and management in a multi-agent context leads to additional issues and questions. For example, multi-agent chance discovery may be computationally more complex than chance discovery by a single, ideal, agent having full knowledge of the domain. Similar issues have recently been considered in the case of multi-agent diagnosis [10.40], where each agent had only a partial view of the domain. However, this work did not consider that agents may have possibly-differing objectives, which is likely in any real-world application of multi-agent chance discovery. We hope to consider such issues in future work.

References

10.1 S. Abramsky. Semantics of interaction: an introduction to game semantics. In A. M. Pitts and P. Dybjer, editors, *Semantics and Logics of Computation*, pages 1–31. Cambridge University Press, Cambridge, UK, 1997.

10.2 L. Amgoud, N. Maudet, and S. Parsons. Modelling dialogues using argumentation. In E. Durfee, editor, *Proceedings of the Fourth International Conference on Multi-Agent Systems (ICMAS 2000)*, pages 31–38, Boston, MA, USA, 2000. IEEE Press.

10.3 L. Amgoud and S. Parsons. Agent dialogues with conflicting preferences. In J-J. Meyer and M. Tambe, editors, *Pre-Proceedings of the Eighth International Workshop on Agent Theories, Architectures, and Languages (ATAL 2001)*, pages 1–14, Seattle, WA, USA, 2001.

10.4 L. Amgoud, S. Parsons, and N. Maudet. Arguments, dialogue, and negotiation. In W. Horn, editor, *Proceedings of the Fourteenth European Conference on Artificial Intelligence (ECAI 2000)*, pages 338–342, Berlin, Germany, 2000. IOS Press.

10.5   Aristotle. *Topics*. Clarendon Press, Oxford, UK, 1928. (W. D. Ross, Editor).

10.6   K. J. Arrow. *Social Choice and Individual Values*. Wiley, New York, 1951.

10.7   J. L. Austin. *How To Do Things with Words*. Oxford University Press, Oxford, UK, 1962. Originally delivered as the William James Lectures at Harvard University in 1955.

10.8   T. J. M. Bench-Capon, P. E. Dunne, and P. H. Leng. Interacting with knowledge-based systems through dialogue games. In *Proceedings of the Eleventh International Conference on Expert Systems and Applications*, pages 123–140, Avignon, France, 1991.

10.9   T. J. M. Bench-Capon, T. Geldard, and P. H. Leng. A method for the computational modelling of dialectical argument with dialogue games. *Artificial Intelligence and Law*, 8:233–254, 2000.

10.10 J. Bohman and W. Rehg, editors. *Deliberative Democracy: Essays on Reason and Politics*. MIT Press, Cambridge, MA, USA, 1997.

10.11 P. R. Cohen and H. J. Levesque. Rational interaction as the basis for communication. In P. R. Cohen, J. Morgan, and M. E. Pollack, editors, *Intentions in Communication*, pages 221–255. MIT Press, Cambridge, MA, USA, 1990.

10.12 P. R. Cohen and C. R. Perrault. Elements of a plan-based theory of speech acts. *Cognitive Science*, 3:177–212, 1979.

10.13 F. H. van Eemeren, R. Grootendorst, F. S. Henkemans, J. A. Blair, R. H. Johnson, E. C. W. Krabbe, C. Plantin, D. N. Walton, C. A. Willard, J. Woods, and D. Zarefsky. *Fundamentals of Argumentation Theory*. Lawrence Erlbaum Associates, Mahwah, NJ, USA, 1996.

10.14 A. Finkelstein and H. Fuks. Multi-party specification. In *Proceedings of the Fifth International Workshop on Software Specification and Design*, Pittsburgh, PA, USA, 1989. ACM Sigsoft Engineering Notes.

10.15 FIPA. Communicative Act Library Specification. Technical Report XC00037H, Foundation for Intelligent Physical Agents, 10 August 2001. Available from: www.fipa.org.

10.16 FIPA. Dutch Auction Interaction Protocol Specification. Technical Report XC00032F, Foundation for Intelligent Physical Agents, 10 August 2001.

10.17 FIPA. English Auction Interaction Protocol Specification. Technical Report XC00031F, Foundation for Intelligent Physical Agents, 10 August 2001.

10.18 J. Forester. *The Deliberative Practitioner: Encouraging Participatory Planning Processes*. MIT Press, Cambridge, MA, USA, 1999.

10.19 J. Habermas. *The Theory of Communicative Action: Volume 1: Reason and the Rationalization of Society*. Heinemann, London, UK, 1984. Translation by T. McCarthy of: *Theorie des Kommunikativen Handelns, Band I, Handlungsrationalitat und gesellschaftliche Rationalisierung*. Suhrkamp, Frankfurt, Germany. 1981.

10.20 C. L. Hamblin. *Fallacies*. Methuen, London, UK, 1970.

10.21 D. Hitchcock, P. McBurney, and S. Parsons. A framework for deliberation dialogues. In H. V. Hansen, C. W. Tindale, J. A. Blair, and R. H. Johnson, editors, *Proceedings of the Fourth Biennial Conference of the Ontario Society for the Study of Argumentation (OSSA 2001)*, Windsor, Ontario, Canada, 2001.

10.22 J. Hulstijn. *Dialogue Models for Inquiry and Transaction*. PhD thesis, Universiteit Twente, Enschede, The Netherlands, 2000.

10.23 L. Hunsberger and M. Zancanaro. A mechanism for group decision making in collaborative activity. In *Proceedings of the Seventeenth National Conference on Artificial Intelligence (AAAI-2000)*, pages 30–35, Menlo Park, CA, USA, 2000. AAAI Press.

10.24 Y. Labrou, T. Finin, and Y. Peng. Agent communication languages: The current landscape. *IEEE Intelligent Systems*, 14(2):45–52, March/April 1999.

10.25 J. A. Levin and J. A. Moore. Dialogue-games: metacommunications structures for natural language interaction. *Cognitive Science*, 1(4):395–420, 1978.

10.26 P. Lorenzen and K. Lorenz. *Dialogische Logik*. Wissenschaftliche Buchgesellschaft, Darmstadt, Germany, 1978.

10.27 J. D. MacKenzie. Question-begging in non-cumulative systems. *Journal of Philosophical Logic*, 8:117–133, 1979.

10.28 J. Mayfield, Y. Labrou, and T. Finin. Evaluating KQML as an agent communication language. In M. J. Wooldridge, J. P. Müller, and M. Tambe, editors, *Intelligent Agents II*, Lecture Notes in Artificial Intelligence 1039, pages 347–360, Berlin, Germany, 1996. Springer.

10.29 P. McBurney, R. M. van Eijk, S. Parsons, and L. Amgoud. A dialogue-game protocol for agent purchase negotiations. *Journal of Autonomous Agents and Multi-Agent Systems*, 2002. *In press*.

10.30 P. McBurney and S. Parsons. Chance discovery using dialectical argumentation. In T. Terano, T. Nishida, A. Namatame, S. Tsumoto, Y. Ohsawa, and T. Washio, editors, *New Frontiers in Artificial Intelligence: Joint JSAI 2001 Workshop Post Proceedings*, Lecture Notes in Artificial Intelligence 2253, pages 414–424. Springer, Berlin, Germany, 2001.

10.31 P. McBurney and S. Parsons. Representing epistemic uncertainty by means of dialectical argumentation. *Annals of Mathematics and Artificial Intelligence*, 32(1–4):125–169, 2001.

10.32 P. McBurney and S. Parsons. Games that agents play: A formal framework for dialogues between autonomous agents. *Journal of Logic, Language and Information*, 11(3):315–334, 2002.

10.33 P. McBurney, S. Parsons, and M. Wooldridge. Desiderata for agent argumentation protocols. In C. Castelfranchi and W. L. Johnson, editors, *Proceedings of the First International Joint Conference on Autonomous Agents and Multi-Agent Systems (AAMAS 2002), Bologna, Italy*, pages 402–409, New York City, NY, USA, 2002. ACM Press.

10.34 B. Meyer. *Introduction to the Theory of Programming Languages*. Prentice Hall, New York City, NY, USA, 1990.

10.35 P. Mittelstaedt. *Quantum Logic*. D. Reidel, Dordrecht, The Netherlands, 1979.

10.36 Y. Ohsawa. Chance discoveries for making decisions in complex real world. *New Generation Computing*, 20(2), 2002.

10.37 M. J. Osborne and A. Rubinstein. *A Course in Game Theory*. MIT Press, Cambridge, MA, USA, 1994.

10.38 P. Panzarasa and N. R. Jennings. The organisation of sociality: a manifesto for a new science of multi-agent systems. In *Modelling Autonomous Agents in a Multi-Agent World: Proceedings of the Tenth European Workshop on Multi-Agent Systems (MAAMAW-01)*. Annecy, France, 2001.

10.39 C. Reed. Dialogue frames in agent communications. In Y. Demazeau, editor, *Proceedings of the Third International Conference on Multi-Agent Systems (ICMAS-98)*, pages 246–253. IEEE Press, 1998.

10.40 N. Roos, A. ten Teije, A. Bos, and C. Witteveen. An analysis of multi-agent diagnosis. In C. Castelfranchi and W. L. Johnson, editors, *Proceedings of the First International Joint Conference on Autonomous Agents and Multi-Agent Systems (AAMAS 2002), Bologna, Italy*, pages 986–987, New York City, NY, USA, 2002. ACM Press.

10.41 J.-P. Sartre. *No Exit and Three Other Plays*. Vintage International, USA, 1989.

10.42 J. Searle. *Speech Acts: An Essay in the Philosophy of Language*. Cambridge University Press, Cambridge, UK, 1969.

10.43 C. E. Shannon. The mathematical theory of communication. In C. E. Shannon and W. Weaver, editors, *The Mathematical Theory of Communication*, pages 29–125. University of Illinois Press, Chicago, IL, USA, 1963. Originally published in the *Bell System Technical Journal*, October and November 1948.

10.44 D. N. Walton and E. C. W. Krabbe. *Commitment in Dialogue: Basic Concepts of Interpersonal Reasoning.* SUNY Series in Logic and Language. State University of New York Press, Albany, NY, USA, 1995.

10.45 M. J. Wooldridge. Semantic issues in the verification of agent communication languages. *Journal of Autonomous Agents and Multi-Agent Systems*, 3(1):9–31, 2000.

10.46 M. J. Wooldridge. *Introduction to Multiagent Systems.* John Wiley and Sons, New York, NY, USA, 2002.

10.47 M. J. Wooldridge and N. R. Jennings. Intelligent agents: Theory and practice. *Knowledge Engineering Review*, 10(2):115–152, 1995.

11. Logics of Argumentation for Chance Discovery

Simon Parsons[1,2] and Peter McBurney[2]

[1] Department of Computer and Information Science,
Brooklyn College, City University of New York,
2900 Bedford Avenue,
Brooklyn, NY 11 210, USA
email:parsons@sci.brooklyn.cuny.edu

[2] Department of Computer Science,
University of Liverpool,
Chadwick Building, Peach Street,
Liverpool L69 7ZF, UK
email:p.j.mcburney@csc.liv.ac.uk

Summary.

If multiple autonomous entities – agents – are involved in chance discovery and management, then the agents involved may disagree as to what constitutes a chance event, and what action, if any, to take in response. One approach to agent communication in this situation is to insist that agents not only send messages, but also support them with reasons why those messages are appropriate. This is argumentation-based communication. In this chapter, we review some of our work on argumentation-based communication, discussing the issues we consider to be important in developing systems for argumentation-based communication between agents in chance discovery and management domains.

11.1 Introduction

When we humans engage in any form of dialogue it is natural for us to do so in a somewhat skeptical manner. If someone informs us of a fact that we find surprising, we typically question it. Not in an aggressive way, but what might be described as an inquisitive way. When someone tells us 'X is true' (where X can range across statements from 'It is raining outside.' to 'the Dow Jones index will continue falling for the next six months.', we want to know 'where did you read that?' or 'what makes you think that?'. Typically we want to know the basis on which some conclusion was reached. In fact, this questioning is so ingrained that we often present information with some of the answer to the question we expect it to provoke already attached – 'It is raining outside, I got soaked through.', 'The editorial in today's Guardian suggests that consumer confidence in the USA is so low that the Dow Jones index will continue falling for the next six months.'. This is exactly argumentation-based communication. It is increasingly being applied to the design of agent communication languages and frameworks, for example Dignum et al. [11.7, 11.8], Grosz and Kraus [11.12], Parsons et al. [11.24, 11.25], Reed [11.27], Schroeder et al. [11.30], and Sycara [11.34]. Indeed, the idea that it is useful for agents to explain what they are doing is not just confined to research on argumentation-based communication [11.28].

Apart from its naturalness, there are two major advantages of this approach to agent communication. One is that it ensures that agents are *rational* in a certain sense. As we shall see, and as is argued at length in [11.20], argumentation-based communication allows us to define a form of rationality in which agents only accept statements which they are unable to refute (the exact form of refutation depending on the particular formal properties of the argumentation system they use). In other words agents will only accept things if they don't have a good reason not to. The second advantage builds on this and, as discussed in more detail in [11.4], provides a way of giving agent communications a *social semantics* in the sense of Singh [11.32, 11.33]. The essence of a social semantics is that agents state publicly their beliefs and intentions at the outset of a dialogue, so that future utterances and actions may be judged for consistency against these statements. The truth of an agent's expressions of its private beliefs or intentions can never be fully verified [11.37], but at least an agent's consistency can be assessed, and, with an argumentation-based dialogue system, the reasons supporting these expressions can be sought. Moreover, these reasons may be accepted or rejected, and possibly challenged and argued-against, by other agents.

An example shows how these two advantages are especially important in the domain of chance discovery and management. Consider, for instance, a network of geographically distributed software agents, each responsible for water monitoring and control in a local domain of the catchment area of a major river [11.21]. One agent may identify, based on its local water level readings, that a flood is a strong possibility in the near future, and that preventative action should be taken. This action may require a second agent in the system, downstream of the first, to release water from a dam in its local domain. But suppose that the second agent has no evidence, in its own domain, of any increased water-levels. If the agents have some degree of relative autonomy, then the first agent cannot simply *order* the second to take the preventative action. Instead, the first agent may need to *persuade* the second, on the basis of the relevant evidence available. This will involve the giving of reasons by the first agent to the second, and, perhaps, the rational challenging by the second agent of these reasons. In other words, where the participants in a system are autonomous and where action is required for chance management, then there will be a need for argumentation-based communications between the participants[1].

This chapter sketches the state of the art in argumentation-based agent communication. We will do this not by describing all the relevant work in detail, but by identifying what we consider to be the main issues, for chance discovery and management domains, in building systems that communicate in this way by briefly describing how our work has addressed them.

---

[1] Multi-agent diagnosis is considered in [11.29], although not from an argumentation perspective.

11.2 Philosophical Background

Our work on argumentation-based dialogue has been influenced by a model of human dialogues due to argumentation theorists Doug Walton and Erik Krabbe [11.35]. Walton and Krabbe set out to analyze the concept of commitment in dialogue, so as to "provide conceptual tools for the theory of argumentation" [11.35, p. ix]. This led to a focus on persuasion dialogues, and their work presents formal models for such dialogues. In attempting this task, they recognized the need for a characterization of dialogues, and so they present a broad typology for inter personal dialogue. They make no claims for its comprehensiveness. Their categorization identifies six primary types of dialogues and three mixed types. As defined by Walton and Krabbe, the six primary dialogue types are:

Information-seeking dialogues: one participant seeks the answer to some question(s) from another participant, who is believed by the first to know the answer(s).

Inquiry dialogues: the participants collaborate to answer some question or questions whose answers are not known to any one participant.

Persuasion dialogues: one party seeks to persuade another party to adopt a belief or point of view he or she does not currently hold.

Negotiation dialogues: the participants bargain over the division of some scarce resource in a way acceptable to all, with each individual party aiming to maximize his or her share [2].

Deliberation dialogues: participants collaborate to decide what course of action to take in some situation.

Eristic dialogues: participants quarrel verbally as a substitute for physical fighting, with each aiming to win the exchange.

Chance discovery – the identification of rare events – between agents in a system would typically involve inquiry, information-seeking and persuasion dialogues. Chance management – actions taken to prevent, mitigate, or facilitate a rare event, or to deal with its consequences – would typically involve deliberation, and possibly negotation dialogues. This framework can be used in a number of ways. First, we have increasingly used this typology as a framework within which it is possible to compare and contrast different systems for argumentation. For example, in [11.3] we used the classification, and the description of the start conditions and aims of participants given in [11.35], to show that the argumentation system described in [11.3] could handle persuasion, information-seeking, and inquiry dialogues. Second, we have also used the typology as a means of classifying particular argumentation systems, for example identifying the system in [11.24] as including elements of deliberation (it is about joint action) and persuasion (one agent is attempting to persuade the other to do something different) rather than negotiation as it was originally billed. Third, we can use the typology as a means of distinguishing the focus

---

[2] Note that this definition of negotiation is that of Walton and Krabbe. Arguably negotiation dialogues may involve other issues besides the division of scarce resources.

(and thus the detailed requirements for) systems intended to be used for engaging in certain types of dialogue as in our work to define locutions to perform inquiry [11.22], chance discovery [11.21], and deliberation [11.14] dialogues.

The final aspect of this work that is relevant, in our view, is that it stresses the importance of being able to handle dialogues of one kind that include embedded dialogues of another kind. Thus a deliberation dialogue about the appropriate action to take to prevent a flood might include an embedded information-seeking dialogue (to discover if water levels are rising everywhere), and an embedded persuasion dialogue (about the value of a particular flood-prevention action). This has led to formalisms in which dialogues can be combined, such as [11.23, 11.27].

11.3 Argumentation and Dialogue

The focus of attention by philosophers to argumentation has been on understanding and guiding human reasoning and argument. It is not surprising, therefore, that this work says little about how argumentation may be applied to the design of communication systems for artificial agents. In this section we consider some of the issues relevant to such application.

11.3.1 Languages and Argumentation

Considering two agents that are engaged in some dialogue, we can distinguish between three different languages that they use. Each agent has a *base language* that it uses as a means of knowledge representation, a language we might call L. This language can be unique to the agent, or may be the same for both agents. This is the language in which the designer of the agent provides the agent with its knowledge of the world, and it is the language in which the agent's beliefs, desires, and intentions (or indeed any other mental notions with which the agent is equipped) are expressed. Given the broad scope of L, it may in practice be a set of languages – for example separate languages for handling beliefs, desires, and intentions – but since all such languages carry out the same function we will regard them as one for the purposes of this discussion.

Each agent is also equipped with a *meta-language* ML which expresses facts about the base language L. Agents need meta-languages because, amongst other things, they need to represent their preferences about elements of L. Again ML may in fact be a set of meta-languages and both agents can use different meta-languages. Furthermore, if the agent has no need to make statements about formulae of L, then it may have no meta-language (or, equivalently, it may have a meta-language which it does not make use of). If an agent does have a separate meta-language, then it, like L, is *internal* to the agent.

Finally, for dialogues, the agents need a shared communication language (or two languages such that it is possible to seamlessly translate between them). We will call this language CL. We can consider CL to be a 'wrapper' around statements in L

and ML, as is the case for KQML [11.9] or the FIPA ACL [11.10], or a dedicated language into which and from which statements in L or CL are translated. CL might even be L or ML, though, as with ML, we can consider it to be a conceptually different language. The difference, of course, is that CL is in some sense *external* to the agents – it is used to communicate between them. We can imagine an agent reasoning using L and ML, then constructing messages in CL and posting them off to the other agent. When a reply arrives in CL, it is turned into statements in L and ML and these are used in new reasoning.

Argumentation can be used with these languages in a number of ways. Agents can use argumentation as a means of performing their own internal reasoning either in L, ML, or both. Independently of whether argumentation is used internally, it can also be used externally, in the sense of being used in conjunction with CL – this is the sense in which Walton and Krabbe [11.35] consider the use of argumentation in human dialogue and is much more on the topic of this chapter.

11.3.2 Inter-agent Argumentation

External argumentation can happen in a number of ways. The main issue, the fact that makes it argumentation, is that the agents do not just exchange facts but also exchange additional information. In persuasion dialogues, which are by far the most studied type of argumentation-based dialogues, these reasons are typically the reasons why the facts are thought to be true. Thus, if agent A wants to persuade agent B that p is true, it does not just state the fact p, but also gives, for example, a proof of p based on information (grounds) that A believes to be true. If the proof is sound then B can only disagree with p if either it disputes the truth of some of the grounds or if it has an alternative proof that p is false. The intuition behind the use of argumentation here is that a dialogue about the truth of a claim p moves to a dialogue about the supporting evidence or one about apparently conflicting proofs. From the perspective of building argumentative agents, the focus is now on how we can bring about either of these kinds of discussion.

There are a number of aspects, in particular, that we need to focus on. These include:

– Clearly communication will be carried out in CL, but it is not clear how arguments will be passed in CL. Will arguments form separate locutions, or will they be included in the same kind of CL locution as every other piece of information passed between the agents?
– Clearly the exchange of arguments between agents will be subject to some protocol , but it is not clear how this is related, if at all, to the protocol used for the exchange of other messages. Do they use the same protocol? If the protocols are different, how do agents know when to move from one protocol to another?
– Clearly the arguments that agents make should be related to what they know, but it is not clear how best this might be done. Should an agent only be able to argue what it believes to be true? If not, what arguments is an agent allowed to make?

One approach to constructing argumentation-based agents is the way suggested in [11.31]. In this work CL contains two sets of illocutions. One set allows the communication of facts (in this case statements in ML that take the form of conjunctions of value/attribute pairs, intended as offers in a negotiation). The other set allows the expressions of arguments. These arguments are un-related to the offers, but express reasons why the offers should be acceptable, appealing to a rich representation of the agent and its environment: the kinds of argument suggested in [11.31] are threats such as: "If you don't accept this I will tell your boss.", promises like: "If you accept my offer I'll bring you repeat business.", and appeals such as: "You should accept this because that is the deal we made before.".

There is no doubt that this model of argumentation has a good deal of similarity with the kind of argumentation we engage in on a daily basis. However, it makes considerable demands on any implementation. For a start, agents which desire to argue in this manner need very rich representations of each other and their environments (especially compared with agents which simply desire to debate the truth of a proposition given what is in their knowledge base). Such agents also require an answer to the second two points raised above, and the very richness of the model makes it hard (at least for the authors) to see how the third point can be addressed.

Now, the complicating factor in three points raised above is the need to handle two types of information – those that are argument-based and those that aren't. One way to simplify the situation is to make all communication argument-based, and that is the approach that we have been following of late. In fact, we go a bit further than even this suggests, by considering agents that use argumentation both for internal reasoning and as a means of relating what they believe and what they communicate. We describe this approach in the next section.

11.3.3 Argumentation at All Levels

In more detail what we are proposing is the following. First of all, every agent carries out internal argumentation using L. This allows it to resolve any inconsistency in its knowledge base (which is important when dealing with information from many sources since such information is typically inconsistent) and to establish some notion of what it believes to be true (though this notion is defeasible since new information may come to light that provides a more compelling argument against some fact that there previously was for that fact). The upshot of this use of argumentation, however it is implemented, is that every agent can not only identify the facts it believes to be true but can supply a rationale for believing them.

This feature then provides us with a way of ensuring a kind of rationality of the agents – rationality in communication. It is natural that an agent which resolves inconsistencies in what it knows about the world uses the same technique to resolve inconsistencies between what it knows and what it is told. In other words the agent looks at the reasons for the things it is told and accepts these things provided they are supported by more compelling reasons than there are against the things. If agents are only going to accept things that are backed by arguments, then it makes sense for agents to only say things that are also backed by arguments. Both of us, separately in [11.20] and [11.4], have suggested that such an argumentation-based approach is a suitable form of rationality, and it was implicit in [11.3][3].

The way that this form of rationality is formalized is, for example, to only permit agents to make assertions that are backed by some form of argument, and to only accept assertions that are so backed. In order words, the formation of arguments becomes a precondition of the locutions of the communication language CL, and the locutions are linked to the agents' knowledge bases.

Although it is not immediately obvious, this gives argumentation-based approaches a *social semantics* in the sense of Singh [11.32, 11.33]. The naïve reason for this is that since agents can only assert things that in their considered view are true (which is another way of putting the fact that the agents have more compelling reasons for thinking something is true than for thinking it is false), other agents have some guarantee that they are true. However agents may lie, and a suitably sophisticated agent will always be able to simulate truth-telling. A more sophisticated reason is that, assuming such locutions are built into CL, the agent on the receiving end of the assertion can always challenge statements, requiring that the reasons for them are stated. These reasons can be checked against what that agent knows, with the result that the agent will only accept things that it has no reason to doubt. This ability to question statements gives argumentation-based communication languages a degree of verifiability that other semantics, such as the original modal semantics for the FIPA ACL [11.10], lack.

11.3.4 Dialogue Games

Dialogues may be viewed as games between the participants, called *dialogue games* [11.17]. In this view, explained in greater detail in Chap. 10 by McBurney and Parsons, each participant is a player with an objective they are trying to achieve and some finite set of moves that they might make. Just as in any game, there are rules about which player is allowed to make which move at any point in the game, and there are rules for starting and ending the game.

As a brief example, consider a persuasion dialogue. We can think of this as being captured by a game in which one player initially believes p to be true and tries to convince another player, who initially believes that p is false, of that fact. The game might start with the first player stating the reason why she believes that p is true, and the other player might be bound to either accept that this reason is true (if she can find no fault with it) or to respond with the reason she believes it to be false. The first player is then bound by the same rules as the second was – to find a reason why this second reason is false or to accept it – and the game continues until one of the players is forced to accept the most recent reason given and thus to concede the game.

---

[3] This meaning of rationality is also consistent with that commonly given in philosophy, see e.g. [11.15].

11.4 A System for Argumentation-Based Communication

In this section we give a concrete instantiation of the rather terse description given in Sect. 11.3.3, providing an example of a system for carrying out argumentation-based communication of the kind first suggested in [11.24].

11.4.1 A System for Internal Argumentation

We start with a possibly inconsistent finite knowledge base Σ with no deductive closure. We assume that Σ contains formulae of a propositional language which we call $\mathcal{L}$, as well as formulae such as $B_i(p)$ and $I_j(q)$ for any p and q which are formulae of $\mathcal{L}$. This extended propositional language is the base language L of the argumentation-based dialogue system we are describing. $B_i(\cdot)$ denotes a belief of agent i and $I_j(\cdot)$ denotes an intention of agent j. Since we are only interested in syntactic manipulation of beliefs and intentions here, we will give no semantics; suitable ways of dealing with the semantics are given elsewhere (e.g. [11.25, 11.36]). The symbol $\vdash$ denotes classical inference and $\equiv$ denotes logical equivalence. An argument is a proposition and the set of formulae from which it can be inferred:

Definition 11.4.1. *An* argument *is a pair $A = (H, h)$ where h is a formula of $\mathcal{L}$ and H a subset of Σ such that:*

1. *H is consistent;*
2. *$H \vdash h$; and*
3. *H is minimal, so no subset of H satisfying both 1. and 2. exists.*

H is called the support *of A, written $H = Support(A)$ and h is the* conclusion *of A written $h = Conclusion(A)$.*

We talk of h being *supported* by the argument (H, h).

In general, since Σ is inconsistent, arguments in $\mathcal{A}(\Sigma)$, the set of all arguments which can be made from Σ, will conflict, and we make this idea precise with the notions of rebutting, undercutting, and attacking.

Definition 11.4.2. *Let A_1 and A_2 be two distinct arguments of $\mathcal{A}(\Sigma)$. A_1 undercuts A_2 iff $\exists h \in Support(A_2)$ such that h attacks $Conclusion(A_1)$.*

Definition 11.4.3. *Let A_1 and A_2 be two distinct arguments of $\mathcal{A}(\Sigma)$. A_1 rebuts A_2 iff $Conclusion(A_1)$ attacks $Conclusion(A_2)$.*

Definition 11.4.4. *Given two distinct formulae h and g of $\mathcal{L}$ such that $h \equiv \neg g$, then, for any i and j:*

- *h attacks g;*
- *$B_i(h)$ attacks $B_j(g)$; and*
- *$I_i(h)$ attacks $I_j(g)$.*

In other words, an argument is rebutted in three cases: if there is another argument which has as its conclusion the negation of the conclusion of the first, and either both are not in the scope of a belief or intention operator, or both are in the scope of the same kind of operator. Thus we recognize "Peter intends that this paper be written by the deadline." and "Simon intends this paper not to be written by the deadline." as rebutting each other, along with "Peter believes God exists." and "Simon does not believe God exists.", but does not recognize "Peter intends that this paper will be written by the deadline." and "Simon does not believe that this paper will be written by the deadline." as rebutting each other. Undercutting occurs in exactly the same situations, except that it holds between the conclusions of one argument and an element of the support of the other[4].

Note that this notion of attack is a generalization of that in [11.2], and, while related to that in [11.25], both extends it (in allowing 'attacks' between things other than intentions) and is less extensive than it (by not allowing 'attacks' between second-order intentions).

To capture the fact that some facts are more strongly believed and intended than others, we assume that any set of facts has a preference order over it. We suppose that this ordering derives from the fact that the knowledge base Σ is stratified into non-overlapping sets $\Sigma_1, \ldots, \Sigma_n$ such that facts in Σ_i are all equally preferred and are more preferred than those in Σ_j where $j > i$. The preference level of a non-empty subset H of Σ, $level(H)$, is the number of the highest-numbered layer which has a member in H.

Definition 11.4.5. *Let A_1 and A_2 be two arguments in $\mathcal{A}(\Sigma)$. A_1 is preferred to A_2 according to $Pref$ iff $level(Support(A_1)) \leq level(Support(A_2))$.*

By $\gg^{Pref}$ we denote the strict pre-order associated with $Pref$. If A_1 is strictly preferred to A_2, we say that A_1 is *stronger* than A_2. We can now define the argumentation system we will use:

Definition 11.4.6. *An argumentation system (AS) is a triple $\langle \mathcal{A}(\Sigma), Undercut/ Rebut, Pref \rangle$ such that:*

– $\mathcal{A}(\Sigma)$ *is a set of the arguments built from Σ,*
– *Undercut/Rebut is a binary relation capturing the existence of an undercut or rebut holding between arguments, $Undercut/Rebut \subseteq \mathcal{A}(\Sigma) \times \mathcal{A}(\Sigma)$, and*
– *Pref is a (partial or complete) pre-ordering on $\mathcal{A}(\Sigma) \times \mathcal{A}(\Sigma)$.*

The preference order makes it possible to distinguish different types of relation between arguments:

Definition 11.4.7. *Let A_1, A_2 be two arguments of $\mathcal{A}(\Sigma)$.*

– *If A_2 undercuts A_1 then A_1 defends itself against A_2 iff $A_1 \gg^{Pref} A_2$. Otherwise, A_1 does not defend itself.*

---

[4] Note that attacking and rebutting are symmetric but not reflexive or transitive, while undercutting is not symmetric, reflexive, or transitive.

– *A set of arguments* $\mathcal{S}$ *defends* A *iff:* $\forall$ B *such that* B *undercuts or rebuts* A *and* A *does not defend itself against* B *then* $\exists$ $C \in \mathcal{S}$ *such that* C *undercuts* B *and* B *does not defend itself against* C.

Henceforth, $C_{Undercut/Rebut,Pref}$ will gather all non-undercut and non-rebut arguments along with arguments defending themselves against all their undercutting and rebutting arguments. Amgoud and Cayrol[11.1] showed that the set $\underline{S}$ of acceptable arguments of the argumentation system $\langle \mathcal{A}(\Sigma), Undercut/Rebut, Pref \rangle$ is the least fixpoint of a function $\mathcal{F}$:

$$\mathcal{F}(\mathcal{S}) = \{(H,h) \in \mathcal{A}(\Sigma) | (H,h) \text{ is defended by } \mathcal{S}\},$$

where $\mathcal{S} \subseteq \mathcal{A}(\Sigma)$.

Definition 11.4.8. *The set of* acceptable *arguments of an argumentation system* $\langle \mathcal{A}(\Sigma), Undercut, Pref \rangle$ *is:*

$$\underline{S} = \bigcup \mathcal{F}_{i \geq 0}(\emptyset) = C_{Undercut/Rebut,Pref} \cup \left[\bigcup \mathcal{F}_{i \geq 1}(C_{Undercut/Rebut,Pref}) \right].$$

An argument is acceptable if it is a member of the acceptable set.

If the argument (H, h) is acceptable, we talk of there being an acceptable argument for h. An acceptable argument is one which is, in some sense, proven since all the arguments which might undermine it are themselves undermined.

Note that while we have given a language L for this system, we have given no language ML. This particular system does not have a meta-language (and the notion of preferences it uses is not expressed in a meta-language). It is, of course, possible to add a meta-language to this system – for example, in [11.5] we added a meta-language which allowed us to express preferences over elements of L, thus making it possible to exchange (and indeed argue about, though this was not done in [11.5]) preferences between formulae.

11.4.2 Arguments Between Agents

Now, this system is sufficient for internal argumentation within a single agent, and the agent can use it to, for example, perform non-monotonic reasoning and to deal with inconsistent information. To allow for dialogues, we have to introduce some more machinery. Clearly part of this will be the communication language, but we need to introduce some additional elements first. These elements are data structures which our system inherits from its dialogue game ancestors as well as previous presentations of this kind of system [11.3, 11.6].

Dialogues are assumed to take place between two agents, P and C. Each agent has a knowledge base, Σ_P and Σ_C respectively, containing their beliefs. In addition, following Hamblin [11.13], each agent has a further knowledge base, read-accessible to both agents, containing commitments made in the dialogue. These commitment stores are denoted $CS(P)$ and $CS(C)$ respectively, and in this dialogue system (unlike that of [11.6] for example) an agent's commitment store is just a subset of its knowledge base. Note that the union of the commitment stores can be viewed as the state of the dialogue at a given time, since it expresses the current commitments of all the participants. Each agent has read- and write-access to their own private knowledge base and read-access to both commitment stores; each agent has only write-access to its own commitment store, with entries made to the store only as a result of utterances in the dialogue. Thus P can make use of

$$\langle \mathcal{A}(\Sigma_P \cup CS(C)), \, Undercut/Rebut, \, Pref \rangle$$

and C can make use of

$$\langle \mathcal{A}(\Sigma_C \cup CS(P), \, Undercut/Rebut, \, Pref \rangle.$$

All the knowledge bases contain propositional formulae and are not closed under deduction, and all are stratified by degree of belief as discussed above.

With this background, we can present the set of dialogue moves that we will use, the set which comprises the locutions of CL. For each move, we give what we call rationality rules, dialogue rules, and update rules. These locutions are those from [11.26] and are based on the rules suggested in [11.19] which, in turn, were based on those in the dialogue game DC introduced by MacKenzie [11.18]. The rationality rules specify the preconditions for making the move. Unlike those in [11.3, 11.6], these rules are not absolute, but are defined in terms of the agent attitudes discussed below, and these provide the social semantics for the locutions. The update rules specify how commitment stores are modified by the move.

In the following, player P addresses the move to player C. We start with the assertion of facts:

assert(p) where p is a propositional formula.

> rationality: the usual assertion condition for the agent.
> update: $CS_i(P) = CS_{i-1}(P) \cup \{p\}$ and $CS_i(C) = CS_{i-1}(C)$.

Here p can be any propositional formula, as well as the special character $\mathcal{U}$, discussed in the next subsection.

assert(S) where S is a set of formulae representing the support of an argument.

> rationality: the usual assertion condition for the agent.
> update: $CS_i(P) = CS_{i-1} \cup S$ and $CS_i(C) = CS_{i-1}(C)$.

The counterparts of these moves are the acceptance moves:

accept(p) p is a propositional formula.

> rationality: the usual acceptance condition for the agent.
> update: $CS_i(P) = CS_{i-1}(P) \cup \{p\}$ and $CS_i(C) = CS_{i-1}(C)$.

accept(S) S is a set of propositional formulae.

> rationality: the usual acceptance condition for every $s \in S$.
> update: $CS_i(P) = CS_{i-1}(P) \cup S$ and $CS_i(C) = CS_{i-1}(C)$.

There are also moves which allow questions to be posed.

challenge(p) where p is a propositional formula.

> rationality: $\emptyset$
> update: $CS_i(P) = CS_{i-1}(P)$ and $CS_i(C) = CS_{i-1}(C)$.

A challenge is a means of making the other player explicitly state the argument supporting a proposition. In contrast, a question can be used to query the other player about any proposition.

question(p) where p is a propositional formula.

> rationality: $\emptyset$
> update: $CS_i(P) = CS_{i-1}(P)$ and $CS_i(C) = CS_{i-1}(C)$.

We refer to this set of moves as the set $\mathcal{M}'_{\mathcal{DC}}$ since they are a variation on the set $\mathcal{M}_{\mathcal{DC}}$ from [11.3] – the main difference from the latter is that there are no 'dialogue conditions'. Instead we explicitly define the protocol for dialogues below. These locutions are the bare minimum to carry out a dialogue, and, as we will see below, require a fairly rigid protocol with a lot of aspects implicit. Further locutions, such as those discussed in [11.23], would be required to be able to debate the beginning and end of dialogues or to have an explicit representation of movement between embedded dialogues.

Clearly this set of moves/locutions defines the communication language CL, and hopefully it is reasonably clear from the description so far how argumentation between agents takes place; a prototypical persuasion dialogue is as follows:

1. P has an acceptable argument (S, p), built from Σ_P, and wants C to accept p. Thus, P asserts p.
2. C has an argument $(S', \neg p)$ and so can not accept p. Thus, C asserts $\neg p$.
3. P can not accept $\neg p$ and challenges it.
4. C responds by asserting S'.
5. P has an argument $(S'', \neg q)$, where $q \in S'$, and asserts $\neg q$.
6. C challenges $\neg q$.
7. ...

At each stage in the dialogue agents can build arguments using information from their own private knowledge base, and the propositions made public (by assertion into commitment stores).

11.4.3 Rationality and Protocol

The final part of the abstract model we introduced above was the use of argumentation to relate what an agent 'knows' (in this case what is in its knowledge base and the commitment stores) and what it is allowed to 'say' (in terms of which locutions from CL it is allowed to utter). We make this connection by specifying the rationality conditions in the definitions of the locutions and relating these to what arguments an agent can make. We do this as follows, essentially defining different types of rationality [11.26].

Definition 11.4.9. *An agent may have one of three* assertion *attitudes.*

- *a* confident *agent can assert any proposition p for which there is an argument* (S, p).
- *a* careful *agent can assert any proposition p for which there is an argument* (S, p) *if no stronger rebutting argument exists.*
- *a* thoughtful *agent can assert any proposition p for which there is an acceptable argument* (S, p).

Thus a thoughtful agent will only put forward propositions which, so far as it knows, are correct. A careful agent will only put forward propositions which aren't directly rebutted. A confident agent won't stop to make either of these checks[5].

Of course, defining when an agent can assert propositions is only one-half of what is needed. The other part is to define the conditions on agents accepting propositions. Here we have the following [11.26].

Definition 11.4.10. *An agent may have one of three* acceptance *attitudes.*

- *a* credulous *agent can accept any proposition p for which there is an argument* (S, p).
- *a* cautious *agent can accept any proposition p for which there is an argument* (S, p) *if no stronger rebutting argument exists.*
- *a* skeptical *agent can accept any proposition p for which there is an acceptable argument* (S, p).

In order to complete the definition of the system, we need only to give the protocol that specifies how a dialogue proceeds. This we do below, providing a protocol (which was not given in the original) for the kind of example dialogue given in [11.24, 11.25]. As in those papers, the kind of dialogue we are interested in here is a dialogue about joint plans, and, in order to describe the dialogue, we need an idea of what one of these plans looks like:

Definition 11.4.11. *A* plan *is an argument* (S, p) *such that there is some* $I_i(p)$. $I_i(p)$ *is known as the* subject *of the plan.*

Thus a plan is just an argument for a proposition that is intended by some agent. If we consider thoughtful/skeptical agents, then the detail of 'attacks' ensures that an agent will only be able to assert a plan if there is no intention which is preferred to the subject of the plan so far as that agent is aware, and there is no conflict between any elements of the support of the plan and what it knows. Equally an agent will only accept a plan if there is no intention that it prefers to the subject of the plan, and it knows nothing that conflicts with any elements of the support of the plan. Similar conditions hold for agents with other attitudes. We then have the following protocol, which we will call $\mathcal{D}$ for a dialogue between agents A and B.

1. A *asserts* a plan (S, p) for some $I_A(p)$.

---

[5] Note that, as a first step, we define these agent attributes uniformly; in later work, we will consider agents which assert or accept propositions in a context-dependent manner.

2. B *accepts* the plan if possible. If the plan is accepted, the dialogue terminates.
3. If the plan is not accepted, then B *asserts* an argument (S', q) which undercuts or rebuts (S, p).
4. A *asserts* either (S''', p), which does not undercut or rebut (S', q), or the statement $\mathcal{U}$. In the first case, the dialogue returns to step 2; in the second case, the dialogue terminates.

The utterance of a statement $\mathcal{U}$ indicates that an agent is unable to add anything to the dialogue, and so the dialogue terminates whenever either agent asserts this.

Note that in B's response it need not assert a plan (A is the only agent which has to mention plans). This allows B to disagree with A on matters such as the resources assumed by A ('No, I don't have the car that week.'), or the tradeoff that A is proposing ('I don't want your Megatokyo T-shirt, I have one like that already.'), even if they don't directly affect the plans that B has.

As it stands, the protocol is rather minimalist but suffices to capture the kind of interaction in [11.24, 11.25]. One agent makes a suggestion which suits it (and may involve the other agent). The second looks to see if the plan prevents it achieving any of its intentions, and if so has to put forward a plan which clashes in some way (we could easily extend the protocol so that B does not have to put forward this plan, but can instead engage A in a persuasion dialogue about A's plan in a way that was not considered in [11.24, 11.25]). The first agent then has the chance to respond by either finding a non-clashing way of achieving what it wants to do or suggesting a way for the second agent to achieve its intention without clashing with the first agent's original plan.

There is also much that is implicit in the protocol, for example: that the agents have previously agreed to carry out this kind of dialogue (since no preamble is required); that the agents are basically co-operative (since they accept suggestions if possible); and that they will end the dialogue as soon as a possible agreement is found or it is clear that no progress can be made (so neither agent will try to filibuster for its own advantage). Such assumptions are consistent with Grice's co-operative maxims for human conversation [11.11].

One advantage of such a minimal protocol is that it is easy to show that the resulting dialogues have some desirable properties. The first of these is that the dialogues terminate:

Proposition 11.4.1. *A dialogue under protocol $\mathcal{D}$ between two agents G and H with any acceptance and assertion attitudes will terminate.*

If both agents are thoughtful and skeptical, we can also show that:

Proposition 11.4.2. *Consider a dialogue under protocol $\mathcal{D}$ between two thoughtful/skeptical agents G and H, where G starts by uttering a plan with the subject $I_G(p)$.*

– *If the dialogue terminates with the utterance of $\mathcal{U}$, then there is no plan with the subject $I_G(p)$ in $A(\Sigma_G \cup CS(H))$ that H can accept.*
– *If the dialogue terminates without the utterance of $\mathcal{U}$, then there is a plan with the subject $I_G(p)$ in $A(\Sigma_G \cup \Sigma_H)$ that is acceptable to both G and H.*

Thus if the agents reach agreement, it is an agreement on a plan which neither of them has any reason to think problematic. In [11.24, 11.25] we called this kind of dialogue a negotiation, but from the perspective of Walton and Krabbe's typology it isn't a negotiation – it is closer to a deliberation with the agents discussing what they will do.

11.5 Argument Aggregation

What happens when agents disagree about some claim or some proposed action, even after a persuasion or deliberation dialogue? One approach we have explored is aggregation across arguments [11.22]. Thus, two or more agents may present the arguments for and against some claim, and then the status of the claim may be determined by the nature and extent of the arguments for and against it. For example [11.16], we could define a set of status labels for claim θ at time t as follows:

- If there have been no arguments uttered for or against θ up to time t, then the claim is *open*.
- If there has been at least one argument uttered for θ up to time t, then the claim is *supported*.
- If there has been at least one argument whose premises are consistent uttered for θ up to time t, then the claim is *plausible*.
- If there has been at least one argument whose premises are consistent uttered for θ up to time t, and no undercutting or attacking arguments have been uttered against θ by this time, then the claim is *probable*.
- If there has been at least one argument whose premises are consistent uttered for θ up to time t, and any undercutting or attacking arguments uttered against θ by this time have themselves been attacked or undercut, then the claim is *accepted*.

The motivation here is that the more and stronger are the arguments for a claim, then the more support it has, and so the greater is the likelihood that it is true. Note that, as with any real-life argument of issues of importance, the status of a claim may change over time, as new arguments are presented for it or against it. Thus claims may become less likely or more likely or both over time. In [11.22], we considered this dynamic aspect to multi-agent arguments, and showed that, under certain conditions, inquiry dialogues would eventually converge to the truth with a high probability.

11.6 Summary

Argumentation-based approaches to interagent communication are becoming more widespread, and there are a variety of systems for argumentation-based communication that have been proposed. Many of these address different aspects of the communication problem, and it can be hard to see how they relate to one another.

This chapter has attempted to put some of this work in context by describing in general terms how argumentation might be used in interagent communication, and then illustrating this general model by providing a concrete instantiation of it, finally describing all the aspects required by the example first introduced in [11.24]. We believe these approaches have great potential for chance discovery and management in multi-agent domains, where agents may need to persuade one another of the possibility of chance events and/or appropriate actions to take in response.

Acknowledgements. We thank Leila Amgoud and Nicolas Maudet for their contribution to the development of many of the components of the argumentation system described here.

References

11.1 Amgoud L, Cayrol C (1998) On the acceptability of arguments in preference-based argumentation framework. *Proc. 14th Conf. Uncertainty in AI*, pp.1–7

11.2 Amgoud L, Cayrol C (2002) A reasoning model based on the production of acceptable arguments. *Annals of Mathematics and AI*, 34:197–215

11.3 Amgoud L, Maudet N, Parsons S (2000) Modelling dialogues using argumentation. In E. Durfee, editor, *Proc. 4th Intern. Conf. on Multi-Agent Systems*, pp.31–38, Boston, MA, IEEE Press, New York, NY

11.4 Amgoud L, Maudet N, Parsons S (2002) An argumentation-based semantics for agent communication languages. In *Proc. 15th European Conf. on AI*

11.5 Amgoud L, Parsons S (2001) Agent dialogues with conflicting preferences. In J.-J. Meyer and M. Tambe, editors, *Proc. 8th Intern. Workshop on Agent Theories, Architectures and Languages*, pp.1–15

11.6 Amgoud L, Parsons S, and Maudet N (2000) Arguments, dialogue, and negotiation. In W. Horn, editor, *Proc. 14th European Conf. on AI*, pp. 338–342, Belin, Germany, IOS Press, Amsterdam, The Netherland

11.7 Dignum F, Dunin-Kęplicz B, and Verbrugge R (2001) Agent theory for team formation by dialogue. In C. Castelfranchi and Y. Lespérance, editors, *Intelligent Agents VII*, pp.141–156, Berlin, Germany, Springer Verlag, Heidelberg, Germany

11.8 Dignum F, Dunin-Kęplicz B, and Verbrugge R (2001) Creating collective intention through dialogue. *Logic Journal of the IGPL*, 9(2):305–319

11.9 Finin T, Labrou Y, and Mayfield J (1997) KQML as an agent communication language. Invited chapter in J. Bradshaw, editor, *Software Agents*. MIT Press, Cambridge, MA

11.10 FIPA. Communicative Act Library Specification. Technical Report XC00037H, Foundation for Intelligent Physical Agents, 10 August 2001.

11.11 Grice HP (1975) Logic and conversation. In P. Cole and J. L. Morgan, editors, *Syntax and Semantics III: Speech Acts*, pp.41–58. Academic Press, New York, NY

11.12 Grosz BJ, Kraus S (1999) The evolution of SharedPlans. In M. J. Wooldridge and A. Rao, editors, *Foundations of Rational Agency*, volume 14 of *Applied Logic*. Kluwer, The Netherlands

11.13 Hamblin CL (1970) *Fallacies*. Methuen, London, UK

11.14 Hitchcock D, McBurney P, Parsons S (2001) A framework for deliberation dialogues. In H. V. Hansen, C. W. Tindale, J. A. Blair, and R. H. Johnson, editors, *Proc. 4th Biennial Conf. Ontario Soc. Study of Argumentation (OSSA 2001)*, Windsor, Canada

11.15 Johnson R (2000) *Manifest Rationality: A Pragmatic Theory of Argument*. Lawrence Erlbaum Associates, Mahwah, NJ

11.16 Krause P, Ambler S, Elvang-Gørannson M, Fox J (1995) A logic of argumentation for reasoning under uncertainty. *Computational Intelligence*, 11 (1):113–131

11.17 Levin JA, Moore JA (1978) Dialogue-games: metacommunications structures for natural language interaction. *Cognitive Science*, 1(4):395–420

11.18 MacKenzie JD (1979) Question-begging in non-cumulative systems. *J. Philosophical Logic*, 8:117–133

11.19 Maudet N, Evrard F (1998) A generic framework for dialogue game implementation. In *Proc. 2nd Workshop on Formal Semantics and Pragmatics of Dialogue*, University of Twente, The Netherlands

11.20 McBurney P (2002) *Rational Interaction*. PhD thesis, Department of Computer Science, University of Liverpool

11.21 McBurney P, Parsons S (2001) Chance discovery using dialectical argumentation. In T. Terano, T. Nishida, A. Namatame, S. Tsumoto, Y. Ohsawa, and T. Washio, editors, *New Frontiers in Artificial Intelligence: Joint JSAI 2001 Workshop Post Proceedings*, LNAI 2253, pp.414–424. Springer, Berlin, Germany

11.22 McBurney P, Parsons S (2001) Representing epistemic uncertainty by means of dialectical argumentation. *Annals of Mathematics and Artificial Intelligence*, 32(1–4):125–169

11.23 McBurney P, Parsons S (2002) Games that agents play: A formal framework for dialogues between autonomous agents. *J. Logic, Language, and Information*, 11(3):315–334

11.24 Parsons S, Jennings NR (1996) Negotiation through argumentation – a preliminary report. In *Proc. 2nd Intern. Conf. on Multi-Agent Systems*, pages 267–274

11.25 Parsons S, Sierra S, Jennings NR (1998) Agents that reason and negotiate by arguing. *Logic and Computation*, 8(3):261–229

11.26 Parsons S, Wooldridge M, Amgoud L (2002) An analysis of formal interagent dialogues. In C. Castelfranchi and W. L. Johnson, editors, *Proc. First Intern. Joint Conf. on Autonomous Agents and Multi-Agent Systems (AAMAS 2002)*, pp.394–401, New York, NY, ACM Press, New York, NY

11.27 Reed C (1998) Dialogue frames in agent communications. In Y. Demazeau, editor, *Proc. 3rd Intern. Conf. on Multi-Agent Systems*, pp.246–253. IEEE Press, New York, NY

11.28 Riley R, Stone P, Veloso M (2001) Layered disclosure: Revealing agents' internals. In C. Castelfranchi and Y. Lespérance, editors, *Intelligent Agents VII*, pp.61–72, Berlin, Germany, Springer Verlag, Heidelberg, Germany

11.29 Roos N, Teije A, Bos A, Witteveen C (2002) An analysis of multi-agent diagnosis. In C. Castelfranchi and W. L. Johnson, editors, *Proc. First Intern. Joint Conf. on Autonomous Agents and Multi-Agent Systems (AAMAS 2002), Bologna, Italy*, pp.986–987, New York City, NY, USA ACM Press, New York, NY

11.30 Schroeder M, Plewe DA, Raab A (1998) Ultima ratio: should Hamlet kill Claudius. In *Proc. 2nd Intern. Conf. on Autonomous Agents*, pp.467–468

11.31 Sierra C, Jennings NR, Noriega P, Parsons S (1998) A framework for argumentation-based negotiations. In M. P. Singh, A. Rao, and M. J. Wooldridge, editors, *Intelligent Agents IV*, pp.177–192, Berlin, Germany, Springer Verlag, Heidelberg, Germany (1998)

11.32 Singh MP (1998) Agent communication languages: Rethinking the principles. In *IEEE Computer 31*, pp.40–47

11.33 Singh MP (1999) A social semantics for agent communication languages. In *Proc. IJCAI'99 Workshop on Agent Communication Languages*, pp.75–88

11.34 Sycara K (1989) Argumentation: Planning other agents' plans. *Proc. 11th Joint Conf. on AI*, pp.517–523

11.35 Walton DN and Krabbe ECW (1995) *Commitment in Dialogue: Basic Concepts of Interpersonal Reasoning*. SUNY Press, Albany, NY

11.36 Wooldridge MJ (2000) *Reasoning about Rational Agents*. MIT Press, Cambridge, MA

11.37 Wooldridge MJ (2000) Semantic issues in the verification of agent communication languages. *J. Autonomous Agents and Multi-Agent Systems*, 3(1):9–31

Part III
Key 2 – Perceptions for Context Shifting

The second key is a human's perception of information, to involve oneself in a new context where the current chance event can be significant. This might be an analogical matching of one's own experience with the current rare situation, or an imagination of a scene or a story in the future situation. This part show methods and theories for context shifting for realizing clear awareness of chances.

12. Awareness and Imagination of Hidden Factors and Rare Events

Yasufumi Takama[1,2]

[1] Tokyo Metropolitan Institute of Technology, Hino, Tokyo 191-0065, Japan
[2] PRESTO, Japan Science and Technology Corporation, 2-2-11 Tsutsujigaoka, Miyagino-ku, Sendai, Miyagi 983-0852, Japan
email: ytakama@cc.tmit.ac.jp

Summary.

In this chapter, we consider the Web as the environment, on which how to help a human notice and evaluate topics is discussed. As the Web provides us with new topics constantly, it is difficult to grasp the trends or change of topics on the Web. Although the hugeness of the Web as well as its dynamic nature is a burden for the users, it will also bring them a chance for business and research if they can notice the trends or movement of the real world from the Web, which can not be found from a single document but from a sequence of document sets.

12.1 Introduction

Awareness of the rare events in the environments that will lead us to a new idea or innovation is one of the important steps in the chance discovery process. However, every rare event does not always lead to a chance, and its importance should be evaluated by imagining its hidden factors. Therefore, we have to notice the rare events and evaluate their importance in terms of our own chance discovery, through an interaction with environments. This process does not only involve the objective (statistical) analysis but also the subjective evaluation, which makes it difficult to be solved by the conventional technologies or theories.

In this chapter, we consider the Web as the environment, on which how to help a human notice and evaluate topics is discussed. As the Web provides us with new topics constantly, it is difficult for us to grasp the trends or change of topics on the Web. In particular, there are so many online news sites on the Web, and they constantly release up-to-date news articles of various topics. Although the hugeness of the Web as well as its dynamic nature is a burden for the users, it will also bring them a chance for business and research if they can notice the trends or movement of the real world from the Web, which can not be found from a single document but from a sequence of document sets.

Information-visualization systems [12.5, 12.14, 12.11, 12.16, 12.18] are promising approaches to help the user notice the trends of topics on the Web. However, what the conventional systems are concerned with has been to process a single set of documents, and no sequential nature of document sets is considered.

In this chapter, an information-visualization method based on document set-wise processing is introduced to find the context through a sequence of document sets.

The plastic clustering method [12.11, 12.12, 12.14, 12.15] is employed, which is one of the essential foundations for realizing the Web information-visualization systems. One of the characteristic features of the plastic clustering method is the generation of a *keyword map* as well as document clustering. Furthermore, the model of a *memory cell* is incorporated into the plastic clustering method, so that it can find a topic stream from a sequence of document sets.

Section 12.2 discusses the interaction with environments in terms of chance discovery, followed by the characteristics of the Web as the environments in Sect. 12.3. Topic visualization on the Web is discussed in Sect. 12.4, and the plastic clustering method is introduced in Sect. 12.5. The experimental results of applying the proposed method to a sequence of document sets are shown in Sect. 12.6.

12.2 Interaction with Environment for Chance Discovery

In our daily life, we always get information from environments (including other people, organizations, and mass media) based on which we make decisions and behave in the environments. As the process of chance discovery is found within the cycle of getting information, decision making, and reaction to environments, the interaction with environments is essential for our chance discovery.

The environments as the interaction partners include other actors (individuals, companies, governments, administrations, and so on) that make decisions by themselves, and mass media such as TV and newspapers. Through the interaction with such environments, our decision making process and behaviors are affected by environments, while the environments are also changed by our behavior. The interaction with other actors is called micro-level interaction, whereas that with mass media is called macro-level interaction. Recently, the Internet has grown as the object that includes both types of interaction, as discussed in Sect. 12.3. Rogers [12.7] discusses in his study of the diffusion process of innovations that mass-media channels (i.e. macro-level interaction) are relatively more important at the knowledge stage, whereas interpersonal channels (micro-level interaction) are relatively more important at the persuasion stage. Such a co-operation of both levels of interactions is not only important for the innovation-diffusion process, but also for the chance discovery process.

The point of the chance discovery is that one has to get a chance by oneself. In other words, no chance can be given by others. There are several reasons for that, as follows:

– A chance is person-specific and subjective. The chance for someone is usually not so important for others.
– It is dangerous for us to accept someone's alluring talks on faith.

In the study of communication [12.3], a communication is divided into three aspects from the viewpoint of their roles:

1. Communication aimed for persuasion.

2. Communication aimed for sharing reality with partners.
3. Communication with no intention, but giving us a reality.

In the case of the first type of communication, we seldom accept someone's proposal unless he is reliable. For example, will you purchase expensive products (such as cars and houses) at a strange shop even though they offer a much cheaper price than other shops? In order for someone to be reliable for us, we usually have to communicate with him frequently in order to share enough contexts (reality) with him. Therefore, the communication of type 2 (3) is very important for the chance discovery process. This discussion also applies to the interaction with mass media.

In general, we have to find reliable information resources as the prestage of chance discovery. However, as our abilities of information collection, calculation (decision making), and acting on environments are limited (they are also referred to as limited rationality) [12.9], heuristics are often used to filter information. The heuristics used for gathering information on the Web is shown in Sect. 12.3.

It is said that out social activities form a micro–macro loop [12.4], i.e. micro-level behaviors result in macro-level information, which works as the constraints for micro-level behaviors. For example, driving an individual car causes various traffic behaviors such as a traffic jam, whereas the traffic information affects a driver's decision making about the route or taking a rest. As the integration of micro-level behaviors leads to the macro-level constraints, we can notice the relation (difference) between others and ourselves from the macro-level information, which will lead to the chance discovery.

12.3 Web as Information Environment for Chance Discovery

As the Web is often viewed as a huge, distributed database, it is also fundamental as the information resource for chance discovery. Moreover, the following characteristics are also important from the viewpoint of chance discovery.

– New information is continuously released on the Web.
– Various communication channels can be available on the Internet (including the Web).
– Everyone can not only collect information, but also issue their message on the Web.

A micro–macro loop as noted in Sect. 12.2 can also be formed on the Web, thanks to these characteristics. Furthermore, the loops on the Web can be turned around more quickly than the usual loops found on the (physical) environments, because we can easily put the information as the result of our activities on the Web, which can spread any information all over the world within a second. Through the micro–macro loop on the Web, it is possible for us to get the feedback of our behaviors quickly.

On the other hand, there exist several difficulties in making the most of the Web for the chance discovery, which are as follows.

1. There is no guarantee of the uniqueness of information.
2. Everyone can upload information on the Web without inspection.
3. Difficulty in distinguishing a communication channel of a certain type from those of different types.

Because of these, we often find a contradiction between the information issued from the different sites, among which it is difficult to select the correct one. We can hardly distinguish a reliable information resource (site) from suspicious ones. Moreover, if we search for information about unfamiliar topics, it is difficult to recognize the importance of incoming information.

Because we have a limited rationality as noted in Sect. 12.2, we often use the heuristics, such as those following, to judge the reliability or importance of an information resource on the Web.

1. Page design: well-organized pages or those containing many audio and visual effects look reliable.
2. Publicity: the information on a famous site seems to be reliable.
3. Hyper-link structure: the pages linked from the reliable pages might also be reliable (e.g. PageRank of Google).
4. Context: the importance of information can not be evaluated independently, but relatively within the surrounding information space.

It should be noted that these heuristics are not always true. Even though a certain heuristic is found to be effective, some sites may make use of it for cheating us. For example, after Google, which uses the Web's hyper-link structure for ranking the retrieved pages, has grown as one of the major search engines in the world, it is observed that the pages (sites) of a certain group make hyper-links to each other, in order to improve their ranking scores in Google's ranking list (it is also considered as one of the effects of the micro–macro loop). Still, we have to use heuristics to tackle the huge volume of the Web. The most important point will be not to define the static measure for a certain heuristic, but to use the heuristics for improving the quality of interaction with environments (the Web).

In this chapter, we propose to visualize the topic distribution on the document space as the context information, which corresponds to the fourth heuristic listed above. In particular, the sequential nature of the information, such as a sequence of online news articles and a series of information-retrieval processes, is handled.

12.4 Visualizing Topic Distribution on Web

When we examine the incoming information, we always try to understand it from the viewpoint that comes from our background knowledge. However, when we are searching for the topics of an unfamiliar field, we do not have enough background knowledge about the field, leading to the difficulty of forming appropriate viewpoints. Such situations often happen on the Web information retrieval, and the important topics tend to be missed easily by the user.

Concerning one of the typical natures of the information on the Web, it is often provided as a sequence of document sets. For example, online news sites on the Web constantly release up-to-date news articles of various topics day by day. A series of user's retrieval processes also provides a sequence of document sets. As a sequence of document sets should have some contexts, the topics found in the previous document sets are expected to be good clues to grasp the topic distribution on the current document set. In particular, the following types of topics are focused on.

Mainstream topic: the topics that survive through several document sets correspond to the main topics of the user's current retrieval task.

Missing topic: when the topics, which have missed by the user in the early stage of browsing, appear again after several document sets, the user can re-evaluate such topics based on the background knowledge obtained during the browsing processes.

As related work, the TDT (topic detection and tracking) project [12.1, 12.17] aims to detect new events from a stream of news stories. In the TDT project, for example, 'the Eruption of Mt. Pinatubo on June 15, 1991', 'the earthquake in Kobe', etc., are considered as events, whereas 'volcanic eruptions', 'airplanes crash', etc., are not events but a more general class of events. Differently from the TDT project, a topic mentioned in this paper should include the latter one (general class of events) as well as the former one, because both of them are expected to be useful as clues to grasp the topic distribution on the document set. While the general visualization systems mainly focus on how to present the data space, the information-visualization systems have to consider what to present as the information space as well. In the following section, the topic streams are extracted by the plastic clustering method equipped with memory cells, and represented on the keyword map [12.5, 12.10, 12.14, 12.15, 12.16] along with the corresponding document clusters. Compared with the ordinary information-visualization methods including a keyword map that handles only a single set of documents, the approach proposed in this paper generates the keyword map of the current document set while considering the keyword maps generated from previous document sets.

12.5 Plastic Clustering Method Based on Immune Network Model

12.5.1 Plastic Clustering Method

A plastering clustering method [12.11, 12.12, 12.14, 12.15] has been proposed to generate a keyword map as well as document clusters. On the keyword map, the keywords relating to the same topic are assumed to gather and form a cluster. The plastic clustering method extracts a representative keyword, called a *landmark*, from each cluster. As the border of keyword clusters on the keyword map is usually not obvious, the constraints for extracting a landmark are adopted from the viewpoint of document clustering. That is, when documents containing the same landmark are classified into the same cluster, there should not exist overlapping among the clusters. The algorithm of the plastic clustering method is as follows:

1. Extraction of keywords (nouns) from a document set using the morphological analyzer[1] and the stop-word list. In this paper, only the keywords contained in more than two documents are extracted.
2. Construction of the keyword network by connecting the extracted keywords k_i to other keywords k_j or documents d_j:
   • Connection between k_i and k_j: (D_{ij} indicates the number of documents containing both keywords.)

   Strong connection (SC): $D_{ij} \geq T_k,$ $\hspace{3cm}$ (12.1)

   Weak connection (WC): $0 < D_{ij} < T_k.$ $\hspace{2.5cm}$ (12.2)

   • Connection between k_i and d_j: (TF_{ij} indicates the term frequency of k_i in d_j.)

   SC: $TF_{ij} \geq T_d,$ $\hspace{4.5cm}$ (12.3)

   WC: $0 < TF_{ij} < T_d.$ $\hspace{3.5cm}$ (12.4)

3. Calculation of keywords' activation values on the constructed network, based on the immune network model ((12.5)–(12.9)).
4. Extraction of the keywords that activate much more highly than others as landmarks after the convergence.
5. Generation of document clusters according to the landmarks.

As for the immune network model in step 3, one of the simplest models proposed in the field of computational biology [12.2, 12.6, 12.8] is adopted.

$$\frac{dX_i}{dt} = s + X_i \left(f \left(h_i^b \right) - k_b \right),$$ $\hspace{2cm}$ (12.5)

$$h_i^b = \sum_j J_{ij}^b X_j + \sum_j J_{ij}^g A_j,$$ $\hspace{2cm}$ (12.6)

$$\frac{dA_i}{dt} = \left(r - k_g h_i^g \right) A_i,$$ $\hspace{2cm}$ (12.7)

$$h_i^g = \sum_j J_{ji}^g X_j,$$ $\hspace{2.5cm}$ (12.8)

$$f(h) = p \frac{h}{h + \theta_1} \frac{\theta_2}{h + \theta_2}.$$ $\hspace{2cm}$ (12.9)

Here X_i and A_i are the concentration (activation) values of B-cell i and antigen i, respectively. s is a source term modeling a constant cell flux from the bone marrow and

---

[1] As the current system is implemented to handle Japanese documents, the Japanese morphological analyzer *Chasen* (http://chasen.aist-nara.ac.jp/) is used to extract nouns.

r is a reproduction rate of the antigen, while k_b and k_g are the decay terms of the antibody and antigen, respectively. J_{ij}^b and J_{ij}^g ($\in \{0, WC, SC\}$) indicate the strength of the connectivity between the antibodies i and j, and that between antibody i and antigen j, respectively. The influence on antibody i by other connected antibodies and antigens is calculated by the proliferation function (12.9), which has a log-bell form with the maximum proliferation rate p. In step 5, it is observed through the experiments that the highly activated keywords have much higher (about 100 times higher) activation values than others [12.11]. A convergence means that the same set of keywords always becomes active, which is observed after at most 1000 times calculation in most of the experiments. Applying a non-monotonic activation mechanism of the immune network model enables us to satisfy the contradictory conditions for a landmark, that is, a landmark should form a keyword cluster with a certain number of connected keywords, while there should not exist any connection between landmarks.

Experiments are performed based on questionnaires [12.14], and the results show that the quality of clusters generated by the plastic clustering method is better than the k-means clustering method in many cases.

Furthermore, as the landmark suppresses the related keywords on the constructed keyword network, such relationship is also useful as the metaphor to improve the understandability of the keyword map [12.14, 12.15] (Fig. 12.4). While the ordinary keyword map uses only the distance information, the immune network metaphor can emphasize the keyword cluster of which the representative is a landmark.

12.5.2 Introduction of Memory Cell for Context Preservation

The information-visualization systems based on the document clustering method assume that documents contained in the same document set should have a relation to each other from a certain viewpoint. Furthermore, it is assumed here that the document sets that are given sequentially have a certain relation to each other. In particular, we aim to find the topic stream as noted in Sect. 12.4 from a sequence of document sets.

To find mainstream topics and missing topics through a sequence of document sets, the landmark that has once been extracted from a certain document set should be preferentially extracted from the subsequent document sets. By analogy with the immune system, such preferential keywords can be realized with the property of a *memory cell*. As the immune system in nature is viewed as an order-sensitive learning system [12.8], which usually remembers all invasions of antigens forever, employing an immune network model is suitable for our purpose of finding topic streams, in particular missing topics. This order-sensitive nature can be viewed as one of the heuristics. To incorporate the property of a memory cell into the plastic clustering method, several types of memory cells are proposed in [12.13]. Among them, the parameters of three types are shown in Table 12.1.

The memory cell of type A has a lower decay term than a normal antibody, while those of type B have the broad activation region $R_{\mathrm{act}}^i = \{h_i^{\mathrm{b}} \mid f(h_i^{\mathrm{b}}) > k_{\mathrm{b}}\}$ by adjusting the values of θ_1 and θ_2.

Table 12.1. Parameter settings for memory cells of several types

Type	k_{b}	p	θ_1	θ_2	R_{act}^i
Normal	0.4	1.06	10^3	10^6	$(607, 1.65\times10^6)$
Type A	0.3	1.06	10^3	10^6	$(395, 2.53\times10^6)$
Type B2	0.4	1.03	625	1.6×10^6	$(397, 2.52\times10^6)$
Type B3	0.4	1.15	741	1.35×10^6	$(395, 2.53\times10^6)$

The experiments are performed to show which type of memory cell is more robust than others in terms of activation, using the following measure, where S_{A} indicates the set of highly activated keywords (i.e. landmarks), S_{M} indicates the set of memory-cell keywords, and S indicates the set of all keywords.

$$\mathrm{Eval} = \frac{|S_{\mathrm{A}} \cap S_{\mathrm{M}}|}{|S_{\mathrm{M}}|} \bigg/ \frac{|S_{\mathrm{A}} - S_{\mathrm{M}}|}{|S - S_{\mathrm{M}}|}. \tag{12.10}$$

Figure 12.1 shows the comparison results, in which the average values of five trials are shown. The leading articles that were released from Yahoo! Japan News[2] on August 8, 2001 are used for the experiments. The number of news articles is 40, from which 115 keywords are extracted.

It is confirmed from Fig. 12.1 that the memory cells of types A, B2, and B3 are given priority to activation compared with normal keywords. However, the memory cell of type B2 can have less priority, because its activation speed that is determined by $(p - d)$ is slightly slower than those of types A and B3.

12.6 Experimental Results

The plastic clustering method equipped with a memory cell is applied to a sequence of online news article sets. In the experiments, a keyword that has once been extracted as a landmark is considered as a memory cell. Regarding the memory-cell model, type A proposed in the previous section is adopted because of its simplicity. The parameter values used in the experiments are shown in Table 12.2, where $k_{\mathrm{b}}^{\mathrm{M}}$ indicates the decay term for a memory cell. These values are empirically determined based on the values used in [12.2, 12.6, 12.8].

Experiments are performed on two sequences of news article sets that were collected from Yahoo! Japan News. The news articles in the entertainment category are used for the experiments, because (1) the entertainment category contains various kinds of topics, and (2) the keywords used in this category are easier to extract without a specific dictionary than those in other categories. The first sequence

[2] http://news.yahoo.co.jp/headlines/

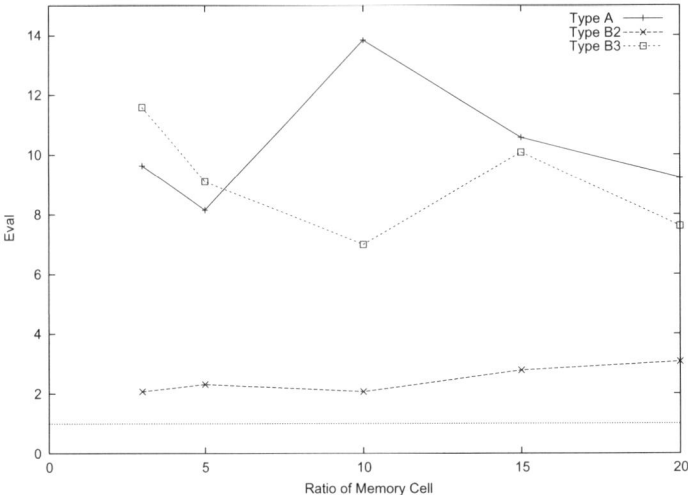

Fig. 12.1. Comparison of memory cell and normal cell keywords in terms of activation ten- - dency

(named *short sequence*) contains the news articles issued from September 17, 2001 to September 21, 2001, while the second (named *long sequence*) contains the news articles issued from January 30, 2002 to February 14, 2002. It is noticed that the results are translated into English, because the news articles used in the experiments are written in Japanese.

Table 12.2. Parameter settings used in the experiments

Parameter	s	r	k_g	k_b	k_b^M	θ_1	θ_2
Value	10	0.01	10^{-4}	0.4	0.3	10^3	10^6
Parameter	$X_i(0)$	$A_i(0)$	T_k	T_d	SC	WC	p
Value	10	10^5	3	3	1.0	10^{-3}	1.0

12.6.1 Topic Stream Extraction from Short Sequence

Figure 12.2 shows the landmarks extracted from the short sequence. The landmarks extracted by both methods (with/without memory cell) are indicated with dotted texture. In Fig. 12.2, one landmark is re-used more than once among 18 landmarks (5.56%) when no memory cell is used. On the contrary, three landmarks are re-used among 17 landmarks (17.6%) using a memory cell. As the news articles within the short sequence were issued just after the tragedy in New York., many articles contain the related topics. Two landmarks, 'Performance', 'Spate', which are re-used more than once, concern the topics related to that New York tragedy. That is, 'Spate' literally concerns the spate of simultaneous terrorism occurred in New York. After

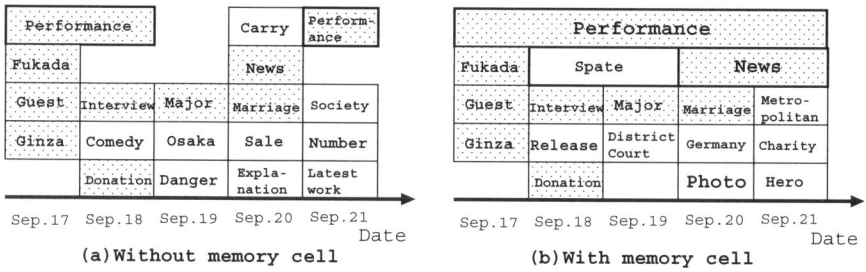

Fig. 12.2. Landmarks extracted from short sequence

that disaster, many concerts, events, etc., were performed/held for the purpose of charity, donation, or encouragement for the victims. The landmark 'Performance' mainly focuses on this kind of topic.

Furthermore, the plastic clustering method with a memory cell can find the topics related to the New York disaster from all the document sets in the short sequence, while the method without a memory cell can not find such topics from the document set of September 19, 2001.

12.6.2 Topic Stream Extraction from Long Sequence

Table 12.3. Distribution of landmarks extracted from long sequence

# of occurrences	1	2	3 or more	Total
Without memory cell	44 (83%)	7 (13%)	2 (4%)	53 (100%)
With memory cell	23 (61%)	9 (24%)	6 (16%)	38 (100%)

Table 12.3 shows the distribution of the landmarks that are extracted from the long sequence, according to the number of occurrences. It is shown in the table that the ratio of landmarks that are extracted more than once in all the landmarks by using a memory cell (39%) is about 2.3 times as much as that without a memory cell (17%). Regarding the landmarks that are extracted more than twice, their streams over the long sequence are shown in Fig. 12.3. In the figure, a solid line indicates the landmark extracted using a memory cell, while a dotted line indicates those extracted without a memory cell. A thin solid line indicates that the same landmark is also extracted without a memory cell.

Except for 'Singer' extracted from the article set of January 31, 2002 and 'Release' extracted from the set of February 5, 2002 the plastic clustering method with a memory cell never misses the landmarks that are extracted without a memory cell. That is, once a keyword becomes a memory cell, it can be a landmark whenever the same one is extracted without a memory cell. It is also confirmed that this fact is applicable to the landmarks extracted more than once.

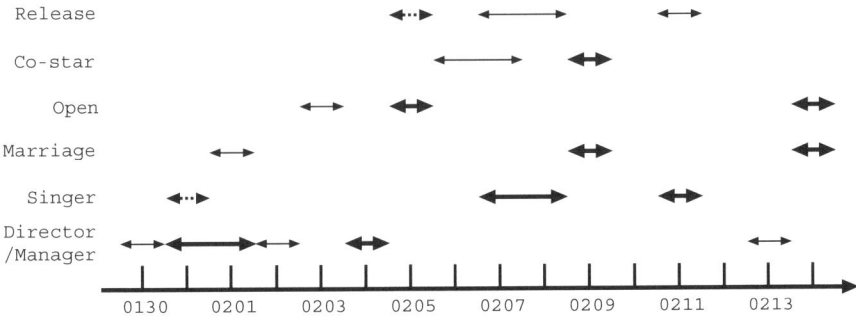

Fig. 12.3. Topic stream extracted from long sequence

Although the long-sequence does not contain the specific topic such as the New York tragedy of the short sequence in Sect. 12.6.1, it can be seen in Fig. 12.3 that there exist mainstream topics concerning dramas ('Co-star' and 'Director/Manager'), music ('Singer'), the release of something new such as a CD ('Release'), the opening of some events or ceremonies ('Open'), and the marriage of a celebrities ('Marriage'). In particular, the topic indicated by the landmark 'Marriage' can also be viewed as the missing topic, because there exists a gap between its first occurrence and the second one.

Figure. 12.4 shows the keyword map generated from the news article set on February 5, 2002, by incorporating *the immune network metaphor* [12.14, 12.15] into the spring model [12.16]. Figure 12.4(a) shows a keyword map without a memory cell, and (b) shows that with a memory cell, in which a landmark is indicated in white color, while a dark-colored one is the keyword suppressed by a landmark. It can be seen that the arrangement of keywords is drastically changed by incorporating memory cells into the plastic clustering method. That is, the landmark 'McCartney' suppresses the keywords 'Paul', 'Show', and 'Presentation' in Fig. 12.4(a). This keyword cluster concerns the topic about the event or tour in which Paul McCartney himself takes part. On the contrary, in Fig. 12.4(b), as the keyword 'Open' has been a memory cell since February 3, 2002, it forms a keyword cluster that concerns the topic about the opening of some events, tours, etc., in which not only the events related to McCartney but also others are included. As the event of Paul McCartney is only a part of the topics represented by 'Open', the keyword 'McCartney' is not arranged so close to the cluster of 'Open'.

We also perform the experiment in which the memory cells found in the short sequence are used as *the initial memory cells* when processing the long sequence. This experiment can be considered as pseudo-longer-period experiments. The experiment that uses memory cells but no initial memory cell is called the experiment with the usual memory cell and that using the initial memory cell as well is called the experiment with the initial memory cell hereinafter.

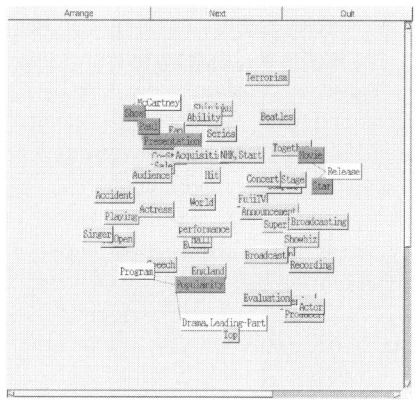
(a)Keyword map without memory cell

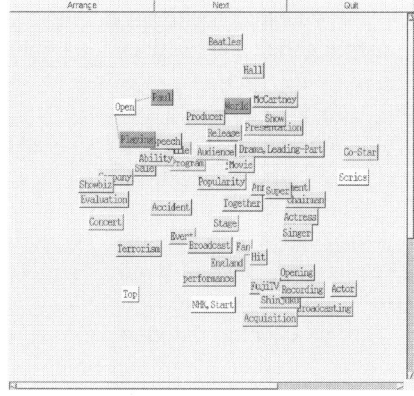
(b)Keyword map with memory cell

Fig. 12.4. Keyword map generated from document set of February 5, 2002

In the experiments with the initial memory cell, seven memory cells in the short sequence ('Interview', 'District Court', 'Marriage', 'Photo', 'Performance', 'Release', and 'Major') are also found in the long sequence. In particular, 'Performance', which has not been found as a topic stream in the experiment with the usual memory cell, can be found three times and forms a major topic stream.

Another interesting result is obtained with 'District Court'. In the data set on February 2, 2002, the first trial hearing of a famous Japanese entertainer that was held in the Tokyo district court is reported in two articles.

One of these two articles also mentions his comeback to the field of show business. Therefore, this entertainer relates to several topics, such as the comeback and crime, among which the topic (landmark) should be selected based on both the topic streams found in the previous document sets and the related documents in the current document set. In the data set on February 2, 2002 the other two documents concern the comeback of actresses, and another document reports that a famous actress was indicted over four different crimes. In the case of the experiment with the usual memory cell, 'Comeback' has become a memory cell in the previous data set, leading to the selection of 'Comeback' as a landmark. On the contrary, as 'District Court' has been one of the initial memory cells in the experiment with the initial memory cell, it is selected as a landmark.

12.7 Conclusion

Awareness of rare events as well as imagining their hidden factors is the essential point for chance discovery, which should be performed through an interaction with environments. This process does not only involve the objective (statistical) analysis but also the subjective evaluation, which makes it difficult to be solved by the conventional technologies or theories.

This chapter considers the Web as the environment, and discusses how to help a human notice and evaluate the topics discussed on the Web. In particular, its dynamic nature, such as a sequence of online news articles constantly released day by day, is focused on, and a method to find a topic stream from a sequence of document sets is introduced. The plastic clustering method that generates a keyword map as well as document clusters is combined with the memory-cell model, in order to find the relation among the keyword maps generated from different document sets. The method is applied to a sequence of online news article sets, and the results show that a memory cell can play a role to find a topic stream through a sequence of document sets.

In this chapter, the characteristic of memory cells is employed to handle the heuristic based on a human's order-sensitive processing (learning, modeling, etc.). There are certainly various heuristics that will help us to tackle the huge volume of the Web, which should be incorporated into the intelligent interface in order to support our chance discovery process.

References

12.1 Allan J, Papka R, Lavrenko V (1998) On-line New Event Detection and Tracking, Proc. 21st Annual International ACM SIGIR Conference on Research and Development in Information Retrieval, pp.37–45

12.2 Anderson RW, Neumann AU, Perelson AS (1993) A Cayley Tree Immune Network Model with Antibody Dynamics, Bulletin of Mathematical Biology, 55(6): 1091–1131

12.3 Ikeda K (2000) Communication, The University of Tokyo Press. (in Japanese. Series on Models of Social Sciences)

12.4 Iwanaga S, Namatame A (2002) Local and Heterogeneous Decisions and Their Collective Phenomena, Journal of Information Processing Society of Japan (IPSJ), 43(5):1528–1537

12.5 Lagus K, Honkela T, Kaski S, Kohonen T (1996) Self-Organizing Maps of Document Collection: A New Approach to Interactive Exploration, 2nd Int'l Conf. on Knowledge Discovery and Data Mining, 238–243

12.6 Neumann AU, Weisbuch G (1992) Dynamics and Topology of Idiotypic Networks, Bulletin of Mathematical Biology, 54(5):699–726

12.7 Rogers E (1983) Diffusion of Innovations 3rd Ed., The Free Press, New York, NY

12.8 Smith DJ, Forrest S, Perelson AS (1996) Immunological Memory is Associative, Int'l Workshop on the Immunity-Based Systems (IBMS'96), pp. 62–70

12.9 Simon H (1945) *Administrative behavior: a study of decision-making processes in administrative organizations* (1st edition 1945, 4th edition 1997, Simon & Schuster Inc., New York, NY

12.10 Sumi Y, Nishimoto K, Mase K (1996) Facilitating Human Communication in Personalized Information Spaces, AAAI-96 Workshop on Internet-Based Information Systems, pp.123–129

12.11 Takama Y and Hirota K (2000) Application of Immune Network Model to Keyword Set Extraction with Variety, 6th Int'l Conf. on Soft Computing (IIZUKA2000)}, pp.825–830

12.12 Takama Y and Hirota K (2001) Employing Immune Network Model for Clustering with Plastic Structure, 2001 IEEE Int'l Symp. on Computational Intelligence in Robotics and Automation (CIRA2001), pp.178–183

12.13 Takama Y and Hirota K (2001) Finding Topic Distribution from A Sequence of Document Sets, Proc. of 2nd Vietnam-Japan Symposium on Fuzzy Systems and Applications (VJFUZZY'2001), pp.132-139

12.14 Takama Y and Hirota K (2002) Immune Network-based Clustering for WWW Information Gathering/Visualization, Motoda, H., ed., *Active Mining* IOS Press, Amsterdam, The Netherlands, pp.21 – 29

12.15 Takama Y and Hirota K (2002) Proposal of Topic Stream Visualization from Sequence of WWW Document Sets, Proc. of Workshop of Japanese Society of Artificial Intelligence, SIG-KBS-A201}, 111-116, (in Japanese)

12.16 Takasugi K and Kunifuji SA (1999) Thinking Support System for Idea Inspiration Using Spring Model, J. of Japanese Society for Artificial Intelligence, 14(3): 495-503 (in Japanese)

12.17 Yang Y, Carbonell J, Brown R, Pierce T, Archibald B, Liu X (1999) Learning Approaches for Detecting and Tracking News Events, IEEE Intelligent Systems, 14(4): 32-43

12.18 Zamir O, Etzioni O (1999) Grouper: A Dynamic Clustering Interface to Web Search Results, Proc. 8th Int'l WWW Conference, pp.1361–1374

13. Effects of Scenic Information

Yasufumi Takama[1] and Yukio Ohsawa[2,3]

[1] Tokyo Metropolitan Institute of Technology, Hino, Tokyo 191-0065, Japan
[2] PRESTO, Japan Science and Technology Corporation, 2-2-11 Tsutsujigaoka, Miyagino-ku, Sendai, Miyagi 983-0852, Japan
[3] Graduate School of Business Sciences, University of Tokyo, 3-29-1 Otsuka, Bunkyo-ku, Tokyo 112-0012, Japan
email: ytakama@cc.tmit.ac.jp, osawa@gssm.otsuka.tsukuba.ac.jp

Summary.

Scenic information is a source of chances, playing a significant role as the 'key 2' for chance discovery mentioned in Chap. 1. It stimulates the contextual shifts of a human, which carries him/her into deeper concerns with actions on chances to be discovered. In other words, scenic information guides a human to the imagination of new environments with new chances. This chapter shows the role of scenic information in up-to-date application domains.

13.1 Introduction: Scenic Information for Chance Discovery

We can not recall everything kept in our mind. Because of our limited rationality [13.3], we can not always derive the best solution, even though we have enough knowledge and information. Scenic information is essential for handling such situations as a trigger that activates our thought and imagination. Let us consider several examples in which scenic information is essential.

13.2 Situation in Answering Questionnaires

Consider the situation when we answer questionnaires about some products. The purpose of the questionnaires is usually to know what the users think about the products, such as the following:

- Advantages of the product against the competitors.
- Inconvenience of the product, which the user wants to improve in new products.
- The situation and purpose for the user.

When we are using the products, we certainly feel something good, or have some complaints. However, such comments about products tend to be lost easily, and we seldom recall what we think about the product when we have the questionnaires in front of us.

Of course, we can answer the questionnaires while using the products. Furthermore, a video that records how we used them is also effective for us to remember what we think when we use the products. This video information is one type of scenic information.

13.3 Making Diagram for Idea Generation

Making diagrams while reading/writing documents, or discussing in a meeting, is useful for the creative thinking process [13.5, 13.4]. As the relationship between documents, opinions, etc., is complicated, it is difficult to translate directly into a linear-order structure like text. A diagram (Fig. 13.1) can include several relationships between objects, which can be formed gradually during reading processes [13.4], or a meeting. As noted later in this chapter, a diagram also activates a group discussion, because a temporarily generated diagram that is formed co-operatively based on the current discussion will invoke the succeeding discussion.

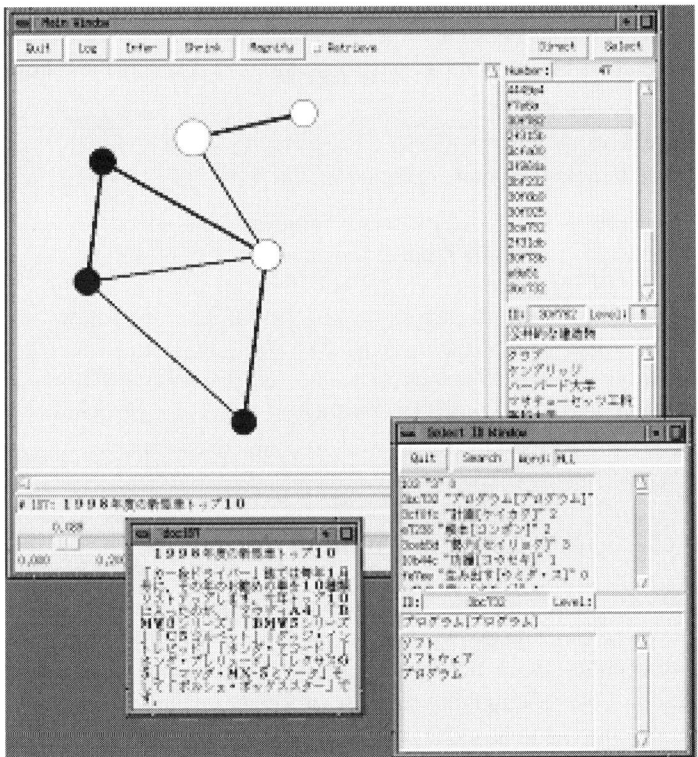

Fig. 13.1. Diagram of FishView system (document-ordering support system)[13.4]

Although a diagram can include a lot more context information than text (document)-style organization, such context information tends to be lost as time goes on, and the readability of a diagram decreases rapidly. The same reason also leads to the fact that it is difficult to read the diagram generated by other people (group). Not only a diagram, but also a memo taken on a bit of paper or on the back of an envelope, easily loses its readability. In order to derive the proper information from a diagram (a memo), the context information in which it was generated (taken) should be required. What gives us such context information is also one of the types of scenic information.

13.4 Scenic Information on the Web

Today, the Web has become one of the important environments for companies and shops to exhibit their activities as well as to sell products and services. They can submit much more information to many more customers than their shops and showrooms in the real world . However, it seems that the virtual shops on the Web have a fatal disadvantage against the actual shops, i.e. the customers can not touch the products in the virtual shops. It does not only mean that, for example, we can not examine the feel of clothes, but also that we have difficulties in knowing the following things:

– How many similar products exist?
– How well does a certain product match another product?

In these cases, it is helpful for customers if scenic information as secondary information about surroundings of a certain product is available.

On the contrary, scenic information is also helpful for companies and shops. As noted in [13.1], ecommerce is an ideal style for retailers, if they can follow customers' behaviors, i.e. what they looked at, what they put in their shopping carts, etc. Although the privacy problem is serious and significant, such detailed observation can not be obtained in the real world. The customers' behaviors are considered as scenic information, which most companies want to obtain at any cost.

13.5 Problem of Losing Scenic Information

As discussed in the previous section, scenic information is very important for our recall ability or creative thinking process. Without scenic information, it is difficult for us to recall what we keep in our memory when needed. Furthermore, the diagrams or memos as the result of the past thinking processes also require the accompanying scenic information in order not to lose their readability.

However, the results of questionnaires are often processed by statistical approaches. The fields that often employ questionnaires, such as data mining and sociology, make use of statistical methods. As noted in Chap. 12 by Takama, a statistical method is one of the major approaches to generate macro-information from the collection of micro (individual)-information. However, the statistical method generally models the objects as the variance from a standard (i.e. average), which means that the information specific to individuals is lost. Although the macro-information generated by the statistical approach may reflect the behavior/opinion of the majority, the actual context information that exists behind the individual's behavior disappears, even though the questionnaire can retrieve the users' honest opinions.

Visual information is also the essential media for transmitting scenic information. Because of the rapid growth of communication channels and recording media (HDD, DVD-ROM, etc.) as well as hardware (computer and display devices) and software, a high-quality, long stream is available. However, the problem is the gap between the virtual reality that can be visualized and the real thing. For example, children can enjoy a 3D video game, in which they can move and beat (shoot) the enemy in the high-quality, virtual environment . It is often said that what children feel in the game is only the pleasure, and they can not know what others feel, such as pain. Therefore, the visual media is a double-edged sword. The visual media is further considered in the next section.

13.6 Media Technologies for Scenic Information

13.6.1 Integration of Real and Information Environments

Integrating the information from information space, such as computers, databases, and the Internet, into the scene in the real world is one of the focused research areas. Typical approaches are virtual reality (VR), augmented reality (AR), and mixed reality (MR). Compared with virtual reality, focusing on the construction of a virtual environment, in which a human feels as if he or she is in the real world, augmented reality aims to enhance the user's view of the real world with visual and text information from a computer. Particular AR systems employ 'see-through' devices, usually worn on the head, which overlay graphics and text on the user's view of his or her surroundings.

Mixed reality, which was proposed in 1994 by Milgram and Kishino [13.2], consists of not only AR, but also augmented virtuality (AV), which aims to enhance the virtual world within computers with the information in the real world. That is, MR tries to connect completely real environments to completely virtual ones.

Currently, several types of devices that are used as the basis for VR, AR, and MR are being developed, which are expected to play an important role to give us the scenic information that is beyond what can be represented by either the real-world view or virtual information. For example, the information from virtual environments will be able to add the context information that is recorded along with time series to the real world information. However, we should remember that the visual media is a double-edged sword, as discussed in the previous section.

13.6.2 Simulation and Visualization

We have a limited rationality, as noted in Chap. 12 by Takama, and can not always predict how things go, or to what result our decision will lead. To cope with this matter, a computer simulation is expected to complement our imagination. Of course, a simulation is based on a certain model that usually consists of the past observations/experiences. Therefore, simulation results can not always predict futures correctly, and it is dangerous to blindly rely on the computer simulation.

Visualization by computer systems gives expressive power to computer simulations. The primary output of computers is a numerical one, which is inconvenient for a human to understand intuitively. Human beings outperform computer systems in terms of pattern-recognition ability. For example, the National Museum of Emerging Science and Evolution in Japan (www.miraikan.jst.go.jp) exhibits a global display 'Geo-cosmos', of which the diameter is 6.5 m. The 'Geo-cosmos' has $951,040$ LEDs on the surface, and constantly displays the image data from NASA, which is taken by a satellite. It also displays various kinds of data about the Earth, such as the surface temperature and the concentration of carbon monoxide. Furthermore, it can also display the simulation result of the global warming from 1890 to 2100. Watching how the Earth ('Geo-cosmos') becomes red (a red color indicates high temperature) shocks the audiences, and they will be conscious of the seriousness of global warming.

References

13.1 Gurak JL (2001) chapter 6. Privacy and Copyright in Digital Space, in cyberliteracy, Yale University Press
13.2 Milgram P and Kishino F (1994) A Taxonomy of Mixed Reality Visual Displays, IEICE Trans. on Information Systems, E77-D(12):1321–1329
13.3 Simon H (1945) *Administrative behavior: a study of decision-making processes in administrative organizations* (1st edition 1945, 4th edition 1997, Simon & Schuster Inc., New York, NY
13.4 Takama Y, Ishizuka M (2000) Fisheye Matching: viewpoint-sensitive feature generation based on concept structure, Knowledge-Based Systems, 13:199–206
13.5 Takasugi K, Kunifuji S (1999) A Thinking Support System for Idea Inspiration Using Spring Model, J. of Japanese Society for Artificial Intelligence, 14(3):495–503 (in Japanese)

14. The Storification of Chances

Corporate Training with Life-Like Characters in a Virtual Social Environment

Helmut Prendinger and Mitsuru Ishizuka

Department of Information and Communication Engineering,
Graduate School of Information Science and Technology,
University of Tokyo,
7-3-1, Hongo, Bunkyo-ku, Tokyo 113-8656, Japan
email: {helmut,ishizuka}@miv.t.u-tokyo.ac.jp

Summary.

   This chapter presents a view of how to use a virtual social environment inhabited by life-like characters to train the awareness of chances. Here, a user is immersed into a virtual story world where he or she interacts with animated agents and can make decisions that affect the future development of the story, eventually leading to positive or negative consequences. Our *storification of chances* approach to chance discovery relies on existing real-world stories that are either 'mistake stories' or 'success stories'. That is, it constitutes a method for realizing the scenic information of the last chapter. The Web-based interaction scenarios serve as a training environment for users striving to acquire practical knowledge that is typically tacit (that is, not explicit). Storified chances can be considered as valuable additions to corporate memories.

14.1 Introduction

The term 'chance' is characterized as information about an event or situation that has significant influence on the decision-making process of a human or artificial agent. Depending on whether this information suggests a desirable or undesirable future development , a chance is perceived as an opportunity or as a risk. Chance discovery can be described as the process of becoming aware of an opportunity or of a risk. After the situation or event is identified as a chance, it can 'be taken' (or ignored) and constitute an essential part of the decision-making process [14.22].

   Chances, or knowledge about chances, do not necessarily have to be 'new' in the sense of an opportunity or risk never encountered before. Often, chance knowledge is part of the experience of humans who managed to turn an opportunity into success or were able to avoid some risk. In organizations, this kind of knowledge is known as *tacit* knowledge, and is typically contrasted with *explicit* knowledge [14.21]. While explicit knowledge is formal knowledge that is documented in reports, manuals, patents, pictures, videos, software, etc., tacit knowledge refers to personal knowledge based on individual experience that is typically exchanged in direct human–human communication. The importance of tacit knowledge about corporate success has been discussed extensively in the area of knowledge management [14.3]. Since tacit knowledge is essentially practical knowledge on 'how to do things' or 'how to decide', it is an important candidate for inclusion in *corporate (or organizational) memories.* Specifically, corporate memories should contain process knowledge related to problem-solving and decision-making activities, and allow for different points of view on relevant issues (in order to prevent the delusion of 'objectivity' often conveyed in explicit knowledge sources).

In this chapter, we will propose *interactive stories* as a technology to communicate knowledge in a corporate context. Our approach is influenced by Lawrence and Thomas [14.16], who suggest story telling in order to build up a (corporate) storybase. They show how social power, possible risk of telling a story , and collaboration (where the audience interrupts with additions, questions, and comments) influence the way a story is told. Most notably, the usefulness of 'mistake stories' is pointed out in the business context. Storybases in their approach essentially refer to textual information resources. Instead, we will suggest the use of life-like characters inhabiting a virtual social environment to deliver stories. Rather than simply telling a story – as virtual story tellers – life-like characters may illustrate events by performing in specified roles of the story. Virtual environments even allow the user to participate as a character in the story. Thereby, the user may influence its development and learn to cope with difficult situations or, in other words, train his or her awareness of chances (risks or opportunities). Similarly, Mott et al. [14.20] motivate stories (or narratives) in the context of learning environments where students are to be actively involved in 'story-centric' problem-solving activities. Their fundamental hypothesis is that "[...] by enabling learners to be co-constructors of narratives, narrative-centered learning environments can promote the deep, connection-building meaning-making activities that define constructivist learning." [14.20, p. 80]. The authors argue that stories lend themselves to active exploration of a domain through challenging and enjoyable problem-solving activities, which is essential for constructivist learning. As a particular case, we will employ interactive stories as a technology to sharpen a user's sense ('awareness') of opportunities and risks in real-life situations, which we call the *storification of chances* and propose as a novel approach to chance discovery.

Some clarification on the meaning of the terms 'narrative' and 'story' is in order. By *narrative* we mean a certain type of artifact that satisfies artistic criteria (in a broad sense), instantiated in various forms such as novels, theater, movies, games, and so forth. A *story* refers to the succession of events that constitute a certain type of narrative (see [14.33] and references therein). Without any deep concerns about the nature of a 'good' story, we briefly introduce some properties of narrative discussed in Bruner's theory [14.4] (as reported in [14.27]).

– *Narrative diachronicity.* A basic property of narrative is diachronicity, which means that events are understood by the way they relate over time rather than by their moment-by-moment significance.
– *Intentional stance entailment.* This property says that what happens in a story is less important than what the involved characters feel about it. It is suggested that characters explicitly express the reasons for their actions and the emotions that trigger their actions.

- *Canonicity and breach.* A narrative is pointless when everything happens as expected. There must be some problem to be resolved, some unusual situation, some difficulty, someone behaving unexpectedly. However, norm deviations can themselves be highly scripted.

Although the mentioned properties do not directly lend themselves to suggest a certain implementation of story systems, they clearly point out issues that deserve consideration in the design of such systems.

The remainder of the chapter is organized as follows. The following section depicts a motivating example for the storification of a chance. Next, we will illustrate two different ways in which a story can be experienced, observation and immersion. After that, we will briefly describe life-like characters as embodied agents that may convey social cues by speech and gesture, and hence perform as actors in story environments. The following section provides a comprehensive overview of existing plot-based and character-based approaches to interactive story systems. We will then demonstrate the storification of chances with a Web-based scenario that allows the user to interact with life-like characters in a corporate setting. The final section summarizes and concludes the chapter.

14.2 Motivating Example

The following is the beginning part of Kenshi Hirokane's [14.14] story 'Suspicion' of the *Division Chief Kosaku Shima* series:

> Shima-san: I'm Shima of the advertising division. I believe we've entered an era in which enterprises must tackle environmental problems head-on. I propose that we at Hatsushiba declare our intention to 'Protect the Earth.'.
>
> Nakazawa-san: What specific actions would that involve?
>
> Shima-san: There are any number of approaches. Chief among them, the disposal of discarded products.
>
> Participant of the meeting: Disposal?
>
> Shima-san: Yes. The streets overflow with discarded appliances. In times of material abundance, who bothers to repair old appliances? It's become quite a social problem. You often see appliances left out as oversized trash. Why don't we, the manufacturers, take the responsibility for their disposal?
>
> Another participant of the meeting: That's absurd. Think of the cost.
>
> Shima-san: The cost would have to be passed on in the prices of products.
>
> Kawauchi-san: Chairman!
>
> Nakazawa-san: Yes, Mr. Kawauchi?

Kawauchi-san: I object to Mr. Shima's proposal. How can we compete if we have to jack up our prices? This is sheer idealism. It's just not realistic.

Shima-san: We'll need our technicians to help us re-consider materials, for one thing. Construct TV cabinets from recyclable resins, for example. Or use inexpensive, easily incinerated substances.

Further participants of the meeting: And where are the R&D funds to come from?

Yeah! Why do we have to be the first?

Hirai-san: Chairman!

Nakazawa-san: Laboratory Chief Hirai?

Harai-san: I agree with Mr. Shima. Profits will decrease. But protection of the environment is a social mission all enterprises must accept from here on. We at Hatsushiba can lead the way for companies the world over. Isn't this of the utmost significance?

In this story excerpt, members of a company discuss Mr. Shima's idea to consider environmental issues in product development. Some argue against the idea, others are in favor of it. Arguments are brought forward from different viewpoints, such as the cost of development and products, or the reputation of the company.

In the setting of this chapter, the following features of the story (excerpt) are of key importance. *First*, different opinions are 'personified' by individual actors of the story. This allows the audience to associate viewpoints to individuals. A similar idea is realized in the car-presentation scenario introduced by André et al. [14.1], where two animated buyer characters represent different points of view, specifically the pros and cons of a certain car (see the next section). A related approach is taken by Don [14.10] in interface design, who proposes the use of techniques from oral story telling in order to organize information in a knowledge base. A narrative structure suggests to view multi-modal contents as 'events' that can be experienced in temporal sequence (as a 'story') rather than as objects in virtual space, and hence supports users in organizing the information in memory.

Second, the excerpt provides the initial sequence of a story where actors follow their goals and associated plans and will eventually succeed (opportunities turn into success), or fail (risks lead to failures).

In summary, stories provide a compelling training context where the audience may experience chances being taken and chance awareness is instantiated by the dramatis personae of a story, possibly by the user as a protagonist. The following section illustrates two different ways in which a story may be experienced, from the spectator or the protagonist viewpoint.

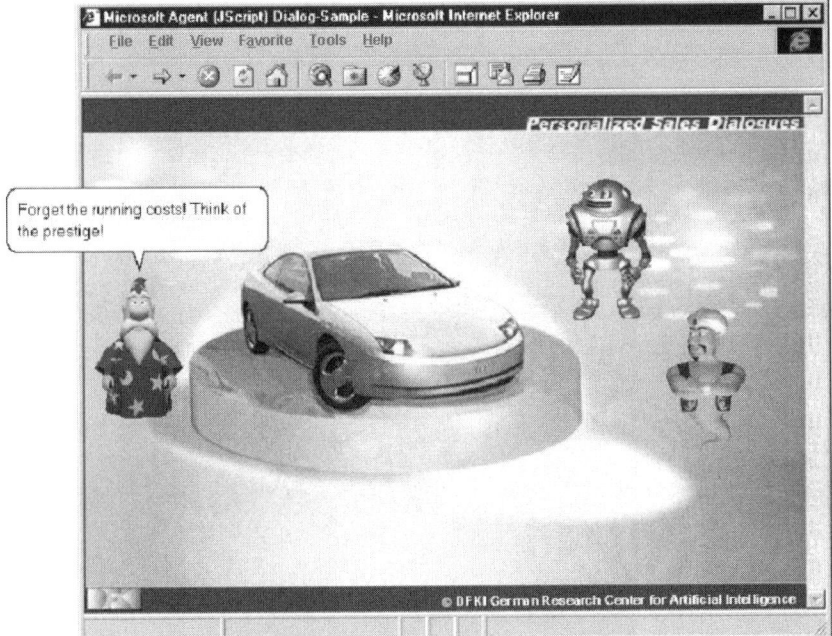

Fig. 14.1. Inhabited market place

14.3 Observation vs. Immersion

There are two major ways in which the audience may perceive story events performed by virtual agents. Users may either passively observe a story as spectators
or they may actively participate in the development of the story, being immersed in
an environment. In the following, we will illustrate both types of involvement with
a story.

14.3.1 Inhabited Market Place

The *inhabited market place* (IMP) is a Web-based system developed at DFKI that
uses multiple animated characters , so-called 'presentation teams', to convey information about products such as cars [14.1]. The IMP can be seen as a virtual
showroom where a seller agent presents products to one or more buyer agents. In
Fig. 14.1 (reproduced from [14.1, p. 229]), the seller agent 'Merlin' presents a car to
two buyer agents, 'Genie' and 'Robby'. All agents shown in the car-sales scenario
come with the Microsoft Agent package [14.19].

Although the IMP does not directly support a narrative interpretation, it provides
a multi-party sales dialogue with different viewpoints taken by the actors involved,
depending on their role (buyer or seller), personality profile (agreeableness, extro-
version), and interests (comfort, cost, sportiness, and so on). The spectator of a

Fig. 14.2. Mission rehearsal exercise

simulated sales dialogue is offered a multi-facetted perspective of a product, with the pros and cons pointed out by animated characters. The IMP designers argue that user–agent interaction is not necessarily the most convenient or effective method for product presentation. However, a development in this direction can be found in their successor work (see e.g. [14.2]).

14.3.2 Mission Rehearsal Exercise

The *mission rehearsal excersise* (MRE) project at the USC Institute for Creative Technologies (ICT) aims to create a virtual reality learning system where participants (Army students) are immersed in a highly stressful environment in order to practice for real-world scenarios [14.32]. The student, e.g. an Army lieutenant, will stand in front of a large curved screen (8.75 feet tall and 31.3 feet unwrapped) and interact, for example, with a sergeant to make decisions in a peacekeeping mission. Figure 14.2 (reproduced from [14.32, p. 412]) shows a shot which was taken during setup of the system.

The kind of envisioned immersion in a MRE-style story and the types of chances encountered are best described in [14.32, p. 409]:

A young Army lieutenant drives into a Balkan village expecting to meet up with the rest of his platoon, only to find that there has been an accident (see Fig. 14.2). A young boy lies hurt in the street, while his mother rocks back and forth, moaning, rubbing her arms in anguish, murmuring encouragement to her son in Serbo-Croatian. The lieutenant's platoon sergeant and medic are on the scene.

The lieutenant inquires, 'Sergeant, what happened here?'

The sergeant, who had been bending over the mother and boy, stands up and faces the lieutenant. 'They just shot out from the side street, sir. The driver couldn't see them coming.'

'How many people are hurt?'

'The boy and one of the drivers.'

'Are the injuries serious?'

Looking up, the medic answers, 'The driver's got a cracked rib, but the kid's ....' Glancing at the mother, the medic checks himself. 'Sir, we've gotta get a medevac in here ASAP.'

The lieutenant faces a dilemma. His platoon already has an urgent mission in another part of town, where an angry crowd surrounds a weapons inspection team. If he continues into town, the boy may die. On the other hand, if he doesn't help the weapons inspection team their safety will be in jeopardy. Should he split his forces or keep them together? If he splits them, how should they be organized? If not, which crisis takes priority? The pressure of the decision grows as the crowd of local civilians begins to form around the accident site. A TV cameraman shows up and begins to film the scene.

Student users of the MRE system have to make decisions in order to overcome dilemmas such as the one described above by instructing virtual humans, for example the sergeant or medic. The visual experience and audio effects add to the user's immersion to the peacekeeping environment. Obviously, the scenario is highly demanding in terms of the user's chance (especially risk) awareness, where wrong decisions may lead to drastic consequences (in military terms). Unlike the previously introduced inhabited market place, interactions with MRE are driven by a story line that is imposed by a so-called 'story net' [14.32], consisting of nodes that allow for confined interactive 'freeplay' with the characters, and links that are linear sequences of events beyond the user's control and serve to carry on the story (more details are given below).

Our approach to 'immerse' users to a scenario will be less dramatic or impressive, but more accessible as we will focus on Web-based interaction scenarios. In the following section, the characters we envision as interaction partners for users will be described.

14.4 Life-Like Characters

Life-like characters are intended to communicate like real people and be able to engage naturally in conversation with humans and other agents. The synthetic characters we will use are cartoon-style 2D animations that run in a Web-page-based JavaScript interface. We also use MPML [14.15], a tool that facilitates scripting the behavior of agents controlled by the Microsoft Agent package [14.19]. The package comes ready with controls for triggering animation sequences, speech recognition, and a text-to-speech (TTS) engine. MPML provides an interface to a system called SCREAM that supports autonomously generated affective behavior, depending on the character's emotional state, personality, and other parameters of the social inter-action context [14.24].

The life-likeness of animated characters derives from their ability to convey so-cial cues. When humans communicate, they employ a variety of signals in combina-tion with verbal utterances, such as body posture, gestures, facial expressions, and gaze. In a similar way, animated characters may their bodies to convey meaning and regulate communication. The most extensive study of non-verbal behaviors for synthetic characters, especially gestures, can be found in Cassell's work on embod-ied conversational agents [14.6].

Emblematic gestures are culturally specified gestures, e.g. signalling 'okay' by a 'thumb-and-index-finger' ring gesture. An example of a *propositional* gesture is the use of both hands to measure the size of an object in symbolic space while saying 'there is a big difference'. There are four types of gestures that support the conveyance of communicative intent (so-called 'co-verbal' gestures [14.6]).

– *Iconic* gestures illustrate some feature of an object or action, e.g. mimicking to hold a phone while saying that someone has been called.
– *Metaphoric* gestures represent a concept without physical form, e.g. a rolling hand gesture while saying 'let's go on now'.
– *Deictic* gestures locate physical space relative to the speaker, e.g. by pointing to an object.
– *Beat* gestures are small baton-like movements to emphasize speech. A special form of a beat gesture is the *contrastive* gesture [14.7] that depicts a 'on the one hand . . . on the other hand' relationship if two items are being contrasted (see Fig. 14.3).

An important class of gestures (including facial gestures) serves the expression of an agent's *emotional state* such as 'suddenly raising arms with widely open eyes and mouth' to signal surprise (see Fig. 14.4). Although the face may express emotions most succinctly [14.11], we will rely on signals involving the whole body, as the size of the characters displayed in the Web browser is relatively small.

Gestures also realize *communicative functions* including conversation initiation, turn taking, back channelling ('nodding'), and breaking away from conversation [14.5]. The communicative behavior corresponding to the (communicative) func-tion of 'giving turn' is typically realized by looking at the interlocutor with raised eyebrows, followed by silence, whereas 'taking turn' is signalled by glancing away and starting to talk.

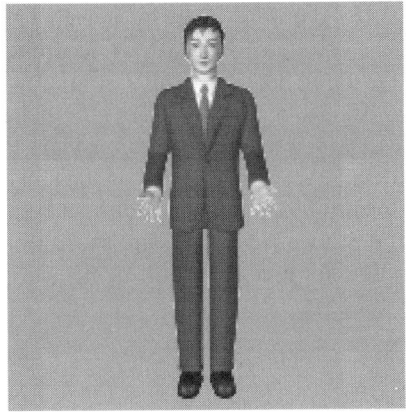

Fig. 14.3. 'Contrast'

Fig. 14.4. 'Surprised'

14.5 Approaches to Interactive Story Systems

Paradigms of interactive story telling can be classified into two categories, depending on the way the story is created.

- In *plot-based story systems* a plot manager is assumed that controls the succession of events (scenes) and specifies what the characters can do (e.g. Sgouros et al. [14.28] or Mateas and Stern [14.17, 14.18]).
- In *character-based story systems* the plot results from the behavior of autonomous characters (e.g. Hayes-Roth and van Gent [14.13] or Cavazza et al. [14.8, 14.9]).

Most of the existing interactive story-telling systems, however, follow a hybrid approach that integrates features of both categories. In the following sections, some representative systems will be introduced.

14.5.1 Story-Morphing

Story-morphing is a technique to achieve a simple form of a plot-based (interactive) story system [14.12]. Story-morphing relies on a given fixed base plot structure (a 'script') that allows us to generate numerous distinct stories (*story-morphs*) by varying the affective models of the involved characters. Optionally, the user may change the evolution of events by interacting with the characters. Story-morph 'tags' refer to emotionally meaningful units in the story, e.g. that a character likes or dislikes a certain activity which leads to different appraisals of events. For instance, depending on a character's interest in racing, it will react differently when losing a race.

In line with Bruner's property of 'intentional stance entailment' [14.4], the authors argue that what makes stories interesting is not solely based in what happens but in the way characters feel about the events in the story. In [14.24, 14.25], the story-morphing method is employed to generate morphs of a Black Jack game and a Japanese comic scenario, where the 'story' is given by the rules of the game and the make-up of the comic. In the Casino scenario the user is guided by an animated advisor to play the Black Jack game whereby the advisor's reactions vary according to its goals and personality profile as well as the user's decisions ('hit', 'stand') and the outcome of the game. In the second scenario, an animated version of a Japanese comic, the user may control an avatar to interact with a female character, and try to guess her wishes correctly. The character's affective reactions depend on the user's choices and her personality. However, the downside of easy creation of distinct event sequences in those scenarios is that the resulting 'stories' depict the same succession of scenes. The approaches discussed below allow for a more flexible composition of scenes.

14.5.2 Plot Control in Interactive Stories

Sgouros et al. [14.28, 14.29, 14.30] develop a framework for plot-based interactive story systems. The main component is a *plot manager* (PM) that controls the behavior of the characters and determines what the protagonist (the user) can do. The plot manager comprises three modules:

– A set of rules for social action.
– A specification of the role of each character in the story.
– The user interface manager.

As input, the PM assumes a story map consisting of a set of 'points', e.g. the destinations of a travel story. Each point (location) in the map has an associated *(local) plot structure* that specifies the cast of characters, their roles, and relations. The arcs between points of the story map are *transfer plans*, a set of conditions under which the user can move between points.

The role specification of characters defines their 'motives', the types of goals they pursue, the types of norms they try to uphold, and what kind of interference they undertake, divided into favorable interference (helping) and unfavorable interference (causing loss). For instance, a character in the *Judge* role tries to enforce legitimate power and watches out for transgressions of norms (laws). Furthermore, social action rules are applied to drive the interaction between characters. Social action includes co-operation, (positive and negative) reciprocation, group performance, and exchange (of resources).

For each interaction situation (point of the story map), the PM prepares a set of possible actions for the user, based on the protagonist's role and the activity of the other characters. Sgouros' approach follows an Aristotelian notion of plot, where the emphasis is on the concept of 'conflict' between antagonistic roles. After the climax is reached through a series of conflicts, a solution is presented at the end that answers all open questions.

14.5.3 Interactive Drama with User-Controlled Character

Szilas [14.33] proposes another plot-based approach to interactive drama. Since the focus of this work is strong emotional involvement of the user, it is argued that characters should be designed to comply with narrative constraints in order to guarantee an engaging story. The 'conflict' of the protagonist, i.e. some action which is incompatible with its values, is motivated as the core of the dramatic narrative. The responsibility of the *story engine* is to set the stage for a conflict, consisting of four units.

- A goal is a state-of-affairs the character wants to achieve.
- An obstacle is a state-of-affairs or event that prevents the character from achieving the goal.
- An 'overstepping task' are actions the character must perform in order to achieve the goal.
- 'Values' is a set of the norms the character maintains, with respect to which character actions are evaluated.

A conflict occurs when the overstepping task is not compatible with the character's values. The conflict is resolved by either ignoring ('overstepping') the values, thereby achieving the goal, or by deciding not to ignore the values, and not reach the goal. The narrative structure is realized by showing the possibility of 'overstepping' actions, the performance of actions (possibly including 'overstepping' actions), and then showing the consequences of the character's behavior.

The architecture of Szilas' story engine contains the story world, the narrative logic, the narrator, a (rudimentary) user model, and the theater (the interface). The narrative logic, a set of rules, encodes a particular narrative theory. The 'narrator' module is based on the rules in the narrative logic module and decides on the next steps in the story – which rules will 'fire' – by either executing events or prompting the user to choose the next action (from a choice list). The narrator's decision is based on artistic requirements from drama. The interaction between the user and system is realized by the user controlling one (or more) characters of the story world.

Szilas' approach is similar to the previously introduced approach by Sgouros as both rely on some concept of conflict and a narrator (plot manager) to control the story events.

14.5.4 Interactive Drama with Human Player

Mateas and Stern [14.17] discuss a plot-based approach to interactive drama where the autonomy of the characters is governed by restrictions from a plot (drama) manager. In their view, character behavior in a story world should depend on the character's mental state, the current world state, the current story state, and the history of all previous interactions.

The basic building blocks of interactive drama are so-called (dramatic) *beats* which are events that 'turn a value'. A 'value' is characterized as 'a property of an individual or relationship, such as trust, love, hope (or hopelessness), etc.' [14.17, p. 116]. An event is considered as a story event (or dramatic action) if it turns a value. Beats are just action–reaction pairs between characters. *Scenes* are seen as larger units of value changes and composed of beats, such as 'one character confesses his love to another character'. The responsibility of the *drama manager* is to compose a sequence of scenes that generates dramatic action on a larger scale. Formally, a scene consists of preconditions, values that are supposed to be changed (and how they change), and a set of beats that create the scene. The preconditions of a scene test if that scene is adequate given story and character state. Scene selection depends on satisfied preconditions, the list of unused scenes, and which value changes satisfy the intended plot arc.

An important architectural entity in Mateas and Stern's approach are *joint plans* that are intended to guide character behavior within one beat. They describe the co-ordinated behavior of characters as one entity rather than having autonomous characters work out a joint plan (which would require complex reasoning, message passing, and so forth). However, joint plans are still reactive, letting the user interfere with plan execution.

In [14.18], the authors define a language, called ABL (*a behavior language*), that allows us to author believable characters for interactive drama. ABL is a reactive planning language with character behaviors written in a Java-style syntax, that may encode joint plans and other story constructs.

Unlike Szilas' [14.33] approach to interactive drama, the authors let the user experience the story from a first-person perspective. The difference to Sgouros' [14.30] approach is that user interaction is not restricted to specific 'points' in the story.

14.5.5 Story Telling with Anytime User Intervention

Cavazza et al. [14.8, 14.9]) propose an approach to character-based interactive story telling. In this system, characters execute plans encoded as hierarchical task networks (HTNs). In order to ensure narratively relevant agent behavior, characters are described in terms of their role-specific goals and actions. For instance, the top goal of the male character (Ross) described in [14.8] is to take out a female character (Rachel). In order to achieve this goal, subgoals include acquiring information about her, getting her attention, isolating her from other characters, and asking her. The goal of acquiring information about Rachel can itself have multiple (disjunctive) subgoals, such as asking her friend or reading her diary.

Users may interfere with the story's progression at any time, either by voice input or by manipulating the environment (e.g. moving an object). However, 'physical' interaction is limited to so-called 'narrative objects', i.e. objects that have narrative consequences (*dispatchers*). For instance, if the user decides to 'steal' Rachel's diary, Ross has to repair his plan to read it, and resort to ask her friend. In this way, the user can eliminate certain courses of actions (of the main character), and force other routes being taken. Speech intervention in the system of Cavazza et al. maintains the user-as-spectator paradigm and is restricted to giving advice to characters (e.g. 'be nice' to Rachel's friend).

Since Cavazza's approach does not include a plot manager, the responsibility for a coherent story is shifted to the plan contained in the role specifications of characters. However, the plans associated with a particular role essentially encode the structure of the story.

A similar route is taken by Paiva et al. [14.23] where a story line is achieved by giving characters specific roles that correspond to predefined story functions [14.26]. For instance, a character in the role of a villain strives to disturb, damage, or harm a happy family.

14.5.6 Story Nets

One hybrid approach to story creation, that contains features of both (linear) plot-based and character-based approaches, is called *story net* [14.32]. A story net contains two types of concepts, nodes and links. A node refers to a situation where the user may 'freely' interact with the story characters. However, the interaction within a node is typically restricted by the verbal utterances characters are able to process. The main purpose of nodes is to play out a confined task or prompt the user to make a decision. Links, on the other hand, refer to linear sequences of events that form the connection between nodes. The user can not influence the events in a link. Those sequences are used to prepare the next situation or demonstrate the consequences of previous decisions of the user. Hidden from the user, each link has a set of conditions that must be satisfied before a link can be traversed to arrive at an outcome node.

The story net approach allows us to combine linear and non-linear story elements, by using deterministic links and nodes including decision points and free interaction, respectively. The authors also think about employing a 'director' agent that controls the flow of events and ensures that pedagogical or dramatic goals are met.

The concept of story net is closely related to the previously introduced plot manager of Sgouros [14.30], with the exception of the natural language understanding based free play component in [14.32].

14.5.7 Digital Director for Interactive Story Telling

Spierling et al. [14.31] focus on the authoring aspect of interactive story telling and take inspiration from film making, where a whole team works on different levels of the product. Example roles are:

- The *Editor* is responsible for the topic and the content of the story.
- The *Playwright* plans the presentation of the content and engages in scriptwriting of the dialogues.
- The *Director* interprets the script and instructs the actors.
- The *Stage director* handles camera, lights, and props.
- The *Casting director* works out characters and their appearance.
- The *Actors* are responsible for playing a certain character and follow instructions of the director.

Based on the film-making metaphor, the authors propose a four-level modular story-telling system: story engine, scene action engine, character conversation engines, and actor avatar engines. Although not all of those modules are fully developed, the approach is very promising. Spierling and her co-workers adhere to the design philosophy of *scalable (or adjustable) autonomy* which allows the author (story creator) to decide on the degree of semi-autonomy on each level. The scale is between 'predefined' and 'autonomous', depending on whether the author chooses to predetermine each behavior (story development, scene selection, dialogue, character animation) or let the system autonomously select behaviors.

The story concept used by Spierling are Propp's narrative functions [14.26], where scenes are chosen if they satisfy a certain function in the narrative. The system may also handle user interaction by allowing for *polymorphic beats* [14.17] which have different outputs depending on users' choices.

14.6 Interactive Story Telling as a Business Training Environment

This chapter set out with the claim that stories can be used as an environment where users may become aware of opportunities and risks. A salient feature of interactive stories is that users may even (virtually) 'take' chances. The following items summarize the requirements of the 'interactive story telling as chance discovery' approach.

– *Immersion.* The user (student) is part of the story in the role of a main character or protagonist.
– *Decisions.* The user is able to make meaningful decisions and hence influence the further development of the story.
– *Consequences.* The user's choices have (virtual) consequences in the course of the story.
– *Edutainment.* The user may achieve a pedagogical goal (chance awareness) in an entertaining way.

As a case study, we will storify a situation describing a decision-making process in a corporate setting. Figure 14.5 illustrates the interaction environment employing a simple plot manager.

The requirement of *immersion* is met by having the user participate in the story. Following a story requirement proposed by Mateas and Stern [14.17], we use three story characters, two computer-controlled agents and the user. This configuration allows for complex social relationships while avoiding too much readiness for interaction on the side of the user. Rather than having the user be the protagonist of the story, the user may 'side with' (i.e. support the opinion of) one of the characters, assuming that both animated agents have conflicting opinions. In this way, the second requirement (*decision*) is satisfied, and the user can influence the future development of the story. User choices are given by a set of pre defined options. The user may select an option by pressing a button in the browser window or the speech recognizer will handle user input (see the bottom browser frame in Fig. 14.6).

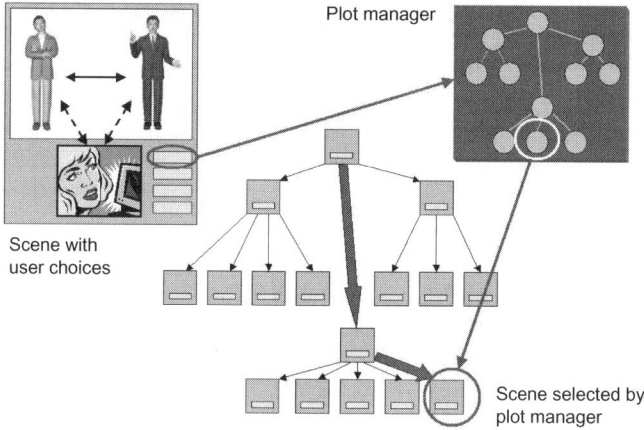

Fig. 14.5. Story telling system

Typically, the outcome of the story is decided in advance, and users – in co-operation with their virtual collaborator – will succeed or fail to achieve their goals. Recall that the chances we want to storify are success stories or 'mistake' stories with real-world counterparts that have known outcomes. It is important that the *consequences* of the user's decisions are clearly demonstrated in the course of the story. The pedagogical aim here is to confront the user with the consequences of his or her actions.

According to the *edutainment* requirement, the story should not only have educational value but also be entertaining *as a story* in an artistic sense. Currently, we use a very simple plot manager to select appropriate scenes. The plot manager contains some rules that decide whether a scene illustrating a certain argument for (or against) a character's opinion should be played. Besides user choices, the plot manager also maintains a history of the interactions in the story. Scenes are realized as joint plans as discussed in [14.18], with no autonomy on the part of the characters. Unlike the concept of 'joint plans' developed by Mateas and Stern, our joint plans are non-reactive. That is, once a joint plan is triggered, a user cannot interfere until the scenario offers a choice to the user.

Figure 14.6 is a screen shot from our implementation of the 'Suspicion' story of the Division Chief Kosaku Shima series [14.14]. Our story is basically a (simplified) animated version and extension of the dialogue given in Sect. 14.2 ('motivating example'). The character on the right-hand side is in the role of Shima from the marketing division, and the character on the left is in the role of the manager of finance. The latter character is supposed to represent all counter-opinions to Shima's proposal in the original story, and eventually grant funding for the proposal or refuse to do so. The participating user is in the role of an expert who is called by Shima to support his proposal. However, the user is free to interfere positively or negatively with Shima's suggestion, thereby influencing Shima's chance of acquiring funding.

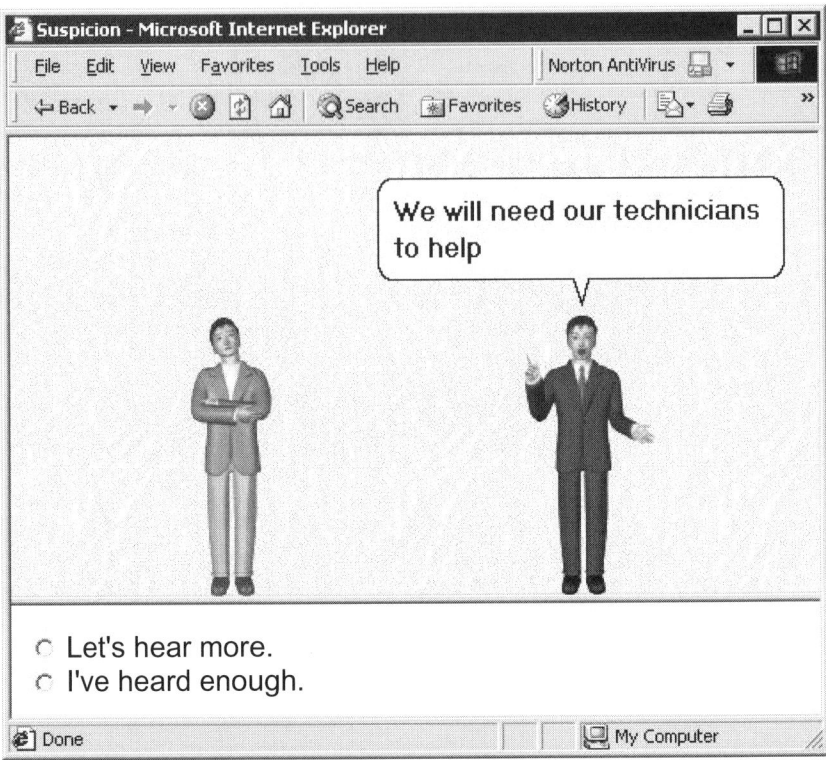

Fig. 14.6. Business training environment

The screen shot depicts the situation where the plot manager decides to let Shima explain more about his proposal. After the user utters 'Let's hear more.', Shima explains his idea and proposes that technicians re-consider materials to construct TV cabinets. The finance manager character displays a 'thinking' gesture while listening to Shima's proposal. In the future course of the story, the manager of finance asks the user whether the company should invest a significant amount of money to realize Shima's idea. The consequences of installing (or not installing) Shima's proposal are shown in a 'Six months later ...' episode (this episode is also shown when the user opts for 'I've heard enough.' in the previously mentioned choice). Here, if the user decides to support the investment, the proposal will turn out as a success as customers accept to pay more for products that are produced in an environment-conscious way, and the company's sales go up. If the user declines the investment, the 'Six months later ...' episode shows a scene with no spectacular differences in the company's development. Consequently, if a chance such as investing in environment-conscious production is not taken, it does not necessarily imply a negative outcome of the story in terms of the company's well-being. On the other hand, if a storified risk situation is ignored, negative consequences will be communicated to the user.

Note that Hirokane [14.14] does not claim that his stories are based on real events, although they can be seen as entertaining introductions to the life in a Japanese company. Moreover, the further development of the story and its consequences are not described in the Shima Kosaku series. Storified chances that are intended to enter a company's corporate memory would have to satisfy more strict criteria than a fictional example story.

14.7 Summary and Conclusion

This chapter presents our approach to chance discovery and awareness of chances. A key assumption of our approach is that interactive story telling featuring life-like characters is an adequate method to sharpen human awareness of situations that may contain chances (i.e. opportunities or risks). While most approaches to chance discovery rely on some technique to extract information essential for decision making from a given situation, we follow an entirely different paradigm that *plays out* situations where chances are present and the user (student) may take chances online, by interacting with virtual agents in a story-like environment. The advantages of this approach can be summarized as follows.

– A significant number of instances of chance discovery are realized in human–human interaction. Environments that simulate those types of communication provide an effective training setting.
– Stories provide the possibility that the consequences of users' decisions can be shown to the user in the course of story development. This is important since, in many cases, the result of a taken (or ignored) chance is only seen in a larger time frame, which can be simulated in virtual environments.
– Interactive stories with animated characters that perform relevant roles offer an entertaining way to immerse users in situations containing chances.

The story-telling systems discussed in this chapter constitute different ways to immerse the user in a narrative experience. Currently, our own training environment is simply an HTML file that embeds the Microsoft Agent package, JavaScript code, and a Java applet containing a basic plot manager. Future versions of the environment will use flash technology to enhance users' interaction experience. Besides speech communication, we also intend to implement pseudo-physical interaction with objects in the environment, such as the 'narrative objects' discussed, for instance, by Cavazza et al. [14.9, p. 21]. Other extensions concern a more elaborate plot manager and a limited form of autonomy for generating characters' affective behavior.

Our approach is intended to make tacit knowledge accessible, which is often seen as complementing the retrieval of explicit knowledge [14.21]. Further studies will show whether storified chances can be beneficially integrated to corporate memories and constitute an effective way to communicate practical knowledge in organizations.

Acknowledgedgments. This research is supported by a Research Grant (1999–2003) for the Future Program ('Mirai Kaitaku') from the Japanese Society for the Promotion of Science (JSPS).

References

14.1 Elisabeth André E, Rist T, van Mulken S, Klesen M, and Baldes S (2000) The automated design of believable dialogue for animated presentation teams. In Cassell J, Sullivan J, Prevost S, and Churchill E (eds) *Embodied Conversational Agents*, The MIT Press, Cambridge, MA, pp.220–255

14.2 Baldes S, Gebhard P, Kipp M, Klesen M, Rist P, Rist T, Schmitt M (2002) The interactive CrossTalk installation: Meta-theater with animated presentation agents. In Prendinger H (ed) *Proceedings International Workshop on Lifelike Animated Agents. Tools, Affective Functions, and Applications, in conjunction with PRICAI-02*, pp.9–15

14.3 Borgkoff UM, Pareschi R (1997) Information technology for knowledge management. *Journal of Universal Computer Science*, 3(8):835–842

14.4 Bruner J (1991) The narrative construction of reality. *Critical Inquiry*, 18(1):1–21

14.5 Cassell J (2000) More than just another pretty face: Embodied conversational interface agents. *Communications of the ACM*, 43(4):70–78

14.6 Cassell J (2000) Nudge nudge wink wink: Elements of face-to-face conversation for embodied conversational agents. In Cassell J, Sullivan J, Prevost S, and Churchill E (eds) *Embodied Conversational Agents*. The MIT Press, Cambridge, MA, pp.1–27

14.7 Cassell J, Vilhjálmsson H, Bickmore T (2001) BEAT: the Behavior Expression Animation Toolkit. In *Proceedings of SIGGRAPH-01*, pp.477–486

14.8 Cavazza M, Charles F, Mead SJ (2002) Interacting with virtual characters in interactive storytelling. In *Proceedings First Conference on Autonomous Agents and Multiagent Systems (AAMAS-02)*, New York, 2002. ACM Press, New York, NY, pp.318–325

14.9 Cavazza M, Charles F, Mead SJ (2002) Character-based interactive storytelling. *IEEE Intelligent Systems*, July-August, pp.18–24

14.10 Don A (1990) Narrative and the interface. In Laurel B (ed) *The Art of Human-Computer Inferface Design*, Addison Wesley, Boston, MA, pp.383–391

14.11 Ekman P, Friesen WV (1969) The repertoire of nonverbal behavior: Categories, origins, usage, and coding. *Semiotica*, 1:49–98

14.12 Elliott C, Brzezinski J, Sheth S, Salvatoriello R (1998) *Story-morphing* in the Affective Reasoning paradigm: Generating stories semi-automatically for use with *emotionally intelligent* multimedia agents. In *Proceedings 2nd International Conference on Autonomous Agents (Agents-98)*, New York, ACM Press, New York, NY, pp.181–188

14.13 Hayes-Roth B, van Gent R (1997) Story-making with improvisional puppets. In *Proceedings First International Conference on Autonomous Agents (Agents-97)*, New York, ACM Press, New York, NY, pp.1–7

14.14 Hirokane K (1995) *Division Chief Kosaku Shima*. Kodansha Bilingual Comics, Tokyo. Translated by Ralph F. McCarthy.

14.15 Ishizuka M, Tsutsui T, Saeyor S, Dohi H, Zong Y, Prendinger H (2000) MPML: A multimodal presentation markup language with character control functions. *Proceedings Agents'2000 Workshop on Achieving Human-like Behavior in Interactive Animated Agents*, pp.50–54

14.16 Lawrence D, Thomas JC (1999) Social dynamics of storytelling: Implications for story-base design. In *Proceedings 1999 AAAI Fall Symposium on Narrative Intelligence*, The AAAI Press, Menlo Park, CA, pp.26–29

14.17 Mateas M, Stern A (2000) Towards integrating plot and character for interactive drama. In *Proceedings AAAI Fall Symposium on Socially Intelligent Agents: The Human in the Loop*, The AAAI Press, Menlo Park, CA, Technical Report FS-00-04, pp.113–118

14.18 Mateas M, Stern A (2002) A behavior language for story-based believable agents. *IEEE Intelligent Systems*, July-August, pp.39–47

14.19 Microsoft (1998) *Developing for Microsoft Agent*. Microsoft Press, Seatle, WA

14.20 Mott BW, Callaway ChB, Zettlemoyer LS, Lee SY, Lester JC (1999) Towards narrative-centered learning environments. In *Proceedings 1999 AAAI Fall Symposium on Narrative Intelligence*, The AAAI Press, Menlo Park, CA, pp.78–82

14.21 Nonaka I, Takeuchi H (1995) *The Knowledge-Creating Company*. Oxford University Press, New York, Oxford

14.22 Ohsawa Y (2003) Modeling the process of chance discovery. This volume, chapter 1

14.23 Paiva A, Machado I, Prada R (2001) Heroes, villains, magicians,...: Dramatis personae in a virtual story creation environment. In *Proceedings International Conference on Intelligent User Interfaces (IUI-2001)*, New York, ACM Press, New York, NY, pp.129–136

14.24 Prendinger H, Descamps S, Ishizuka M (2002) Scripting affective communication with life-like characters in web-based interaction systems. *Applied Artificial Intelligence*, 16(7–8):519–553

14.25 Prendinger H, Ishizuka M (2002) Evolving social relationships with animate characters. In *Proceedings of the AISB-02 Symposium on Animating Expressive Characters for Social Interactions*, pp.73–78

14.26 Propp V (1968) *Morphology of the Folktale*. University of Texas Press, Austin, TX

14.27 Sengers Ph (2000) Narrative intelligence. In Dautenhahn K (ed) *Human Cognition and Social Agent Technology*, John Benjamins Publishing Company, Amsterdam, The Netherlands, pp.1–26

14.28 Sgouros NM Papakonstantinou G, Tsankakas P (1996) A framework for plot control in interactive story systems. *Proceedings 13th National Conference on Artificial Intelligence (AAAI-96)*, pp.162–167

14.29 Sgouros NM (1997) Dynamic, user-centered resolution in interactive stories. In *Proceedings 15th International Joint Conference on Artificial Intelligence (IJCAI-97)*, pp. 990–995

14.30 Sgouros NM (1999) Dynamic generation, management and resolution of interactive plots. *Artificial Intelligence*, 107:29–62

14.31 Spierling U, Grasbon D, Braun N, Iurgel I (2002) Setting the scene: playing digital director in interactive storytelling and creation. *Computer & Graphics*, 26:31–44

14.32 Swartout W, Hill R, Gratch J, Johnson WL, Kyriakakis C, LaBore C, Lindheim R, Marsella S, Miraglia D, Moore B, Morie J, Rickel J, Thiébaux M, Tuch L, Whitney R, Douglas J (2001) Toward the Holodeck: Integrating graphics, sound, character and story. In *Proceedings 5th International Conference on Autonomous Agents (Agents-01)*, New York, ACM Press, New York, NY, pp.409–416

14.33 Szilas N (1999) Interactive drama on computer: beyond linear narrative. In *Proceedings 1999 AAAI Fall Symposium on Narrative Intelligence*, The AAAI Press, Menlo Park, CA, pp.150–156

15. The Prepared Mind: the Role of Representational Change in Chance Discovery

Eric Dietrich[1], Arthur B. Markman[2], C. Hunt Stilwell[2], and Michael Winkley[1]

[1] Philosophy Department, Binghamton University, Binghamton, NY 13 902-6000, USA
email:dietrich@binghamton.edu, mwinkley@binghamton.edu
[2] Psychology Department, University of Texas, Austin, TX 78 712, USA
email:markman@psy.utexas.edu, stilwell@psy.utexas.edu

Summary.

Analogical reminding in humans and machines is a great source for chance discoveries because analogical reminding can produce representational change and thereby produce insights. Here, we present a new kind of representational change associated with analogical reminding called *packing*. We derived the algorithm in part from human data we have on packing. Here, we explain packing and its role in analogy making, and then present a computer model of packing in a micro-domain. We conclude that packing is likely to be used in human chance discoveries, and is needed if our machines are to make their own chance discoveries.

"Chance favors only the prepared mind." – Louis Pasteur

15.1 Introduction

A classic example of a chance discovery happened to the mathematician Henri Poincaré. In the late 1800s, Poincaré was working on what he termed Fuchsian functions (what we now call automorphic functions). One day, taking a break from his work, he took a small trip, and as he was about to board a bus, a new thought occurred to him. He relates "…just as I put my foot on the step, the idea came to me, though nothing in my former thoughts seemed to have prepared me for it, that the transformations I had used to define Fuchsian functions were identical to those of non-Euclidean geometry." [15.16]. This insight turned out to be one of his more important mathematical discoveries.

Poincaré concluded that he had made this propitious discovery because his ideas and concepts had changed and altered in his head. Indeed, in his published works and speeches, he related many similar episodes that happened to him during his mathematical life, and in all of them he describes his ideas as jostling each other, coalescing, and, in the process, changing.

What happened to Poincaré on that fateful day is what modern cognitive scientists call *analogical reminding*. His work on Fuchsian functions reminded him of transformations in non-Euclidean geometry. However, these two domains – Fuchsian functions and transformations in non-Euclidean geometry – are not similar in detail, but instead are *analogous*. We believe that many chance discoveries involve analogical remindings. So, in order to get our computers to make such discoveries, we need to implement in them a theory of analogical reminding. There is a robust and well-supported suite of theories, collectively called *structure mapping theory*, about analogy and analogical reminding in humans which postulates that analogies are high-level, structural isomorphisms between knowledge representations [15.5, 15.6, 15.7]. Many of the theories in this suite have been implemented, and the results of the implementations compare well with human data [15.2, 15.3]. We will assume this suite of theories, which we explain below, because of the wealth of data supporting it, and because it is the dominant theory in the field of analogy research.

Structure mapping theory postulates that representational change occurs after the analogy is made [15.5, 15.8]. But in cases of chance discovery, there are reasons to believe that at least some representational change has to occur before the analogy can be made [15.1]. This sort of change is not currently included within structure mapping theory. In this paper, we extend the theory by incorporating within it a proposed mechanism for how knowledge representations change before an analogical reminding is completed. In our proposal, the chance reminding occurs *because* the representational change has been made. We call this mechanism for representational change *packing* .

15.2 Analogical Reminding and Structure Mapping Theory: Background

Analogical reminding is a common psychological phenomenon in humans. It occurs any time some concept (or representation, we use the two terms interchangeably) in one domain facilitates recall of an analogous concept in another domain. A famous example of analogical reminding occurred when Ernest Rutherford was reminded of comets and their highly elliptical orbits around the Sun by studying the strongly deflected trajectories of alpha particles around nuclei. This analogy eventually led to the "solar-system" model of the atom. Another example is Kepler's analogy between light and the notion of the *anima motrix* (moving spirit). This analogy eventually led to Kepler's more refined analogy between light and the *vis motrix* (motive force), which postulated the Sun as the cause of the planet's orbits around it. The concept of the *vis motrix* was a precursor to the concept of gravity. Kepler knew that more distant planets moved more slowly in their orbits than could be predicted simply from their greater distances, and he had noticed that the planets moved more quickly when they were closer to the Sun and more slowly when they were distant from the Sun – a fact that could only be discerned by viewing the solar system as a heliocentric system. What would cause such a change in velocity? Kepler reasoned that this behavior could be nicely explained if the *anima motrix* emanated from the Sun and behaved like light: weakening with distance (see [15.4], which examines this case in detail).

Before going further, we should state that we adopt the standard working assumption in cognitive science and AI about the composition of concepts. On this view, concepts are built from three main constituents: entities, attributes, and structural relations. Entities represent abstract or concrete objects. Attributes describe specific properties of entities. Relations provide information about the relationships among entities, attributes, or other relations. In mental representations constructed from these elements, attributes are denoted by unary predicates whose argument denotes the item described by that attribute. For example, *bright (x)* denotes that the argument to this attribute is bright. Relations are denoted by two or more place predicates having two or more arguments that specify the items that play particular roles within that relation. For example, in the relation *produce (x, y)*, the first argument is the producer and the second is the product.

Quite a bit is known about analogical reminding and the more inclusive class of similarity-based retrievals to which analogical reminding belongs. Analogical reminding is thought to comprise two processes: *access* and *mapping* [15.3, 15.9]. Access (also simply called 'retrieval') is the process of retrieving some representation from long-term memory based on some active representation that serves as a retrieval cue. In the usual case of analogical retrieval , the retrieving representation is the domain that we want to understand (or understand better). We refer to this domain as the *target*. The retrieved concept illuminates the target. We refer to it as the *base*. In Rutherford's analogy, the strongly deflected orbits of the alpha particles were the target domain, and the highly elliptical orbits of comets were the base domain.

Before the analogical retrieval, the target domain is accessible, but the base domain is not. After retrieval, both the base and target are accessible. The structure of the analogy between them may or may not be clear at this point, but it will be clear that the domains are similar in some way.

Mapping is the process of matching and placing in correspondence the constituents of the two active representations. Hence, mapping is a process of finding *functional counterparts* between concepts. Such functional counterparts are said to be in *structural alignment*. What makes these objects functional counterparts is that they play the same roles within a matching relational structure. It is this identity of role that is the basis of structural alignment[1].

Mappings are found by determining the best match between representations that satisfies the constraints of *parallel connectivity*, *one-to-one mapping*, and *systematicity*. The parallel connectivity constraint ensures that when two elements are placed in correspondence, the arguments of those elements are also placed in correspondence. The one-to-one mapping principle states that each element in a domain can be mapped to at most one element in the corresponding domain. The systematicity constraint requires that, all else being equal, correspondences between *systems* of elements in the domains are preferred to matches between isolated elements. The mapping process is local-to-global. At the beginning of the comparison process individual entities, attributes, and relations are matched, and then the constraints are utilized to determine more global systems of relations. In sum, in making an anal-

---

[1] The mapping process in analogy tends to concentrate and preserve relational information over attribute information. But this happens only in analogy. In cases of ordinary similarity comparisons, attributes are also important [15.7, 15.13].

ogy, the mapping process locates entities in the target domain that play the same roles (by being in identical structural relations) as the entities in the base domain.

Finally, we define two other technical terms of crucial importance. When a similarity or analogy comparison is made, elements (e.g. entities, attributes, and relations) of one concept are placed in correspondence with elements of another concept. Some of the elements of the concepts are identical (of course, as we mentioned above, in an analogy, certain crucial structural relations have to be identical), but some are not. The elements that are not identical come in two major types. *Alignable differences* are non-identical elements in corresponding positions in a system of relations. They are seen as similar – aligned – because of the similarity of the relational structure they participate in, rather than some inherent similarity of the items themselves. *Non-alignable differences* are non-identical elements that are *not* in corresponding positions in any alignable system of relations and are therefore not seen as similar [15.13, 15.14]. In Fig. 15.1, the pig and baby are alignable differences, as are the farmer and mother. The helicopter, barn, and hay in Fig. 15.1a (top) are non-alignable differences relative to the bottom scene.

15.3 Why Representational Change Is Needed to Understand Chance Discovery

Representational change is crucial to chance discovery, and analogical retrieval is responsible for particularly interesting kinds of representational change – kinds that look well suited to the task of chance discovery. Structure mapping theory postulates five kinds of change.

1. *Projection of candidate inferences*, in which structure from the base is transferred to the target, with appropriate substitutions when elements in the base and target correspond [15.5, 15.6, 15.11].
2. *Progressive alignment*, whereby children's knowledge becomes more abstract so that more high-order similarities can be recognized [15.10];
3. *Highlighting*, whereby less-salient conceptual properties are made more salient;
4. *Rerepresentation*, whereby representations or either or both domains are changed to improve the analogical match;
5. *Restructuring*, whereby whole systems of knowledge get changed [15.8].

These kinds of representational change occur as a part of analogical mapping, and hence they happen after the base and target are both active. In the context of chance discovery, however, we are particularly interested in the kinds of representational change that occur in the process of retrieving information.

Previous implemented models of analogical retrieval have not assumed that any representational occurs during the retrieval process (e.g. [15.3, 15.18]). We believe that representational change must be an integral part of the analogical retrieval process based on what we call the *low-probability argument* [15.1]. It seems unlikely that disparate domains that have not been compared previously already are represented using the same structure and yet that parallel structure had not been noticed.

Fig. 15.1. Sample scene pair from the comparison task. The top scene is the *target* and the bottom the *base*. The pig and farmer in the target scene are alignable with the baby and mother, respectively, in the base scene. The helicopter in the target is non-alignable with the base scene

Clearly, there must be some similarity between domains to get a comparison started, but it is implausible that two domains are already represented with highly parallel structures before the target domain retrieves the base. Hence, retrieval models that do not posit representational change do not seem sufficiently powerful to explain the number of analogical remindings that occur both in extraordinary minds like Kepler's as well as in more ordinary ones.

If the high-level relational structures of the two analogues do not match before the analogy (except in very rare cases), but they do match at the time the analogy is made (and of course, afterwards), then some representational change must occur at the time of retrieval that permits an analogical alignment to occur. On this view, analogical retrieval is a constructive process, and not like taking a book off a shelf. The base and target representations may be altered in the process of retrieving an analog. Thus, analogical 'retrieval' is not retrieval in the sense of getting something that is already there.

To illustrate the kind of representational changes the low-probability argument suggests, we consider two examples. The first is derived from experiments on analogical learning [15.6]. The second is about Kepler and his theory of planetary motion.

Assume that the idea that heat flows is new to a subject, S, but that S is well acquainted with flowing water. Suppose S is presented with a small, solid, silver bar sticking into a cup of hot coffee. And suppose that there is an ice cube on the end of the bar sticking out of the coffee. What happens to the ice cube? It rapidly melts. Suppose in contemplating this phenomenon, the following analogy occurs to S. S imagines a large beaker connected to a smaller beaker by some rubber tubing running from the base of the large beaker to the base of the small one. When the large beaker is filled with water, the water will flow from the large beaker into the smaller beaker through the tubing. The analogy is that heat flows from the hot coffee up the bar to the ice just as water flows from the large beaker through the tubing into the small beaker. The target domain is heat flow, and the base domain is water flow. The large beaker is analogous to the hot coffee, the small beaker is analogous to the ice cube, the tubing is analogous to the silver bar, and the flowing water is analogous to the moving heat. The core of the analogy is that the greater temperature of the coffee relative to the ice cube causes heat to flow up the bar just like the greater pressure of the water in the large beaker relative to the pressure in the small beaker causes the water to flow through the tube. [15.6] represented the base domain of water flow like this:

(CAUSE (GREATER
(PRESSURE large-beaker) (PRESSURE small-beaker))
(FLOW water tubing large-beaker small-beaker))

A parallel relational structure was used to represent heat flow:

(CAUSE (GREATER

(TEMPERATURE coffee) (TEMPERATURE ice-cube))
(FLOW heat bar coffee ice-cube))

The two uses of the relation FLOW(w, x, y, z) have to be identical in order for the analogy to be constructed. But according to the low-probability argument, the above two representations must be viewed as *end* products in S's representations of the domains. Before the analogy occurred to S, it is unlikely that the flow of heat was conceptualized by S in the same way as the flow of water. Otherwise, there would have been nothing to discover. Instead, flowing water was probably conceptualized in a manner tied closely to knowledge of liquids. Therefore, in order for the analogy between heat flow and water flow to occur, the concept of FLOW had to become more abstract, losing its relationship to specific properties of liquids. There are several ways that a concept might change to allow such abstraction. We consider two of them here.

The first type of conceptual change involves broadening the range of items to which a particular predicate could apply. For this example, assume that our intrepid scientist, S, did conceive of heat as something that moves. This conception was likely different from S's conceptualization of water. There are probably several different kinds of movement predicates, such as one for describing the movement of stuff like water, and one for other kinds of more abstract movements like the flow of emotion through a person or a group of people or the flow of information through a community. In this case, let us assume that there is a predicate *L-flow* (L for 'liquid') that would have one argument for the substance that is flowing. The value of this argument would be restricted to physical matter in a liquid state, specifically to a unified body of matter of no particular shape with a tendency to flow. This restriction could be implemented in many different ways, the easiest being a data-type restriction on the type of argument the predicate *L-flow* can take. Thus, *L-flow* involves a particular attribute of one of its arguments, and it is these attributes that change.

Because of this restriction, *L-flow* can not apply to heat or to emotion (or to other movements that seem less like flow such as a person walking down the street or a ball thrown through the air). In order to recognize that the movement of heat or emotion is like the movement of liquid, it is necessary to generalize the concept by removing the type restriction on *L-flow*. Even less radical types of flow, such as the flow of granular masses like flour, sugar, or dirt, would require some loosening of the type restriction.

It is possible that comparing types of movement might suggest that a particular type restriction could be relaxed. For example, the physical similarity of water pouring from a glass and sugar pouring from a bowl might be sufficient to recognize that the substance that flows need not be a liquid. In this case, the type restriction is altered. Then, when flow is applied to more abstract concepts like heat or emotion, two things are likely to occur. First, the type restriction on the argument is likely to become even more loose and abstract. Second, the representation of the more abstract concepts may also change to include some attributes that make them more similar to other items that flow. Indeed, early theories of heat conceptualized it as a liquid or a particulate substance (e.g. [15.19]. This type of representational change involving attribute change will be considered further in Sect. 15.5; it is, we claim, the basic kind of packing.

The second case of conceptual change during analogy involves the creation of a new predicate as a result of the juxtaposition of mismatching items. In this case, assume that S has not yet conceptualized heat as something that can flow. Hence, the concept of flow would only be applied to describe the movements of physical substances. The movement of heat would require some other predicate. For example, S might represent heat using a general predicate like 'moves,' which would yield a representation like the following.

(CAUSE (GREATER
(TEMPERATURE coffee) (TEMPERATURE ice-cube))
(MOVE heat bar coffee ice-cube))

The analogy could still go through, however, because there is enough similarity between the above representation and the representation for flowing water. What is crucial about this case is the predicate change from 'moves' to 'flows.' 'Move' is too general. Planets, feet, and money move, but they don't flow. Still, in order for the predicate 'flow' to replace the predicate 'move' in this context, 'flow' must be abstracted; it must lose information. This is because it is used for standard liquids. Changing from 'move' to 'flow' allows S to understand heat as some sort of unified stuff (but unlike matter) that slowly climbs its way up the bar from the hot coffee. Changing from 'move' to 'flow' is a particularly robust representational change. This case will also be considered further in Sect. 15.5; it requires a special kind of packing.

We now return to Kepler. Kepler was an Aristotelian until his dying day, meaning that he had no notion of inertia and believed that things move only because they are pushed; if something is not pushed, it ceases to move. In his first book, the *Mysterium Cosmographicum* (1596), Kepler asked this question concerning the planetary orbits: why did the outer planets move more slowly than the inner planets, even factoring in their great distance? Saturn's orbit is about twice the size of Jupiter's, but Saturn takes substantially more than twice as long to go around once – about two and a half times as long. His answer was:

"...we must choose between two assumptions: either the souls which move the planets are less active the farther the planet is removed from the Sun, or there is only one moving soul in the center of all the orbits, that is the Sun, which drives the planet the more vigorously the closer the planet is, but whose force is quasi-exhausted when acting on the outer planets because of the long distance and the weakening of the force which it entails.".

He chose the second and thereby was analogically reminded of light (it is not known that the analogical reminding occurred right then; in fact it is not known when the analogy occurred to him, but it is reported in the *Mysterium*). In short order, because of his insight, the reason for the planets' movements went from being souls (*anima motrix*) to being a mechanical, physical cause (*vis motrix*). This reconceptualization of planetary motion set the stage for Newton's proposal for the role of a gravitational force in planetary motion.

In making this analogy, however, Kepler had to suppress a lot of what he knew about light. The *vis motrix* had to somehow whip the planets around, veering at a large angle from the direction of emanation from the Sun. But light does not work this way at all. It leaves the Sun and reaches all the planets in a straight line beginning at the Sun and ending at the side of the planet facing the Sun. Indeed this is more of a problem for Kepler's conception of the *vis motrix* than it is for the *anima motrix* because, for the latter, one can assume a modicum of intelligence: the moving soul *knows* it has to push the planets, the moving force doesn't. Furthermore, light from the Sun illuminates everything from planets to the noses on our faces. But the *vis motrix* apparently only operates on planets (and not noses). Again, information crucial to understanding light from the Sun had to be suppressed in order for the analogical mapping to occur; that is, information had to be suppressed or packed.

15.4 Packing: an Overview

Packing is a process of representational change during memory retrieval that suppresses irrelevant information in a structured mental representation. This suppression alters the structure of the representation by making aspects that are not relevant to the correspondence between base and target less accessible. Though we explore packing as it relates to analogical reminding, packing is not restricted to analogical reminding. Rather, it is a general memory process that simplifies structured representations.

We have done some experiments on memory for pairs of complex scenes following comparisons of those scenes, and the data suggest that attributes of objects are packed when those attributes are irrelevant to the relational match among items [15.17]. Though we only have data indicating that irrelevant attribute information is packed, the computer model described below extends packing to handle packing of whole relational predicates, as well as predicate *change*. These types of representational change parallel the two processes described above about heat flow.

Packing is a syntactic process, and is domain-general. The packing process does not care about the semantic content of the representation (such as the labels or inferential connections or the information content, etc.). This contrasts with many representational processes, which are domain-specific and require lots of domain knowledge to be carried out.

The packed representations are more *abstract* than their less abstract versions. By 'abstract,' we mean that the representations with packed components contain less information than their unaltered versions. This usually means that the representations have wider applicability than their unpacked cousins.

Finally, packed information can be *unpacked*. That is, the suppressed information can be recovered, made explicit, and used for other reasoning tasks such as inferencing.

15.5 STRANG: a Computational Model of Packing During Analogy Making

Our model of packing in analogical reminding is called STRANG, short for 'STRing AnaloGizer.' It is implemented in Java, and operates over letter strings (e.g. ab-bccc). STRANG has only limited knowledge of the English alphabet. In particular, it knows the shapes of the letters (in one font only) and their sequence. The shapes are shown in Fig. 15.2. The labels on the segments of the template correspond to attributes that are used to describe the letters. STRANG's knowledge of the sequence of the letters is limited to the ability to produce the successor of a letter presented.

There are two reasons why letter strings are a good arena for an initial exploration of packing. First, by using such a sparse domain, the representations need not be judged on the basis of their match to what people know about the domain. Thus, our representational assumptions can be made explicit. It also prevents us from reading too much into the content of the predicates. It is tempting to see meaningful predicate names in knowledge representations as actually representing what they name, i.e. to assume that a node labeled 'dog' actually represents dogs. It is then also tempting to assume that a knowledge representation actually models the relevant concept, i.e. that a dog-node (or a dog-frame, or whatever) is an accurate model of a human's concept of a dog [15.15]. Both of these assumptions are unjustified and should be avoided. In fact, no one currently knows the actual content or structure of mental representations in the human mind, hence no one is sure how to mirror human representations in computers [15.12]. Second, because these packing processes are complex, a simple domain provides us with a small number of predicates and hence keeps the packing process managable so we are able to evaluate the success of the model. If there were too many predicates in the domains, it might be difficult to determine why the model acts as it does.

Strictly speaking, STRANG makes analogies during a process of analogical retrieval: given a string simple working memory, STRANG retrieves an analogous string from long-term memory. But STRANG is not intended as a model of analogical retrieval. The essential elements of STRANG are those that pack the representations. STRANG has to make retrievals in order to exhibit representational change via packing. The essential parts of STRANG could be included in other, more psychologically accurate, models of analogical retrieval such as MAC/FAC [15.3] or ACME [15.18].

15.5.1 Background on the Operation of STRANG

STRANG has two sets of letter strings. First, there is a collection of letter in long-term memory. There are no connections among strings in long-term memory. Sec-

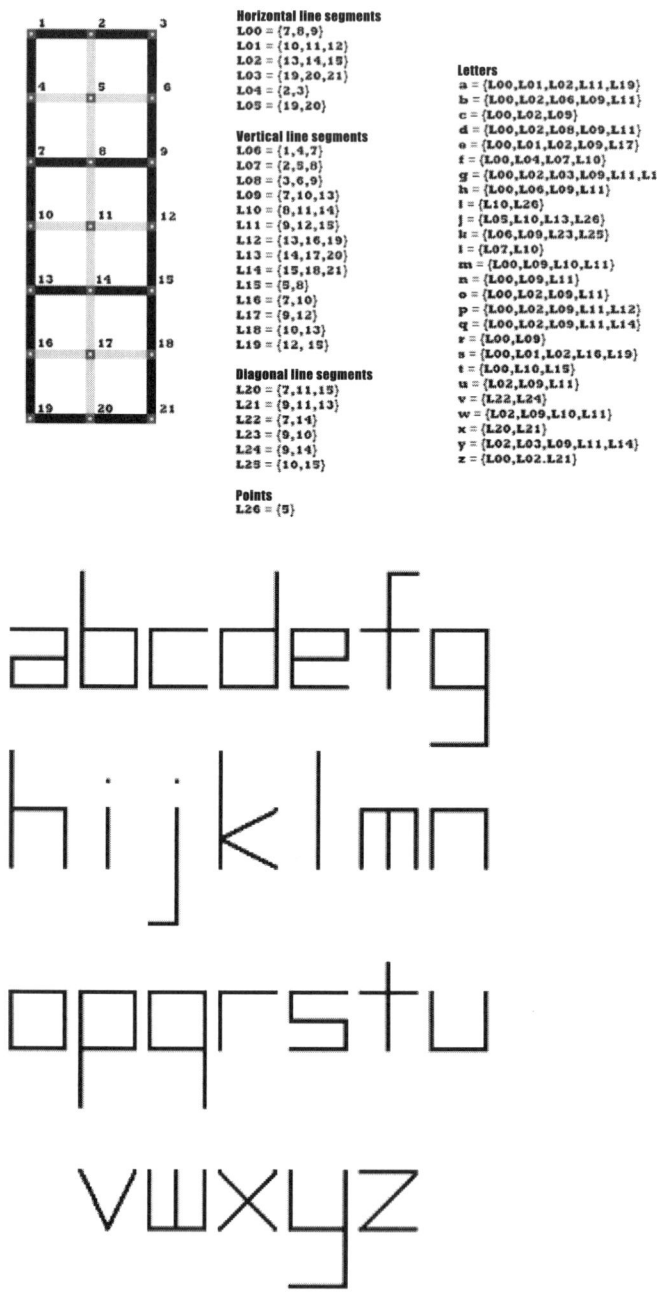

Horizontal line segments
L00 = {7,8,9}
L01 = {10,11,12}
L02 = {13,14,15}
L03 = {19,20,21}
L04 = {2,3}
L05 = {19,20}

Vertical line segments
L06 = {1,4,7}
L07 = {2,5,8}
L08 = {3,6,9}
L09 = {7,10,13}
L10 = {8,11,14}
L11 = {9,12,15}
L12 = {13,16,19}
L13 = {14,17,20}
L14 = {15,18,21}
L15 = {5,8}
L16 = {7,10}
L17 = {9,12}
L18 = {10,13}
L19 = {12, 15}

Diagonal line segments
L20 = {7,11,15}
L21 = {9,11,13}
L22 = {7,14}
L23 = {9,10}
L24 = {9,14}
L25 = {10,15}

Points
L26 = {5}

Letters
a = {L00,L01,L02,L11,L19}
b = {L00,L02,L06,L09,L11}
c = {L00,L02,L09}
d = {L00,L02,L08,L09,L11}
e = {L00,L01,L02,L09,L17}
f = {L00,L04,L07,L10}
g = {L00,L02,L03,L09,L11,L14}
h = {L00,L06,L09,L11}
i = {L10,L26}
j = {L05,L10,L13,L26}
k = {L06,L09,L23,L25}
l = {L07,L10}
m = {L00,L09,L10,L11}
n = {L00,L09,L11}
o = {L00,L02,L09,L11}
p = {L00,L02,L09,L11,L12}
q = {L00,L02,L09,L11,L14}
r = {L00,L09}
s = {L00,L01,L02,L16,L19}
t = {L00,L10,L15}
u = {L02,L09,L11}
v = {L22,L24}
w = {L02,L09,L10,L11}
x = {L20,L21}
y = {L02,L03,L09,L11,L14}
z = {L00,L02.L21}

Fig. 15.2. STRANG's letter grid and letters

ond, there is a current active string that serves as a retrieval cue. Both the strings in long-term memory and the one serving as a cue are represented by structured representations consisting of attributes that describe the letters in the string as well as relations that describe the construal of the string. Before we can describe the operation of STRANG, we must first describe the representational system in more detail.

STRANG represents letters as entities with collections of attributes. Most of these attributes are derived from viewing letters in a grid-like font where each line segment of the grid is given a designation. Figure 15.2 shows the set of segments that describe the letters as well as an image of the letters themselves. Here are the letters we use in the examples:

a - LET, L00, L01, L02, L11, L18
b - LET, L00, L02, L06, L09, L11
c - LET, L00, L02, L09
e - LET, L00, L01, L02, L09, L17
f - LET, L00, L04, L07, L10
g - LET, L00, L02, L03, L09, L11, L14
h - LET, L00, L06, L09, L11
i - LET, L10, L26
j - LET, L05, L10, L13, L26
m - LET, L00, L09, L10, L11
n - LET, L00, L09, L11
o - LET, L00, L02, L09, L11
p - LET, L00, L02, L09, L11, L12
q - LET, L00, L02, L09, L11, L14
r - LET, L00, L09

(where the attribute LET is common to all letters, and designates that the entity is a letter)

All strings (both the cues and those in long-term memory) are parsed by creating a parse tree of descriptions. When analyzing a string, the category name is used as the root node of the parse tree. Parse trees (not necessarily binary) will be represented as lists: (root, leftmost-subtree, next-subtree, . . . , rightmost-subtree).

When parsed, there are five types of categories into which strings are placed:

1. *sequence* (abbreviated: Seq)
2. *identity group* (abbreviated: IG)
3. *repeating sequence* (abbreviated: Rseq)
4. *repeating group* (abbreviated: RG)
5. *string* (abbreviated: ST)

[Note: this category is used when none of the above four correctly describes the vector of letters.]

The main relationship governing the formation of complex strings (i.e. strings with multi-letter substrings as parts) is the relation *Concats-with* (i.e. *letter-string1 concatenates with letter-string2*). This relation is an n-ary relation.

As mentioned, categories have attributes (single-place predicates) modifying them. These attributes are:

1. Constituent = ⟨ string category⟩ (i.e. *The constituents of this string are of these categories*)
2. Card ⟨ integer⟩ (i.e. *this string has cardinality X*)
3. Begins-with ⟨ letter⟩ (i.e. *this string begins with this letter*)

Not all attributes are always present, especially after packing.

All strings are made up of letters in binary relations. The two relations are:

1. IAF ⟨ letter1, letter2⟩
   (i.e. *letter1 immediately alphabetically follows letter2*)
2. IW ⟨ letter1, letter2⟩
   (i.e. *letter1 is identical with letter2*)
3. OW ⟨ letter1 letter2⟩
   (i.e. *letter1 occurs with letter2*)

[Note: this relation was chosen over 'immediately proceeds' because in English we read from left to right, but it also forces the parser to list the strings backwards.]
[Note: this is the default relation letters enter into when the above two are not applicable.]

Finally, in the descriptions of the parses below:

1. Entities are enclosed in quotes,
2. Attributes will be enclosed in angle brackets and are to be read as attributes of the entities they come after,
   and
3. Relations are underlined.

15.5.2 Packing Irrelevant Attribute Information

Suppose STRANG gets as input the repeating sequence *abab* and has in memory the repeating sequence *efef* (and suppose here, for simplicity, that only *efef* is in memory so that it will be retrieved). Here are the representations of these two strings.

abab =

('Rseq' ⟨ (card 2) (Constituent = seq, begins-with 'a', card 2)⟩
(Concats-with
('Seq' ⟨ (card 2) (begins-with a)⟩
(IAF ('b' ⟨ LET⟩ , ⟨ L00⟩ , ⟨ L02⟩ , ⟨ L06⟩ , ⟨ L09⟩ , ⟨ L11⟩)
('a' ⟨ LET⟩ , ⟨ L00⟩ , ⟨ L01⟩ , ⟨ L02⟩ , ⟨ L11⟩ , ⟨ L18⟩)))
('Seq' ⟨ (card 2) (begins-with a)⟩
(IAF ('b' ⟨ LET⟩ , ⟨ L00⟩ , ⟨ L02⟩ , ⟨ L06⟩ , ⟨ L09⟩ , ⟨ L11⟩)
('a' ⟨ LET⟩ , ⟨ L00⟩ , ⟨ L01⟩ , ⟨ L02⟩ , ⟨ L11⟩ , ⟨ L18⟩)))))

efef =
('Rseq' ⟨ (card 2) (Constituent = seq, begins-with 'e', card 2)⟩
(Concats-with
('Seq' ⟨ (card 2) (begins-with 'e')⟩
(IAF ('f' ⟨ LET⟩ , ⟨ L00⟩ , ⟨ L04⟩ , ⟨ L07⟩ , ⟨ L10⟩)
('e' ⟨ LET⟩ , ⟨ L00⟩ , ⟨ L01⟩ , ⟨ L02⟩ , ⟨ L09⟩ , ⟨ L17⟩)))
('Seq' ⟨ (card 2) (begins-with 'e')⟩
(IAF ('f' ⟨ LET⟩ , ⟨ L00⟩ , ⟨ L04⟩ , ⟨ L07⟩ , ⟨ L10⟩)
('e' ⟨ LET⟩ , ⟨ L00⟩ , ⟨ L01⟩ , ⟨ L02⟩ , ⟨ L09⟩ , ⟨ L17⟩)))))

Note how entities and relations alternate. Note also that all the entities at any level of abstraction have attributes. And finally note that the attributes of 'seq' and 'Rseq' are derived from the relational structures they dominate, together with the attributes of those structures. We call this 'promotion of attributes.'

There is plenty of relational structure between the above two representations to give us an analogy between them. Indeed, *abab* and *efef* are intuitively analogous (in string world, anyway; in the ordinary world, they might be regarded as too representationally thin to be true analogies). However, the irrelevant attribute information could get in the way of the analogy, for the representations are quite detailed. To the extent that the attributes in the two representations are taken seriously by STRANG, there really can not be an analogy between them, for the representations do not in fact look alike. Of course, we can see that the representations *do* look alike, but that is just because we are looking past the details: we are *packing away* the irrelevant details. This is exactly what packing is for. For example, being a sequence beginning with 'a' is not like being a sequence beginning with 'e' unless what the sequence starts with doesn't matter.

Now for the central idea behind STRANG – to form the analogy between *abab* and *efef*, the irrelevant attribute information is packed away. We define *irrelevant attribute information* as the information that two objects do *not* have in common during a similarity comparison. The packed information is stored in *packed variables*, denoted by lower-case Greek letters. The packed variables contain the irrelevant attribute information. This now gives:

packing of *abab* relative to *efef* =
('Rseq' ⟨ (card 2) α⟩

(Concats-with
('Seq' ⟨ (card 2) β⟩
(IAF ('b' ⟨ LET⟩ , ⟨ L00⟩ , χ)
('a' ⟨ LET⟩ , ⟨ L00⟩ , ⟨ L01⟩ , ⟨ L02⟩ , δ)))
('Seq' ⟨ (card 2) β⟩
(IAF ('b' ⟨ LET⟩ , ⟨ L00⟩ , χ)
('a' ⟨ LET⟩ , ⟨ L00⟩ , ⟨ L01⟩ , ⟨ L02⟩ , δ)))))

packing of *efef* relative to *abab* =
('Rseq' ⟨ (card 2) κ⟩
(Concats-with
('Seq' ⟨ (card 2) ϕ ⟩
(IAF ('f' ⟨ LET⟩ , ⟨ L00⟩ , γ)
('e' ⟨ LET⟩ , ⟨ L00⟩ , ⟨ L01⟩ , ⟨ L02⟩ , η)))
('Seq' ⟨ (card 2) ϕ ⟩
(IAF ('f' ⟨ LET⟩ , ⟨ L00⟩ , γ)
('e' ⟨ LET⟩ , ⟨ L00⟩ , ⟨ L01⟩ , ⟨ L02⟩ , η)))))

The representations are now more parallel than they were before. The packed variables differ, of course, but they can now be easily excluded from the similarity match because of their special status as packed variables. Given this, the above two representations are now *analogous*. In this case, some of the attributes are retained. In particular, the shared attributes that make up the letters are kept. This seems plausible because when people make analogies, some shared attributes are kept. For example: from Fig. 15.1, a mother yelling at her baby who is making a mess by drawing on the walls and floors and a farmer yelling at her mess-making pig have some attributes in common. Both the mother and the farmer are yelling ('yelling-at' is a relation, but 'yelling' is an attribute), and both the baby and the pig are living things that can make a mess and, in fact, are making a mess. So it seems that these, but only these, attributes are retained.

As another example, consider *abab* and *efgefg*; are they analogous? They are clearly similar (they are both repeating sequences), but do they share enough structure to be analogies?

abab =
('Rseq' ⟨ (card 2) (Constituent = seq, begins-with 'a', card 2)⟩
(Concats-with
('Seq' ⟨ (card 2) (begins-with a)⟩
(IAF ('b' ⟨ LET⟩ , ⟨ L00⟩ , ⟨ L02⟩ , ⟨ L06⟩ , ⟨ L09⟩ , ⟨ L11⟩)
('a' ⟨ LET⟩ , ⟨ L00⟩ , ⟨ L01⟩ , ⟨ L02⟩ , ⟨ L11⟩ , ⟨ L18⟩)))
('Seq' ⟨ (card 2) (begins-with a)⟩
(IAF ('b' ⟨ LET⟩ , ⟨ L00⟩ , ⟨ L02⟩ , ⟨ L06⟩ , ⟨ L09⟩ , ⟨ L11⟩)

('a' ⟨ LET⟩ , ⟨ L00⟩ , ⟨ L01⟩ , ⟨ L02⟩ , ⟨ L11⟩ , ⟨ L18⟩)))))

efgefg =
('Rseq' ⟨ (card 2) (Constituent = 'Seq', begins-with 'e', card 3)⟩
(Concats-with
('Seq' ⟨ (card 3) (begins-with 'e')⟩
(IAF ('g' ⟨ LET⟩ , ⟨ L00⟩ , ⟨ L02⟩ , ⟨ L03⟩ , ⟨ L09⟩ , ⟨ L11⟩ , ⟨ L14⟩)
('f' ⟨ LET⟩ , ⟨ L00⟩ , ⟨ L04⟩ , ⟨ L07⟩ , ⟨ L10⟩))
(IAF ('f' ⟨ LET⟩ , ⟨ L00⟩ , ⟨ L04⟩ , ⟨ L07⟩ , ⟨ L10⟩)
('e' ⟨ LET⟩ , ⟨ L00⟩ , ⟨ L01⟩ , ⟨ L02⟩ , ⟨ L09⟩ , ⟨ L17⟩)
('Seq' ⟨ (card 3) (begins-with 'e')⟩
(IAF ('g' ⟨ LET⟩ , ⟨ L00⟩ , ⟨ L02⟩ , ⟨ L03⟩ , ⟨ L09⟩ , ⟨ L11⟩ , ⟨ L14⟩)
('f' ⟨ LET⟩ , ⟨ L00⟩ , ⟨ L04⟩ , ⟨ L07⟩ , ⟨ L10⟩))
(IAF ('f' ⟨ LET⟩ , ⟨ L00⟩ , ⟨ L04⟩ , ⟨ L07⟩ , ⟨ L10⟩)
('e' ⟨ LET⟩ , ⟨ L00⟩ , ⟨ L01⟩ , ⟨ L02⟩ , ⟨ L09⟩ , ⟨ L17⟩)

Now, upon packing, we get:

packing of *abab* relative to *efgefg* =
('Rseq' ⟨ (card 2) α⟩
(Concats-with
('Seq' ⟨ β⟩
χ)
('Seq' ⟨ β⟩
χ)

packing of *efgefg* relative *abab* =
('Rseq' ⟨ (card 2) δ ⟩
(Concats-with
('Seq' ⟨ ϕ ⟩
γ)
('Seq' ⟨ ϕ ⟩
γ)

We see that the two remaining abstract structures are rather thin, indeed. Even the letters and letter structures have been packed because a sequence of three and a sequence of two are too dissimilar. Basically, there is one relationship governing this similarity comparison: namely that each string is composed of two sequences, i.e. each is a repeating sequence. In this case, the relational nodes that have IAF as their roots were also packed away on the grounds that they are non-alignable differences. Thus STRANG explicitly packs away more than irrelevant attribute information. Here is one place where STRANG goes beyond the experimental data. In the next section, we discuss another, more extreme case.

15.5.3 Packing and Predicate Change

Now let's turn our attention to predicate change. We have no human data on this kind of representational change, so here is where STRANG goes significantly beyond the experimental results.

We consider the analogy between *ababccc* and *mnopqrhijhijhij*. First, here are the relevant representations.

ababccc =
('ST' ⟨ (card 2) (Constituent = 'Rseq'; 'IG')⟩
(Concats-with
('Rseq' ⟨ (card 2) (Constituent = 'Seq', begins-with 'a', card 2)⟩
(Concats-with
('Seq' ⟨ (card 2) (begins-with 'a')⟩
(IAF ('b' ⟨ LET⟩ ⟨ L00⟩ ⟨ L02⟩ ⟨ L06⟩ ⟨ L09⟩ ⟨ L11⟩)
('a' ⟨ LET⟩ ⟨ L00⟩ ⟨ L01⟩ ⟨ L02⟩ ⟨ L11⟩ ⟨ L18⟩)))
('Seq' ⟨ (card 2) (begins-with 'a')⟩
(IAF ('b' ⟨ LET⟩ ⟨ L00⟩ ⟨ L02⟩ ⟨ L06⟩ ⟨ L09⟩ ⟨ L11⟩
('a' ⟨ LET⟩ ⟨ L00⟩ ⟨ L01⟩ ⟨ L02⟩ ⟨ L11⟩ ⟨ L18⟩)))))
('IG' ⟨ (card 3) (begins-with 'c')⟩
(IW ('c' ⟨ LET⟩ ⟨ L00⟩ ⟨ L02⟩ ⟨ L09⟩)
('c' ⟨ LET⟩ ⟨ L00⟩ ⟨ L02⟩ ⟨ L09⟩)
(IW ('c' ⟨ LET⟩ ⟨ L00⟩ ⟨ L02⟩ ⟨ L09⟩)
('c' ⟨ LET⟩ ⟨ L00⟩ ⟨ L02⟩ ⟨ L09⟩)))))

mnopqrhijhijhij =
('ST' ⟨ (card 2) (Constituent = 'Seq'; 'Rseq')⟩
(Concats-with
('Seq' ⟨ (Card 6) (begins-with 'm')⟩
(IAF ('r' ⟨ LET⟩ ⟨ L00⟩ ⟨ L09⟩)
('q' ⟨ LET⟩ ⟨ L00⟩ ⟨ L02⟩ ⟨ L09⟩ ⟨ L11⟩ ⟨ L14⟩))
(IAF ('q' ⟨ LET⟩ ⟨ L00⟩ ⟨ L02⟩ ⟨ L09⟩ ⟨ L11⟩ ⟨ L14⟩)
('p' ⟨ LET⟩ ⟨ L00⟩ ⟨ L02⟩ ⟨ L09⟩ ⟨ L11⟩ ⟨ L12⟩))
(IAF ('p' ⟨ LET⟩ ⟨ L00⟩ ⟨ L02⟩ ⟨ L09⟩ ⟨ L11⟩ ⟨ L12⟩)
('o' ⟨ LET⟩ ⟨ L00⟩ ⟨ L02⟩ ⟨ L09⟩ ⟨ L11⟩))
(IAF ('o' ⟨ LET⟩ ⟨ L00⟩ ⟨ L02⟩ ⟨ L09⟩ ⟨ L11⟩)
('n' ⟨ LET⟩ ⟨ L00⟩ ⟨ L09⟩ ⟨ L11⟩))
(IAF ('n' ⟨ LET⟩ ⟨ L00⟩ ⟨ L09⟩ ⟨ L11⟩)
('m' ⟨ LET⟩ ⟨ L00⟩ ⟨ L09⟩ ⟨ L10⟩ ⟨ L11⟩)))
('Rseq' ⟨ (card 3) (Constituent = 'Seq', begins-with 'h', card 3)⟩
(Concats-with
('Seq' ⟨ (card 3) (begins-with 'h')⟩
(IAF ('j' ⟨ LET⟩ , ⟨ L05, ⟨ L10, ⟨ L13, ⟨ L26⟩)
('i' ⟨ LET⟩ , ⟨ L10⟩ , ⟨ L26⟩))

(<u>IAF</u> ('i' ⟨ LET⟩ , ⟨ L10⟩ , ⟨ L26⟩)
('h' ⟨ LET⟩ , ⟨ L00, ⟨ L06, ⟨ L09, ⟨ L11⟩))
('Seq' ⟨ (card 3) (begins-with 'h')⟩
(<u>IAF</u> ('j' ⟨ LET⟩ , ⟨ L05, ⟨ L10, ⟨ L13, ⟨ L26⟩
('i' ⟨ LET⟩ , ⟨ L10, ⟨ L26⟩)
(<u>IAF</u> ('i' ⟨ LET⟩ , ⟨ L10, ⟨ L26⟩)
('h' ⟨ LET⟩ , ⟨ L00⟩ , ⟨ L06⟩ , ⟨ L09⟩ , ⟨ L11⟩))
(<u>Concats-with</u>
('Seq' ⟨ (card 3) (begins-with 'h')⟩
(<u>IAF</u> ('j' ⟨ LET⟩ , ⟨ L05⟩ , ⟨ L10⟩ , ⟨ L13⟩ , ⟨ L26⟩)
('i' ⟨ LET⟩ , ⟨ L10⟩ , ⟨ L26⟩))
(<u>IAF</u> ('i' ⟨ LET⟩ , ⟨ L10⟩ , ⟨ L26⟩)
('h' ⟨ LET⟩ , ⟨ L00⟩ , ⟨ L06⟩ , ⟨ L09⟩ , ⟨ L11⟩)))
('Seq' ⟨ (card 3) (begins-with 'h')⟩
(<u>IAF</u> ('j' ⟨ LET⟩ , ⟨ L05⟩ , ⟨ L10⟩ , ⟨ L13⟩ , ⟨ L26⟩)
('i' ⟨ LET⟩ , ⟨ L10⟩ , ⟨ L26⟩))
(<u>IAF</u> ('i' ⟨ LET⟩ , ⟨ L10⟩ , ⟨ L26⟩)
('h' ⟨ LET⟩ , ⟨ L00⟩ , ⟨ L06⟩ , ⟨ L09⟩ , ⟨ L11⟩)))))))

Basic STRANG aligns (finds an analogy between) the two repeating sequences, *abab* and *hijhijhij* , from the two original strings (this assumes that *ababccc* is the base domain). Using STRANG's standard packing algorithm, packing the above two representations relative to each other would produce:

packing of *ababccc* relative to *mnopqrhijhijhij* =
('ST' ⟨ (card 2) α⟩
(<u>Concats-with</u>
'Rseq' ⟨ β⟩
χ))

and

packing of *mnopqrhijhijhij* relative to *ababccc* =
('ST' ⟨ (card 2) δ ⟩
(<u>Concats-with</u>
φ
'Rseq' ⟨ γ⟩))

These packings do not preserve much information. That is unfortunate, because there is additional similar structure if one looks more deeply. To get at this structure, STRANG uses an algorithm called the *cookie cutter* – so called because the structure of one representation is used to *re*structure the other like a cookie cutter cutting out cookies in dough. The cookie cutter is only called during certain situations: STRANG measures how much structure is in the packed representations it creates and if that amount is low enough, then in a small percentage of those cases, the cookie cutter is called.

When the cookie cutter is called in the case of *ababccc* and *mnopqrhijhijhij* , then the substrings *abab* and the *mnopqr* are *forced* to have similar structure by altering their predicates. Then *ccc* and the *hijhijhij* naturally align due to their having the same cardinality. Here is what the cookie cutter produces in this case:

packing of *ababccc* relative to *mnopqrhijhijhij* =
('ST' ⟨ (card 2) (Constituent = '?_NewNode' 'β')⟩
(Concats-with
('?_NewNode' ⟨ (card 2)⟩)
(Concats-with
'η'
'φ'
('β' (card 3)))

packing of *mnopqrhijhijhij* relative to *ababccc* =
('ST' ⟨ (card 2) (Constituent = '?_NewNode' 'δ')⟩
(Concats-with
('?_NewNode' ⟨ (card 2)⟩)
(Concats-with
'μ'
'λ'))
('δ' ⟨ (card 3)⟩)))

There are several things to note about these representations. First and foremost, there is the new node, *?_NewNode*. This node is a second kind of packed variable which is an unnamed new entity, i.e. it designates a new category that STRANG doesn't have detailed knowledge of, hence the '?'. Second, note that this category has a cardinality of 2, so the relevant cardinality attribute changed from 6 to 2. This means *mnopqr* has been rerepresented as *(mno)(pqr)*. This is quite a radical restructuring of the representation, and is accomplished by changing the cardinality predicate and packing away the string *mnopqr* as *(mno)(pqr)*. STRANG has no category for a list or group of non-repeating sequences. But because of the re-representation, STRANG has the beginnings of such a category after the analogy. If need be, STRANG also can generate representations of these two strings with more semantics than the above two representations have, thus making the fact that the two constituents are sequences explicit. This is accomplished by attaching to the list of attributes of *?_NewNode* the predicate *(Constituents = 'Seq')*, giving us:

('?_NewNode' ⟨ (card 2) (Constituents = 'Seq'⟩).

Finally, note that the remaining substrings, *ccc* and *hijhijhij*, are aligned because they each have cardinality 3. This means that STRANG has packed away all the other information and has just focused on the fact that these strings have the same number of constituents.

15.6 Chance Discoveries and the Prepared Mind

Our argument for the importance of packing in chance discovery is simple. We showed in Sect. 15.3 that packing is crucial to making at least some analogical remindings. And analogical reminding is crucial to at least some chance discoveries. Hence, packing is likely to be crucial to chance discovery.

It remains to be seen whether similar mechanisms can be used to understand machines. As discussed in the previous section, interesting and useful chance discoveries require seeing two separate domains of knowledge as interestingly analogous at a new abstract level, which is more abstract than the level at which the domains were learned. A mind prepared for chance discoveries, whether natural or artificial, is a mind capable of packing and analogical reminding. Without these capacities, chance discoveries are unlikely to occur.

However, it seems that packing via the cookie cutter is quite powerful: virtually any two strings can be made analogous to each other using this approach. It is legitimate to worry, then, that packing may be *too* powerful. If packing could render any two concepts analogous, then chance discoveries would be virtually impossible, because sifting through the enormous quantities of fruitlessly analogous concepts would swamp any intelligent system. However, the strength of packing is partly an artifact of the domain. Packing is very powerful in the string micro-world precisely because this domain is so simple and therefore produces extremely simple representations – parse trees are nowhere near as complicated as real human concepts or the knowledge representations needed in real-world AI. It is unlikely that packing is as strong a mechanism in humans as it is in STRANG. That said, packing is still a powerful technique for producing analogies via reminding. Even in humans, it does seem as if it licenses analogies between a very large number of potential analogical correspondences. This could still be a serious problem for our theory, for the number of useful, chance discoveries is quite small, and it is not clear how any of them are ever discovered if a large portion of computing time is spent packing representations away in a myriad of different combinations.

To solve this problem, we must delve more deeply into the mechanisms of analogical reminding, and then explore the factors that make analogies useful. Data on human reminding show that the majority of remindings are based on surface similarities (e.g. similarities in the letters in the strings rather than similarities in the relations). Yet analogies based on structural similarities do occur from time to time. These two facts are the basis of one of the most prominent extant models of analogical retrieval, the MAC/FAC model. MAC/FAC stands for 'many are called but few are chosen.'. The first stage of this model involves a cheap similarity computation between semantic vectors composed of the predicates that appear in the representations of a variety of domains. This vector-similarity calculation can be carried out efficiently, even for large knowledge bases. This first stage is used to weed out all but a few possible analogical remindings. Then, the second stage performs the mapping stage of analogy finding using the structure mapping engine. This model tends to produce remindings that share substantial surface similarity with the cue.

We are exploring two different alternatives. The first is that packing is assumed to work independently of this similarity-based reminding process. In the context of chance discovery, packing has to be more opportunistic. Because the process is computationally expensive and produces a variety of potential matches given a pair of domains, it is unlikely to be fruitfully applied to a whole knowledge base. However, in the context of scientific discovery, there are many opportunities for packing to operate on domains that happen to be available for other reasons. For example, Kepler was exposed to William Gilbert's nascent concept of magnetism, and spent considerable time working out the analogy between magnetism and the *vis motrix*. These opportunistic juxtapositions provide cases for the packing mechanism to operate on.

The second is that perhaps STRANG could be made part of MAC/FAC. Because MAC/FAC starts with the computationally cheap calculation and then moves on to the more complex calculation, STRANG could be inserted between them, creating a retrieval program that would do a reduced form of structure mapping with abstraction via packing during retrieval. STRANG would, in effect, do a second, packing based 'retrieval' – a kind of filtering – on the representations retrieved first by MAC, and then output these filtered representations to FAC.

A final important point to address is whether the analogies that emerge from STRANG will be useful. This issue is important, because the representations that emerge from the packing process are guaranteed to be analogous; they are constructed to create parallel relational structures between domains. Whether these matches are important for science (or anything else) depends on several other considerations, some of which have to do with the knowledge possessed by the human (scientist) having the analogy, and some having to do with the nature of the world. Whether an analogy turns out to be useful depends on the candidate inferences and other kinds of knowledge change the scientist can muster. This is purely a matter of what knowledge the system possesses. Kepler was obviously very good at developing analogies and mining them for insights. However, many profound analogies in science don't pan out not because the scientist lacks some requisite knowledge but because the world doesn't conform to the predictions of the analogy. These analogies are seldom reported, but again, Kepler's research provides a good example: his analogy between the *vis motrix* and magnetism was brilliant but incorrect. As it turns out, gravity is not like magnetism. All of this means that *chance* really does play a large role in chance discoveries, which in turn means that the kinds of analogies that result in profound scientific chance discoveries on our model will be rare – which they are.

A very interesting research project would be to add packing and the rest of the associated analogical retrieval algorithms to knowledge discovery and data-mining machines. There is much research to do on optimizing the packing algorithm. Packing could be placed in different parts of the retrieval process. And, finally, packing could be made a background process going on constantly between knowledge representations in long-term store. All of these options remain to be explored.

15.7 Conclusion

We have proposed extending the structure-mapping theory of analogy and analogical retrieval by adding a process of packing parts of representations away, thereby producing more abstract concepts. We have detailed a model, based on human data, that implements packing for analogical retrieval. We have shown that packing can introduce new representational structure with new semantics. And, finally, we have shown where packing fits into the broader theoretical foundation for analogical retrieval.

Really understanding chance discovery will require understanding how representations change. Unfortunately, it has proven difficult to make headway on mechanisms for representational change. We believe that the packing mechanisms described here will help change this situation and greatly enhance reasoning systems that are devoted to making new discoveries. These mechanisms will permit automated systems to make use of the power of analogy to create new knowledge.

References

15.1 Dietrich E (2000) "Analogy and conceptual change, or you can't step into the same mind twice. In E. Dietrich and A. Markman (eds.), *Cognitive Dynamics: Conceptual and representational change in humans and machines*, Lawrence Erlbaum Associates, Mahwah, NJ, pp. 265-294

15.2 Falkenhainer B, Forbus K, Gentner D (1989) The structure-mapping engine: Algorithm and examples, *Artificial Intelligence*, 41(1): 1-63

15.3 Forbus K, Gentner D, Law K (1995) MAC/FAC: A model of similarity-based retrieval. *Cognitive Science*, 19: 141-205

15.4 Gentner D, Brem S, Ferguson R, Markman A, Levidow B, Wolff P, Forbus K (1997) Analogical reasoning and conceptual change: A case study of Johannes Kepler. *Journal of the Learning Sciences*, 6(1): 3-40

15.5 Gentner D (1983) Structure-mapping: A theoretical framework for analogy, *Cognitive Science* 7:155-170

15.6 Gentner D (1989) The mechanisms of analogical learning, In S. Vosniadou & A. Ortony (eds.) *Similarity and analogical reasoning*, Cambridge Univ. Press, Cambridge, UK, pp. 199-241.

15.7 Gentner D, Markman AB (1997) Structure mapping in analogy and similarity. *American Psychologist*, 52(1):45-56

15.8 Gentner D, Wolff P (2000) Metaphor and knowledge change," In Dietrich E, Markman A (eds), *Cognitive Dynamics: Conceptual and representational change in humans and machines*, Lawrence Erlbaum Associates, Mahwah, NJ, pp. 295-342

15.9 Hummel JE, Holyoak KJ (1997) Distributed representations of structure: A theory of analogical access and mapping. *Psychological Review*, 104(3):427-466

15.10 Kotovsky L, Gentner D(1996) Comparison and categorization in the development of relational similarity, Child Development, 67:2797-2822.

15.11 Markman AB (1997) Constraints on analogical inference. *Cognitive Science*, 21(4): 373–418

15.12 Markman AB (1999) *Knowledge Representation*. Lawrence Erlbaum, Mahwah, NJ

15.13 Markman AB, Gentner D (1993) Structural alignment during similarity Comparisons. *Memory and Cognition*, 24(2): 235-249

15.14 Markman AB, Gentner D (1996) Commonalities and differences, *Memory and Cognition*, 24: 235–249

15.15 Palmer SE (1978) Fundamental aspects of cognitive representation. In Rosch E, Lloyd BB (eds) *Cognition and Categorization*, Lawrence Erlbaum Associates, Hillsdale, NJ, pp. 259–302

15.16 Poincare H (1908/1952) *Science and Method*, Dover Publications, New York

15.17 Stilwell CH, Markman AB, Dietrich E (2001). The Fate of Irrelevant Information in Analogy, Unpublished Ms.

15.18 Thagard P, Holyoak KJ, Nelson G, Gochfeld D (1990) Analogical retrieval by constraint satisfaction. *Artificial Intelligence, 46,* 259-310

15.19 Wiser M, Carey S (1983) When heat and temperature were one. In Gentner D, Stevens AL (eds) *Mental Models* (pp. 267-298), Lawrence Erlbaum Associates, Hillsdale, NJ

16. Abduction and Analogy in Chance Discovery

Akinori Abe

NTT MSC R&D,
No. 43 000, jalan APEC, 63 000 Cyberjaya
Selangor Darul Ehsan, Malaysia
email: ave@cslab.kecl.ntt.co.jp

Summary.

In this chapter, we first introduce abduction and analogy as a discovery reasoning. Second, we show a hypothetical reasoning system, Theorist, as an example of computational abductive reasoning. This hypothetical reasoning system can be applied to explanatory reasoning such as design and diagnosis. However, it can not generate new hypotheses. In our explanation of hypothetical reasoning, we also show the possibilities and the limitations of conventional abduction when we use it in the context of chance discovery. Third, we show abductive analogical reasoning (AAR), which can generate new hypotheses. AAR is an extension of hypothetical reasoning that is achieved by combining abduction and analogical mapping. Finally, we show AAR as a tool for chance discovery and explain the roles of abduction and analogy in chance discovery.

16.1 Introduction

Recently, research on discovery science and knowledge discovery has been carried out in various fields. Although these studies deal with real and large data, they involve types of learning that learn tendencies from sets of data of the same or similar categories. Of course, this is important research for the prediction of coming events. However, such learning techniques can not foretell events that are different from the trends. For example, financial engineering seems to be done in inductive or statistical ways. Financial engineering regards the dynamic world as chaos and, in stock option pricing, solves problems by using the Black–Scholes equation (probability differentiation equation). The Black–Scholes model gives a theoretical price of an option in a world where trading is continuous. It assumes stochastic models that follow the geometric Brownian motion process. Thus, in the field of financial engineering, risk management or risk avoidance is usually done by such stochastic models. However, since abnormal values are ignored, models like the Black–Scholes model could not explain the end of the economic bubble phenomenon in Japan [16.19]. When we try to predict unknown events, it is quite unnatural to have a stochastic model assumption that is made only from the known data of the same or similar category and ignores abnormal phenomena. Of course, it is important to make a stochastic model that can predict future trends, but it is more important to find factors that can not be reasoned by a stochastic process and that may cause rare or novel events. That is, finding exceptional rules and exceptional facts is also important for risk management or risk avoidance in the real world.

Peirce wrote that abduction is an operation for adopting an explanatory hypothesis, which is subject to certain conditions, and that in pure abduction there can never be justification for accepting the hypothesis, other than through interrogation [16.14]. In Peirce's sense, abduction is a type of discovery reasoning. In AI, abduction can be formalized as a logical explanation of an observation. Indeed, explanation seems to be different from prediction. Explanation deals with the current problems or past problems, while prediction deals with future problems. If the knowledge base is perfect[1], this characterization is true. However, if all of the current affairs are not known, that is, the knowledge base is not perfect, explanation becomes prediction. Accordingly, abduction under an incomplete knowledge base can be formalized as prediction, since hypotheses generation and testing can be regarded as ways to predict the explanation of the observation.

Gentner defined analogy as a 'structure mapping', that is, a mapping of knowledge from one domain that is already stored in memory (the source domain) into another (the target domain), which conveys that a system of relations that holds among the base objects also holds among the target objects [16.6, 16.7]. Thus, an analogy is a way of focusing on relational commonalities independently of the objects in which those relations are embedded. Actually, analogy is an inference to find the hidden relations among two or more domains. If we can find a hidden relation, we can regard such an inference as a discovery inference in which new knowledge can be generated.

From the above viewpoints, this chapter shows the roles of abduction and analogy in chance discovery that discovers or suggests a chance (the definition is given in the next section) for future events. In fact, abduction, which is an explanatory reasoning, is adopted as a discovery reasoning or a suggestive reasoning. However, normal abduction such as hypothetical reasoning (Theorist [16.15], etc.) can not do such tasks because normal abduction usually requires a perfect knowledge base. Therefore, abductive analogical reasoning [16.3] that can generate unknown hypotheses is adopted to solve the problem.

16.2 Definition of 'Chance'

Ohsawa defined chance (risk) [16.13] as "*a new event/situation that can be conceived either as an opportunity or a risk*". It is naturally understood that chance, which is either known or unknown, includes the possibility of eliciting unfamiliar observations. It can also be said that chance acts as an alarm, such as how an inflation of money supply can affect the mid- or long-term economic recession (Japan, in 1990). We sometimes ignore such critical factors because we can not understand that they are important. This is because the results or the factors are exceptions and rare or novel events. However, it is important to be aware of such exceptions and rare or novel events as important factors and explain unknown or unfamiliar observations by reference to them. Therefore, this chapter regards chance discovery as

---

[1] The meaning of 'perfect' will be explained in Sect. 16.4.1.

an explanatory reasoning for unknown or unfamiliar observations, and thus defines 'chance' as follows:

Definition 1 . *'Chance'*

1. **Chance** *is a set of unknown hypotheses. Therefore, the explanation of an observation is not influenced by it. Accordingly, a possible observation that should be explained can not be explained. In this case, a hypothesis base or a knowledge base lacks the necessary hypotheses. Therefore, it is necessary to generate missing hypotheses. Missing hypotheses are characterized as chance.*

2. **Chance** *itself is a set of known facts, but it is unknown how to use them to explain an observation. That is, a certain set of rules is missing. Accordingly, an observation can not be explained by the facts. Since rules are usually generated in inductive ways, rules that are different from the trend can not be generated. In this case, rules are generated by abductive methods, so trends are not considered. Abductively generated rules are characterized as a chance.*

In fact, chance has the flavor of probabilistic reasoning, but this chapter does not represent chance in an explicitly probabilistic form. Instead, this chapter treats chance in a logical way because a logical inference, as shown later, especially abduction, seems to be a powerful weapon for performing chance discovery as explanatory reasoning. Therefore, in Definition 1, we defined chance by using the terminology of abduction like explanation and hypotheses. Accordingly, this chapter deals with chance discovery in the context of abduction.

16.3 Abduction and Analogy as Discovery

16.3.1 Reversed Deduction as Discovery

Peirce showed three types of human reasoning: deduction, induction, and abduction [16.14]. Peirce characterized deduction as analytic reasoning. Then, deduction can be regarded as an if–then type of reasoning. For example, the inference feature of deduction can be exemplified as follows:

(1)   All human beings die.
(2)   Socrates is a human being.
(3)   Therefore, Socrates dies.

Deduction is a useful inference that is used in daily life. However, there is no chance for discovery. This is because deduction can not work with an incomplete knowledge base. For example, if either knowledge (1) or (2) is missing, we can not conclude (3). Discovery can be thought of as a process for dealing with incompleteness. That is, discovery is a type of process to find new or missing information.

Abduction and induction can be thought of as the reverse of deduction. As Peirce characterized, both are synthetic reasonings that can deal with incomplete knowledge. However, their characterizations are slightly different.

Peirce wrote that abduction is an operation for adopting an explanatory hypothesis, which is subject to certain conditions, and that in pure abduction there can never be justification for accepting the hypothesis, other than through interrogation.

Peirce also wrote that induction is an operation for testing a hypothesis by experiment, and, if it is true, an observation made under certain conditions ought to have certain results. Consequently, if these conditions are fulfilled and the results are favorable, we extend a certain confidence to the hypothesis. Simply speaking, induction can be formalized as the generalization of examples. Induction finds tendencies in examples and generates general rules (hypotheses) from examples and background knowledge.

In Peirce's sense, abduction is a type of discovery reasoning. It discovers new events. On the other hand, in fact, induction is also a type of discovery reasoning; however, it discovers tendencies that are not new events. The following example may be useful in comparing abduction with induction. When we find fossils of sea shells in a certain place, the result from induction (inductive hypothesis) is that fossils of sea shells will be found if we dig there. On the other hand, the result from abduction is that the place used to be under the sea.

As shown above, the role of abduction is to discover something that can not be directly observed. From fossils of sea shells, it is slightly difficult to reason that 'sea' is a new concept.

16.3.2 Analogy as Discovery

Analogical reasoning (analogy) is achieved by analogical mapping from source knowledge to target knowledge. Thus, analogy might be used to generate conjectures about an unfamiliar domain. For example, if we know about the water flow system where water flows from a place with greater pressure to a place with less pressure, we can guess or find the heat flow system where heat flows from a place with greater temperature to a place with less temperature (see the figure on p. 65 of [16.7]). Thus, we usually utilize analogy to guess or supplement an unknown system. Actually, analogical mapping can be regarded as a sort of induction, but analogy can also be used for creation of new things. For example, some designs are created by an analogical reference to an existing design. Sometimes, the new design will be created by analogy from a thing in a different field. Also, we sometimes solve a problem by considering a problem that is analogous to the problem to be solved. The most popular example is Poincaré's case. Dietrich et al. pointed out that Poincaré's work on Fuchsian functions reminded him of transformations in non-Euclidean geometry. Although, these two domains – Fuchsian functions and transformations in non-Euclidean geometry – are not similar in detail, they are analogous. This analogy points out that many chance discoveries involve analogical reminding [16.4].

Gentner defined analogy as a 'structure mapping'. Accordingly, Gentner pointed out that analogical mapping is in general a combination of matching existing predicate structures and importing new predicates (carry-out). Gentner also pointed out that most explanatory analogies are a combination of matching and carry-out

[16.6, 16.7, 16.8]. Importing new predicates seems to work as a chance introduction. Furthermore, a computational implementation of SME (structure-mapping engine) has been achieved by Falkenhainer and Forbus [16.5].

Dietrich et al. showed that analogical reminding is a great source for chance discoveries because it can produce representational change and thereby produce insights. They proposed extending the structure-mapping theory of analogy and analogical retrieval by adding a packing[2] away of parts of representations, thereby producing more abstract concepts [16.4].

Greiner pointed out that the teacher can often express the same chunk of information in many fewer words by using an appropriate analogy. In order to reclaim the information, it is necessary to decode the analogy. The decoding process is analogical inference. From this viewpoint, Greiner proposed an analogical inference system NLAG [16.10] that is explicitly given analogical hints like $A \sim B$ (A is like B) and then performs analogical inference. Let Th denote the theory, PT be the target problem to solve, and $\varphi(A)$ be a proposition. When the inputs are Th, $A \sim B$, and PT, the analogical inference is done as follows:

$$Th \not\models \varphi(A) \quad (unknown), \tag{16.1}$$

$$Th \not\models \neg\varphi(A) \quad (consistent), \tag{16.2}$$

$$Th \models \varphi(B) \quad (common), \tag{16.3}$$

$$Th \cup \{\varphi(A)\} \models PT \quad (useful). \tag{16.4}$$

The output $\varphi(A)$ is a 'new' proposition. Greiner defines the above inference as a useful analogical inference that can seek analogies that help solve the target problem.

Russell proposed DBAR (determination-based analogical reasoning) [16.18] as a solution to the justification and non-redundancy problems. In DBAR, a determination relation is considered as a logical generalization of the different types of dependency relations defined in database theory, and determination rules facilitate a sound rule inference and valid conclusion projected by analogy from a single instance, without implying what the conclusion should be prior to an inspection of the instance.

The determination rule (P determines Q) is
$P(\underline{x}, \underline{y}) \succ Q(\underline{x}, \underline{y})$ *iff*

$$\forall \underline{wyz}[\, P(\underline{w}, \underline{y}) \wedge Q(\underline{w}, \underline{z}) \implies \forall \underline{x}[\, P(\underline{x}, \underline{y}) \Rightarrow Q(\underline{x}, \underline{z})\,]\,], \tag{16.5}$$

where $\underline{x}$ is a set of variables such as $\{a, b, c, \ldots\}$.

Then, the analogical mapping can be shown as follows:

$$\frac{(P(\underline{x}, \underline{y}) \wedge Q(\underline{x}, \underline{z})),\ P(S, A),\ P(T, A),\ Q(S, B)}{Q(T, B)}. \tag{16.6}$$

For example, if we have the following background knowledge:

---

[2] Packing is a process of representational change during memory retrieval that suppresses irrelevant information in a structured mental representation.

```
Nationality(Jack,UK)
Male(Jack)
Height(Jack,6')
....
Nationality(Giuseppe,Italy)
Male(Giuseppe)
Height(Giuseppe,6')
NativeLanguage(Giuseppe,Italian)
......
Nationality(Jill,UK)
Female(Jill)
Height(Jill,5'10'')
NativeLanguage(Jill,English)
.....
```

and if we have a determination rule such as

$$(Nationality(x, n) \wedge \neg Nationality(x, Swiss)) \\ \succ NativeLanguage(x, l),$$

$$(Nationality(x, n) \wedge i_1 Dualcitizen(x, US))^3 \\ \succ (i_2 NeedVisa(x, US) \wedge Maxstay(x, t)),$$

then by using the determination rule and knowledge regarding Jill's source knowledge, `NativeLanguage(Jack,English)` can be inferred[4].

Although the derived solution can be justified, it is unnatural to assume that one determination rule can determine every relation. Nonetheless, if the determination rule can be induced for several relations, it is useful in the analogical mapping procedure, and the result seems to be a new rule.

Goebel has shown that analogical reasoning (mapping) can be explained by the Theorist framework [16.9]:

$$So \cup T \not\models G, \tag{16.7}$$

$$So \cup T \cup M \models G, \tag{16.8}$$

$$So \cup T \cup M \not\models \neg m \ (m \in M), \tag{16.9}$$

where So is source knowledge, T is target knowledge, M is equality assumption, and G is analogical reasoning. Goebel's framework shows that analogical reasoning is the generation of analogical mapping functions M as hypotheses. From this viewpoint, analogy can be discussed in abduction that works as an inference for chance discovery.

From the above studies, analogy offers the possibility to find, supplement, or suggest new or novel systems and to extend knowledge to another domain. Naturally, these mechanisms include chance discovery.

---

[3] $iP(x)$ represents 'either $P(x)$ or not'.
[4] For details of analogical reasoning, see [16.18].

16.4 Computational Abduction

This section shows computerization of abduction (mechanized abduction). The first computerization of abduction was done by Pople [16.16]. This abduction is very primitive and simply the reverse of deduction. After his work, various systems have been proposed. In this section, we review two of these systems.

16.4.1 Hypothetical Reasoning (Theorist)

Abduction is an explanatory reasoning that explains an observation by a set of hypotheses. This type of reasoning is also called hypothetical reasoning, where the reason (set of consistent hypotheses) is found (adopted) in a logical way to explain an observation. One of the systems that mechanize hypothetical reasoning is Theorist [16.15]. The inference mechanism of Theorist that explains an observation (O) by a consistent hypothesis set (h) selected from a set of hypotheses (H) operates as follows.

$$F \nvdash O \qquad (O \text{ can not be explained by only } F), \qquad (16.10)$$

$$F \cup h \vdash O \qquad (O \text{ can be explained by } F \text{ and } h), \qquad (16.11)$$

$$F \cup h \nvdash \Box \qquad (F \text{ and } h \text{ are consistent}), \qquad (16.12)$$

where F is a fact (background knowledge) and $\Box$ is an empty clause.

The inference image of hypothetical reasoning is shown in Fig. 16.1.

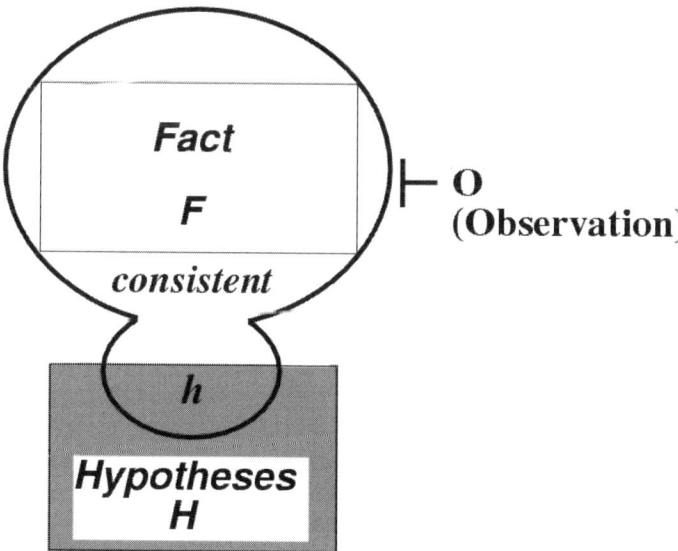

Fig. 16.1. Inference image of Theorist

Let us show an application of Theorist (LSI circuit design [16.12]) to explain the behavior of computational abduction. Let the knowledge used in the application be a set of fact clauses as shown in Fig. 16.2. They show the function of devices, connection rules, and information on inconsistent relationships.

```
fact((equ(out(1,D),IN+1):- sub(D,inc,1,1)&
        equ(in(1,D),IN))).
fact((equ(out(1,D),IN-1):- sub(D,dec,1,1)&
        equ(in(1,D),IN))).
fact((equ(out(1,D),STATE):- sub(D,reg,1,1)&
        equ(state(1,D),STATE))).
fact((equ(n_state(1,D),IN):- sub(D,reg,1,1)&
        equ(in(1,D),IN))).
....
fact((equ(out(N,f1),O):- conn(Node,out(N,f1))&
        equ(Node,O))).
fact((equ(in(N,D),O):- conn(Node,in(N,D))&
        equ(Node,O))).
...
fact((inconsistent:- type(D,S)& type(D,S1)&
        S1\==S)).
....
```

Fig. 16.2. Knowledge base (*fact*)

In hypothetical reasoning, an observation is explained by adopting hypotheses. As such, if hypothetical reasoning is applied to some constraint satisfaction problem (CSP) applications, an observation corresponds to a sort of specification and a set of hypotheses corresponds to a solution.

Possible function blocks and their possible connection information are represented as *hypotheses* (Fig. 16.3), which are included in a hypotheses base. The result of the design is information about connections between function blocks and function-block assignments, which are included in a set of consistent hypotheses adopted from the hypotheses base.

```
hyp (conn (in (N, f1), in (M, D)), (conn (in (N, f1),
in (M, D)) :- D \== f1 )).
hyp (conn (out (N1, D1), in (N2, D2)),
(conn (out (N1, D1),  in (N2, D2)) ) :-
        D2 \== f1 & domain (dev (D1)) &
D1 \== D2 )).
. . . .
hyp (function (D, S), type (D, S)).
hyp (equ (in (X, D), V), (equ (in (X, D), V) :-
D \== f1 & integer (V) & V \== 1)).
. . . .
```

Fig. 16.3. *Hypotheses* base

Therefore, the result is a set of generated (adopted) hypotheses.

If the user gives the input output relationship of the function blocks (devices) and a circuit state transition as shown below (an example of a filter design),

```
((equ (out (1, f1), input+x2+x3+x4) :-
                    equ (in (1, f1), input) &
                    equ (state (1, X2), x2) &
                    equ (state (1, X3), x3) &
                    equ (state (1, X4), x4)) &
(equ (n_state (1, X2), input) :-
                    equ (in (1, f1), input)) &
(equ (n_state (1, X3), x2) :-
                    equ (state (1, X2), x2)) &
(equ (n_state (1, X4), x3) :-
                    equ (state (1, X3), x3))).
```

then the name of function blocks and their connection information are returned as a solution by the hypothetical reasoning system as shown in Fig. 16.4.

By using the above hypothetical reasoning system, some applications like a circuit block design system can be implemented. Basically, hypothetical reasoning works by generating and testing; it is a general inference paradigm that is not specified for special applications. If we make use of a hypothetical reasoning system, it is very easy to give specifications without the need for strict programing.

However, it is obvious that, although variables in hypotheses are unified, it is not the generation (abduction) of new hypotheses but only the restriction of a candidate hypothesis set that may be adopted.

Thus, 'reason' (= explanation) is usually selected from the knowledge (hypotheses) base. Therefore, conventional abduction requires a perfect hypotheses base from which an entire consistent hypothesis set is selected to explain an observation. Here, 'perfect hypotheses base' means that it contains all of the necessary hypotheses. Chance discovery is also characterized as an explanatory reasoning; however,

```
conn(out(1,x5),in(1,x6)) &
conn(out(1,x4),in(1,x5)) &
conn(in(1,f1),in(1,x4)) &
function(x6,reg) & dev(x6,1,1) &
conn(out(1,x6),in(2,x1)) &
function(x5,reg) & dev(x5,1,1) &
conn(out(1,x5),in(2,x2)) &
function(x4,reg) & dev(x4,1,1) &
conn(out(1,x4),in(2,x3)) &
conn(in(1,f1),in(1,x3)) &
function(x3,plus) & dev(x3,2,1) &
conn(out(1,x3),in(1,x2)) &
function(x2,plus) & dev(x2,2,1) &
conn(out(1,x2),in(1,x1)) &
function(x1,plus) & dev(x1,2,1) &
conn(out(1,x1),out(1,f1))
```

Fig. 16.4. Circuit block design system result

as defined in Definition 1, since 'chance' is defined as unknown hypotheses, so techniques that can deal with an empty or an imperfect hypotheses base are required. If so, such an inference mechanism as conventional abduction (hypothetical reasoning, etc.) is not sufficient to achieve Chance discovery. Chance discovery needs an explanatory reasoning that can deal with an empty or imperfect hypotheses base.

16.4.2 Clause Management System (CMS)

Reiter and de Kleer proposed a clause management system (CMS) [16.17], which is a type of deductive database management system. If the database is incomplete, CMS returns necessary (missing) knowledge. From the abductive viewpoint, this is an abduction. If we use CMS as an abduction, the CMS inference mechanism is as follows.

When

$$\Sigma \not\models C \qquad (C \text{ can not be explained by only } \Sigma), \tag{16.13}$$

if propositional clause C (observation) is given, CMS returns a set of minimal clauses S to clause set Σ such that

$$\Sigma \models S \vee C, \tag{16.14}$$

$$\Sigma \not\models S. \tag{16.15}$$

S is called a minimal support clause, and $\neg S$ is a clause missing from clause set Σ that can explain C. Therefore, $\neg S$ can be thought of as an abductive hypothesis according to the abductive point of view.

For example, if

$$\Sigma = \{\{p \vee q \vee r\}, \{p \vee \neg q\}, \{p \vee \neg r\}, \{q \vee r \vee \neg t\}, \{\neg p \vee q \vee s\}\},$$

then minimal supports for $\{s \wedge r\}$ are $\{q\}$, $\{\neg t\}$, $\{\neg r\}$, and minimal supports for $\{q\}$ are $\{s\}$, $\{r \wedge \neg t\}$, $\{\neg q\}$ (Fig. 16.5).

The hypothesis set (negated minimal supports) is not included in the knowledge base ($\Sigma \not\models S$) and minimal clause set. In fact, the hypothesis set is required to make an inference (for example, to explain q), but there is no justification for the hypotheses except that they provide a minimal completion for the abductive puzzle.

The CMS mechanism has the possibility of generating unknown clauses, but if we use CMS as an abduction system, we must solve this justification problem.

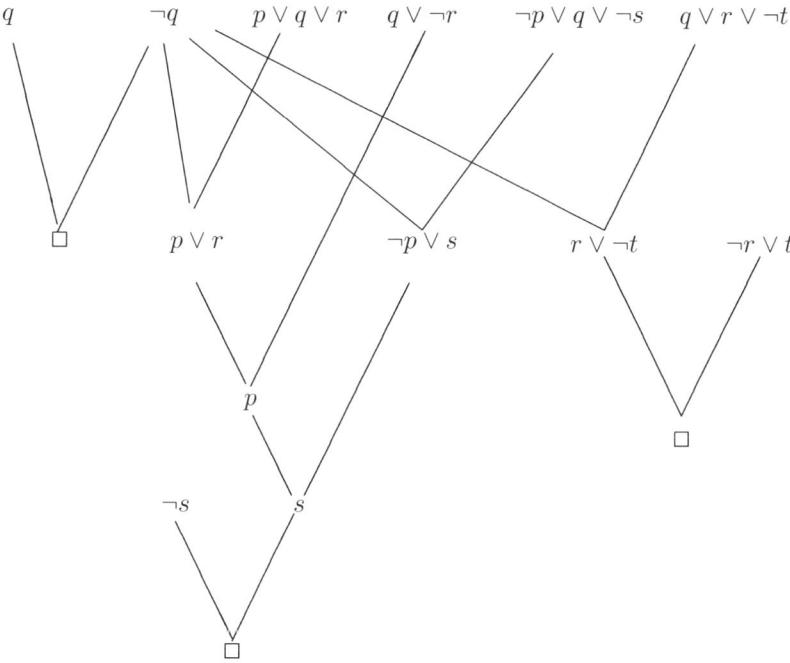

Fig. 16.5. Minimal supports for q

16.5 Abductive Analogical Reasoning (AAR)

Chance discovery can be defined as an abduction that can deal with an empty or imperfect hypothesis base. To perform such abduction, we proposed abductive analogical reasoning (AAR) [16.3] to generate the missing knowledge needed to explain an observation. A brief formalization of AAR[5] is given below.

Definition 2 . **Analogical clause:** *in general, 'A is an analogical clause of B' means that B is analogically mapped from A. In this chapter, analogical mapping is simplified. Let A be $a \vee b \vee \ldots \vee l$. An analogical clause of A (=B) has at least one atom that is similar to that of A and other atoms that are the same as those of A. For example, if a is similar to a', B is $a' \vee b \vee \ldots \vee l$. The similarity between words is calculated by the conceptual base [16.11] or dictionaries that can calculate or show an analogical relation. Actually, this chapter does not intend to discuss the details of analogical mapping, so it only deals with simplified (linear) analogical mapping. Of course, this analogical mapping can be extended to structure mapping [16.6].*

Definition 3 . *Let A be a set of clauses. 'A $\mapsto$ A'' is the relationship between A and A', such that A' is a set of clauses that can derive analogical clauses of the clauses derived from A.*

AAR explains an observation, O, by satisfying the following formulae. If observation O is given, AAR tries to explain it using clauses in the knowledge base Σ. When

$$\Sigma \not\models O \quad (O \text{ can not be explained by only } \Sigma), \tag{16.16}$$

this means that Σ lacks a certain set of clauses to explain O. Consequently, AAR returns a set of minimal clauses S such that

$$\Sigma \models S \vee O, \tag{16.17}$$

$$\Sigma \not\models S. \tag{16.18}$$

$\neg S$ is a missing clause set that is necessary to explain O. In this framework, missing knowledge can be thought of as hypotheses. However, since $\Sigma \not\models S$ ($\neg S \notin \Sigma$), the justification of $\neg S$ is not guaranteed.
N.B. In Theorist, h is selected from the hypothesis base, so the justification of h is guaranteed.

In fact, $\neg S$ can be a candidate for the hypothesis set and can be used as a hypothesis set to explain an observation; however, it would be better to generate a more plausible hypothesis set. Since we have obtained the candidate hypothesis set, our solution is to generate a plausible hypothesis set by referring to the existing clause set that is similar to the candidate hypothesis set. Therefore, in order to find a similar clause set in the knowledge base Σ, clause set S' is generated (selected from the knowledge base) as follows.

$$S \mapsto S' \quad (S' \text{ is analogically transformed from } S), \tag{16.19}$$

$$\Sigma \models S'. \tag{16.20}$$

---

[5] The style of formulae follows that of CMS.

By this transformation, the structure of the clause set may change. A detailed explanation is given in our previous paper [16.3]. Since $\neg S'$ is selected from the knowledge base, it is a guaranteed hypothesis set. However, it can not correctly explain the observation. This is because it is not a real hypothesis set. Anyway, we can utilize this clause set to generate a plausible hypothesis set. Then, by referring to S', clause set S'' is generated.

$$S' \mapsto S'', \tag{16.21}$$

$$\Sigma \models S'' \vee O, \tag{16.22}$$

$$\Sigma \not\models S''. \tag{16.23}$$

Accordingly, O is then explained by $\neg S''$ as a hypothesis set. The inference image of AAR is shown in Fig. 16.6.

A clause set $\neg S''$ is not included in Σ but is newly generated to explain an observation. The generated clause set is logically equivalent[6] to or identical to $\neg S$, because the series of transformations by analogical mapping keeps a logical equivalence. Moreover, a clause set is generated by analogical transformation from the existing clause set $\neg S'$. Therefore, $\neg S''$ is a more plausible set of hypotheses to explain the observation than $\neg S$. Consequently, AAR can guess necessary clauses and generates missing clauses by analogical transformation from the existing clause set in the knowledge base to explain an observation. Thus, AAR can be regarded as an explanatory reasoning that can deal with an empty or imperfect hypothesis base.

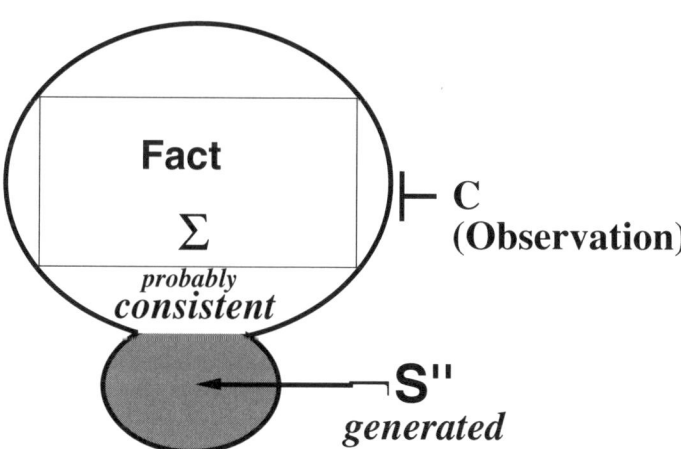

Fig. 16.6. Inference image of AAR

---

[6] In this chapter, 'logically equivalent' means that both results from forward-chained clauses are the same. For example, if A is $\neg a \vee c$ and B is $\neg a \vee \neg b \vee c$ and b, then A is logically equivalent to B.

16.6 Abductive Analogical Reasoning as Chance Discovery

As shown above, AAR can generate missing hypotheses. Missing hypotheses are either leaf hypotheses or hypotheses in the middle of an inference path. Hypotheses in the middle of an inference path are regarded as propositional rules, and some of them are generated by analogical mapping from the known clauses. This analogical mapping is performed to guarantee a justification of generated hypotheses. In this sense, we may impose restrictions that are similar to constraints in hypothetical reasoning. These types of restriction work for generating plausible hypotheses (to avoid generating short-cut hypotheses), although constraints usually work by excluding no-good pairs. Nevertheless, in the context of chance discovery, hypotheses generated by AAR are regarded as 'chance' in the sense defined in Definition 1. Therefore, it is effective to use AAR as a tool for chance discovery to give some suggestions for possible signs of rare or novel observation.

If we use AAR to generate hypotheses (symptoms) that can explain a possible observation (never explained by the current hypothesis set), then AAR enables an inference like chance discovery. This section illustrates two types of chance discovery according to Definition 1 by using the framework of AAR.

16.6.1 Type 1: When Some of the Hypotheses Are Unknown

When some of the necessary hypotheses are unknown (not found), the current observation can not be explained. In the context of chance discovery, chance seems to be a set of unknown hypotheses, and this type of inference can be regarded as pure abduction. For example, when we must predict the occurrence of a serious earthquake, we show that if such events (symptoms) as active faults can be found, a serious earthquake will occur, and then attempts to find the events must be carried out.

Let a knowledge base (Σ) be as follows:

$earthquake :- movement, distortion, oceanic\_plate.$
$faults.$
$distortion.$
$movement.$

By the usual hypothetical reasoning, since *oceanic_plate* is not included in a hypothesis base as the possible hypothesis, the observation (*O*) *earthquake* can not be explained. However, when we want to show a possibility of an earthquake, we must discover and show symptoms (= chance). When we apply AAR to solve this problem, AAR explains an observation *earthquake* by generating a hypothesis ($\neg S$) *oceanic_plate* [7].

$$\Sigma \models \neg oceanic\_plate \lor earthquake. \tag{16.24}$$

---

[7] This abduction is the same as CMS's abductive knowledge management.

Then, in the above example, chance is *oceanic_plate*, which was not found (in the knowledge base). It is logically discovered and suggested by abduction. AAR shows that 'if the hypothesis *oceanic_plate* is justified, the observation will be explained'. In this case, with a suggestion from abduction, the user will be aware of the occurrence of an earthquake, and can assume that *oceanic_plate* is a chance of earthquake; therefore a possible way to predict an earthquake will be to find *oceanic_plate*. Consequently, if we can find *oceanic_plate*, we can predict (explain) the occurrence of a serious earthquake.

16.6.2 Type 2: When Some of the Rules Are Unknown

When some of the necessary rules are unknown, even if we are aware of all the symptoms, the future observation can not be explained (predicted). In this case, reasons are usually shown afterward. For example, when a serious earthquake like the Great Hanshin Earthquake occurs, even if we know that there are a lot of active faults near the Kansai area, we can not predict an earthquake because we did not know the relationship between active faults and the Hanshin Earthquake. Therefore, when we want to predict an earthquake, we must predict unknown rules from future observation (to be predicted) and the current symptoms.

Let a knowledge base (Σ) be the above knowledge base. In this case, as shown, *oceanic_plate* can be shown as a chance. However, sometimes we must be aware of existing things (*faults*) that are similar to the known but non-existing things (*oceanic_plate*). In such cases, we predict or generate new rules as hypotheses. AAR can explain an observation by generating new hypotheses by referring to the known knowledge. If we use this type of inference, the following inference can be achieved. Since the knowledge base has *faults* as a fact, AAR uses *faults* as one of the symptoms (= hypotheses) to explain an observation. When the observation can not be explained by *faults*, AAR, by referring to '*earthquake :- movement, distortion, oceanic_plate.*', generates a new rule '*earthquake :- movement, distortion, faults.*' to explain an observation *earthquake* when *oceanic_plate* is similar to *faults*.

$$\Sigma \models S'' \vee earthquake. \tag{16.25}$$

$$\neg S'' = \neg movement \vee \neg distortion \vee \neg faults \vee earthquake. \tag{16.26}$$

In this case, *faults* is a chance, and by the similar experiences '*earthquake :- movement, distortion, faults.*', a forthcoming observation can be explained (predicted). AAR shows that 'if the rule '*earthquake :- movement, distortion, faults.*' is justified, the observation will be explained by *faults*'[8]. This is a suggestion of the occurrence of an earthquake that makes the user aware of the the occurrence of the earthquake. If the user thinks the newly generated rule is plausible, he/she can

---

[8] In fact, a certain movement of faults is also caused by the movement of oceanic plates, and it would be slightly difficult to find a similarity between faults and oceanic plates in a simple way. Anyway, this scenario is only used for explanation.

decide $faults$ as a chance and predict an earthquake by using the rule. Of course, if there is no similar knowledge, some rules can be generated to explain the observation only in an abductive way. However, in this case, generated rules may be shortcut rules or wrong rules. This is because in AAR, at first, minimal hypotheses are selected. For example, in this case, if the knowledge in the knowledge base can not be accessed, the generated hypothesis is $earthquake :- faults$. This hypothesis seems too minimal and is not correct. Even so, if it is justified by the user, $faults$ can be a chance. Nevertheless, this type of reasoning can be helpful for finding a chance and prediction. The result will be shown later, so the important thing is to show all possibilities that may cause a future observation.

The above example may seem quite simple; however, even if rules in the middle of an inference path are missing, the problem can be solved by using the proposed abduction [16.1].

16.7 The Role of Abduction and Analogy in Chance Discovery

In the previous section, two types of reasoning were shown for chance discovery. The first one is to show (suggest) unseen or unknown events as chance, and the second one is to show (suggest) known events as chance by generating new rules. Both reasonings can be achieved by abduction from the possible observations. Therefore, in this formalization, a chance seems to be a set of abductive hypotheses[9] that is generated in a logical way. Thus, the role of abduction in chance discovery is to logically discover and suggest potential factors that will cause serious events. In addition, AAR is a reasoning that combines abduction and analogical mapping. By only using abduction, as shown above, we sometimes can not guarantee the generated hypothesis set. With analogical mapping, we can generate a more plausible hypothesis set that is supported by the existing clause set. That is, the role of abduction is discovery and suggestion of a chance and the role of analogical mapping is adjustment and confirmation of a chance. By combining abduction and analogical mapping, a more plausible chance can be suggested.

In our framework, since abduction is an inference from an observation to hypotheses, it seems that it is necessary to prepare all possible observations. However, all possible observations do not need to be prepared. We only need to explain or predict the relevant observations. Therefore, if we concerned about the occurrence of an earthquake, the only thing to do is to check whether an earthquake is explained by the current facts or unseen facts. Such a small number of observations can be given by the user.

Anyway, after abduction suggests chance, the final decision will be made by the user. No decision is automatically made by abduction and analogical mapping.

---

[9] In the second reasoning, new rules are generated. Actually, some of them are in the form of a clause (not an atom). However, since they are propositional clauses in our framework, they are also thought of as abductive hypotheses.

16.8 Conclusions

Chance discovery is a promising research field for risk management or risk avoidance as well as fortune making. This chapter defines 'chance' and shows one of the methods for chance discovery as a type of abduction that is an explanatory reasoning with analogical mapping. Then, the roles of abduction and analogy in chance discovery were explained. In addition, this chapter showed that a chance for future observation is suggested as a set of abductive hypotheses by performing AAR from the possible observations. Considering the results, abduction appears to be a promising framework for performing chance discovery.

We also proposed an integration of abduction and induction [16.2], in which induction generates predicate rules (inductive hypotheses) and abduction generates propositional rules (abductive hypotheses). In this integration, of course, exceptional facts do not reflect the generated rules. However, if the proposed inference could be implemented in the integration, then it would produce hopeful results.

References

16.1 Abe A (1999) "Two-sided Hypotheses Generation for Abductive Analogical Reasoning", Proc. of ICTAI99, pp. 145–152
16.2 Abe A (2000) "The Relation between Abductive Hypotheses and Inductive Hypotheses", Abduction and Induction (Flach P. and Kakas T. eds.), Kluwer Academic Press, Hingham, MA, pp. 169–180
16.3 Abe A (2000) "Abductive Analogical Reasoning", Systems and Computers in Japan, 31(1): 11–19
16.4 Dietrich E et al (2002) "The Prepared Mind: The Role of Representational Change in Chance Discovery", Chap 19, in this book
16.5 Falkenhainer B, Forbus KD (1986) "The Structure-Mapping Engine: Algorithm and Examples", Artificial Intelligence, 41: 1–63
16.6 Gentner D (1983) "Structure-Mapping: A Theoretical Framework for Analogy", Cognitive Science, 7: 155–170
16.7 Gentner D (1988) Analogical inference and analogical access. In Prieditis A (ed) Analogica, Pitman, London, UK, pp.63-88
16.8 Gentner D (1989) "The mechanisms of analogical learning", SIMILARITY AND ANALOGICAL REASONING, Cambridge University Press, pp.199–241
16.9 Goebel R (1989) A sketch of analogy as reasoning with equality hypotheses", In Proc. of Int'l Workshop Analogical and Inductive Inference (LNAI-397), pp. 243–253
16.10 Greiner R (1988) Learning by Understanding Analogies, In Prieditis A (ed) Analogica, Pitman, London, UK, pp.1–36
16.11 Kasahara K et al (1996) "Viewpoint-based measurement of semantic similarity between words", Proc. of 5th. International Workshop on Artificial Intelligence and Statistics (LNS 112), pp. 433–442
16.12 Makino T, Ishizuka M (1990) "A Hypothetical Reasoning System with Constraint Handling Mechanism and its Application to Circuit-Block Synthesis", Proc. of PRICAI90, pp. 122–127
16.13 Ohsawa Y (2002) "Chance Discovery for Making Decision in Complex Real World," New Generation Computings, 20(2): 143–163
16.14 Peirce CS (1955) "Abduction and Induction", Chap. 11, Philosophical Writings of Peirce, Dover, Mineola, NY

16.15 Poole D, Goebel R, Aleliunas R (1987) "Theorist: A Logical Reasoning System for Defaults and Diagnosis", Cercone NJ, McCalla G (eds) The Knowledge Frontier: Essays in the Representation of Knowledge , Springer-Verlag, Heidelberg, Germany, pp. 331–352

16.16 Pople HE Jr (1973) "On The Mechanization of Abductive Logic", Proc. of IJCAI73, pp.147–152

16.17 Reiter R, Kleer J (1987) "Foundation of assumption-based truth maintenance systems: preliminary report", Proc. of AAAI87, 1987, pp.183–188

16.18 Russell SJ (1988) The Use of Knowledge in Analogy and Induction, Pitman, London, UK

16.19 Takeuchi K (2000) "Where probabilistic and stochastic society goes?", revue de la pensée d'aujourd'hui, 28(1): 84–99

Part IV
Key 3 – Computer-aided Chance Discoveries

The third key shown in Part I was data mining. In a more advanced sense, we look at the interactive system between a human and a computer which extracts patterns of the appearance of chance events. The algorithms are beyond existing machine-learning methods, for triggering the in-depth understanding by a human of his/her environment.

17. Active Mining with Visual Human Interface

Wataru Sunayama

Osaka University, 1-3 Machikaneyama, Toyonaka, Osaka 560-8531, Japan
email: sunayama@sys.es.osaka-u.ac.jp

Summary.

In this chapter, the criteria for information selection and two-dimensional inter-faces for application are presented. We require multi-sided analysis for shrinking enormous amounts of data, and we should understand visualized outputs by intu-ition for acquiring new ideas. Visualization methods will be heavily relied upon for aiding humans' comprehension by intuition, and promoting the urge for finding biased outputs. Such a system will provide clues for discovering chances.

17.1 Introduction

We seek information that is available for research exploitation and sales promotions from enormous dynamic World Wide Web resources. However, it is difficult for us to obtain new knowledge and ideas simply by seeing those data. Therefore, we'd like to see an information arrangement based on some viewpoints for acquiring latent chances in this world. Thus, the methodology for those are active mining and information visualization.

In this chapter, I point out three facts for discovery as follows:

1. Discovery of new viewpoints.
2. Discovery of communities.
3. Discovery of timing.

First, the discovery of new viewpoints is to hit upon a new interpretation by being aware of new aspects of objects. This means that an object has unknown means and unknown aspects, like a word has various meanings. In addition, we may find something in common among multiple objects which are otherwise unlikely to be related.

These new viewpoints won't appear from whole data but from some restricted data. It is thought that this restriction also becomes another viewpoint. Therefore, a discovery doesn't end with a single viewpoint discovery, but continues to another discovery like links in a chain of the model of chance discovery, which is described in the first chapter of this book.

Secondly, the discovery of communities means that a system finds a large group of people who have the same consciousness in this world. The viewpoints described above correspond to common points in data, but the origin of a community is a common point between people who create that data. As such, it is necessary to analyze people to find their needs and flows in the modern world. As a matter of course, there may be a person who creates all of the needs and communities in the world. However, it is difficult, if not impossible, to succeed only by intuition, and such a person needs to know the world to acquire that intuition.

Finally, discovery of timing means that a system must consider when to give values to information. For example, fruits which are unripe or over-ripe are not good. It is important for us to know not only the most valuable information, but also information that is likely to be valuable or information whose value is decreasing. Related to the flows mentioned in the above paragraph, the state of the world always varies as long as we're alive. We'd like to see the most noteworthy things by grasping this transition.

In any case, although it is a person that discovers viewpoints, communities, and timing, a system aids people by its outputs . That is, a system doesn't output only effective information automatically, but outputs information which is likely to be effective, and the final decision will be made by a person. When a system provides outputs, those outputs should be arranged in a two-dimensional interface on a computer display. The human activity of selection is aided by the information-visualization methods of effectve arranging.

Firstly, the difference between two-dimensional and three-dimensional is explained. A person who wants to find the nature and tendencies of information repeatedly sees and compares both global and local tendencies. Although global information should be viewed on a display, 3D information is compressed onto a 2D display. Therefore, a user must decide a viewpoint and the color shade will appear to be far from the viewpoint; that is, 3D interfaces are not suitable for grasping global information on a 2D screen. Since 3D interfaces have higher representation ability than those of 2D, some specific global information shapes should be shown in 3D. However, trial-and-error is needed for dealing with multi-dimensional and prodigious information, so an output should be make it easy to revise and remember whole output images. Therefore, in this chapter, the features and meanings of 2D interfaces and relationships between 2D information visualization and chance discovery are described.

Chance discovery with a 2D interface is defined by a user discovering unknown original truth from information arranged by known or general relationships. It is impossible to output a correct answer which is unknown to the user. Therefore, we expected that a system outputs information implying unknown relationships, and shows possible candidates for a correct answer.

17.2 Features of Two-Dimensional Interfaces

The meaning of information visualization using a 2D interface is how systems enable humans to find a large amount of information and how systems arrange that information. It is natural that the arrangement varies, depending on the type of information. I then describe the features of 2D interfaces related to arrangement and representation, such as order by directions, order by arrangement methods, and relationships among arranged multiple objects.

17.2.1 Order by Directions

Since people read from left to right and from top to bottom, people see figures and pictures by the same order when searching. Therefore, it is rational for us to arrange objects in 2D from the upper side or left side. In an example expressing vertical order, the higher object tends to be important. Similar to the tournament of the World Cup soccer, the top of the tree indicates championship, so the top of the tree structure represents a concept that includes all. In this case, though, horizontal order is not as important.

17.2.2 Order by Arrangements

From another aspect, if the center of the interface is specified as the most important place, then the edges of the interface are less important. For example, a hyperbolic tree [17.3] is one of those. As the center of a field of view tends to be a gazing point, an object placed at the center will be easily noticed. In cases where the objects are enormous, the point a user wants to see will be moved to the center of an interface. However, this move is regarded as not only the movement of objects, but also the movement of a field of view.

17.2.3 Potential Relationships Among Objects

As a method for expressing relationships by arrangements, the Euclidean distance usually stands for the gap in meaning. Such systems give a relative value to two distant objects, and the values are used to arrange multi-dimensional scaling or spring models. Though this arrangement is easily interpreted by intuition, there are two disadvantages: that the relationship limit is not cleared, and the calculation cost is too high to achieve fast arrangement.

17.2.4 Cleared Relationships Among Objects

To overcome such problems mentioned in the last paragraph, the relationship among objects is expressed by cleared lines (or links) and the relationship is clearly displayed. Such expression methods contain tree structures, networks, and graphs such as *KeyGraph* [17.4]. However, if the objects have complex relationships, the links can not be distinguished by users.

There are clustering methods to classify objects by definite boundaries. Clustering methods make up for the disadvantage of arrangements by distance. The objects inside the boundaries have similar meanings and belong to the same class. If labels are not given to each cluster, a user should give adequate labels. This process means that a user may be able to discover previously unknown common features of the objects in the same class; however, disadvantages are that it is too difficult to give labels and the clustering may not make sense.

The relationships are sometimes expressed by colors, where similar objects are given the same color. Because these colors imply relationships, this coloring method is both a tacit and a clear method.

17.3 Information for Selection

In this section, the criteria for selection are described. By using these criteria, systems select enormous quantities of information in a database and output to the interface. These criteria are also used as the order for arrangement on an interface, meaning the user can select information by this arrangement. It is desired that the information-visualization methods should be examined for each criterion described as follows.

17.3.1 Necessity of Information

Necessity is the most important and basic criterion indicating whether a user currently needs information. However, systems have no way of knowing a user's subjective necessity, so users only select objects that he/she needs. Search keywords are such a directional input tool.

Public necessity may be calculated by text data in the World Wide Web, because information the world needs probably already exists in proportion to the necessity. This doesn't necessarily mean much information always exists. Small amounts of high-quality information may be sufficient to satisfy needs. For example, for questions having only one answer, such as 'What are the capital cities of every country?', the page including all capitals satisfies the question. On the contrary, for questions which have many answers, such as 'Tell me how to cook chicken?', the quantity of the information increases in proportion to the necessity. The concrete number for necessity is given as hit numbers, i.e. the number of pages including the search keywords. To have a high hit number is a necessary condition that the information is very necessary.

17.3.2 Adaption of Information

In this world, there are topics whose necessity is increasing, decreasing, and always required. Adaption is defined as the rate of increasing necessity for information, related to the transition of necessity from time to time. Topics the necessities do not fix are transitory, and topics constantly required are universal. This adaption is calculated by the difference of necessity as time goes by. It can be simply said that a significant increase in the hit number becomes adaption. However, the number of Web pages is generally increasing, because older information also remains on the Web without being deleted. Therefore, as the decrease of information is difficult to notice, the adaption is calulated by the difference from the mean rate of increase.

As it happens, researchers do not acquire information after the increase of adaption, but prospect topics whose adaption is likely to increase. As such, we investigate methods for discovering topics whose necessity will increase from data of topics which have already increased in adaption, and also investigate the tendencies of those topics. In particular, we will examine topics that temporarily increase when they are in fashion, and slightly decrease when they are out of fashion.

This process of decrease compares to that of radioactive-element decay. It is necessary for the researchers to grasp information necessity and be sensitive to the global information stream to succeed in their work.

17.3.3 Effectiveness of Information

Though necessity of information is focused on human beings, effectiveness is focused on information. Effectiveness is concerned with time, not adaption, and the most valuable time is not past nor future but the present time. This is also evaluated by the hit number, because so much information inundates the world. In other words, public necessity corresponds to effectiveness.

17.3.4 Originality of Information

Though large information hit numbers tend to be noticed, some information with small hit numbers is available. Originality is given to combinations of information. Interesting and novel combinations should be evaluated, even if the information exists at low frequency. That is, if each factor for combination exists in a random location, and they are not already combined, the combined factor has originality.

For example, let N denote a whole number of Web pages, A denote the number of pages related to topic A, and B denote the number of pages related to topic B. If topic A and topic B are probabilistically independent, the number of Web pages including both A and B becomes AB/N^2. Therefore, if the number of Web pages including both A and B is smaller than that number, that combination has originality. Originality is very close to novelty. If each factor has adaption to the world, it is thought that the possibility of the combination appearing will increase, and the new combined topic will be remarkable.

17.3.5 Quality of Information

Finally, I describe the quality of Web pages as information resources. This quality evaluation is formed by superficial information, constituting known source texts, and by contents described in those texts. It is difficult for the systems to analyze the contents without human assistance. Therefore, relational feedback methods [17.5] that improve the results by human–computer interaction are suggested. Now, however, superficial evaluation is preferred to reduce a user's workload. In the search engine Google [17.1], each page is measured by the Pagerank method based on the number of connecting links. In addition, quantities and rates of texts, images, tags, commas, and periods are used for this purpose.

17.3.6 Selection Criteria for Chance Discovery

Here, the relationship between selection criteria and chance discovery is summarized. Chances are defined as ideas by humans' subjective thoughts. That is, though such a thought is not known objectively at the time, that must be acknowledged in the near future. The steps for chance discovery are follows:

1. An idea occurs, but it is not known whether the idea will be a chance or a risk.
2. A chance or a risk arises.
3. Act to achieve the chances or avoid the risks.

Though the growth of chances, as in the third step, is an important step in chance discovery, generating a chance is the most fundamental and basic process (the first step). In other words, chance discovery with information-visualization systems is to imply new opportunities in the near future, and to support original idea generation.

The selection criteria described above are summarized to be applied to chance discovery. Information necessity is reflected by a user's subjective input. Even search keywords for a simple question differ from person to person. Therefore, the number of inputs corresponds to the user's subjective intention; if a user wants to get objective information, he/she will give less information such as search keywords, but if the user wants subjective information, he/she gives a lot of information.

Users become aware of trends or fashions in the world by information adaption. This criterion aims to discover objects that will be in fashion. However, if the object has a high possibility of success, that object will be acknowledged around the world. Therefore, the possibility of success and high reward are not compatible. There is a trade-off between success and reward, and the user should decide which they prefer.

Users may find objects that do not yet have high effectiveness, but are recognized as having value now. Though objects that have high effectiveness are objective, they won't be reflected in the user's subjectivity. If one is not careful, one might follow an object which is already established in fashion and will not grow to be a new trend. In practice, creative methods consist of both subjective and objective aspects as seen in brand expansions.

Users discover things previously unseen by information originality. This originality is the user's revealed subjectivity. The ability to make use of a chance is the ability to adapt an original idea to the world.

Users distinguish information resources as of certain or uncertain quality. Information from certain resources is treated as reliable, while the other is used for exploring possibilities. In other words, information will be valuable if it is reliable.

Clues for discovering chances will be seen if these criteria are correctly comprehended and used for the purpose of chance discovery. Although the ability to discover chances differs from the ability to utilize chances, the chance wouldn't be utilized without first being discovered. The criteria described above are combined and utilized for supplying useful information. Such information will lead users to chances and aid chance discovery. In summary, originality expresses a user's subjectivity, effectiveness, and adaption to express world trends and changes, while quality enables users to ascertain information reliability, and guides users to new chance discoveries.

17.4 Information-Visualization Systems for Chance Discovery

In this section, real two-dimensional information-visualization systems are described for corresponding to features of two-dimensional interfaces and criteria for information selection.

17.4.1 Interface 1: Display of Web Structures

Web pages in the Internet are linked with hyper-links. The research hyperbolic tree [17.3] visualizes the hyper-links as the Web structure. This tree has a root at the center of the interface, and the links are spread in all directions around, though the ordinal tree expands from the top to the bottom. The self-organizing map [17.2] expresses existing information as two-dimensional maps, and research [17.6] is output as geographical maps using the SOM.

These are the nature of geographical maps.

– Use of the center: viewpoint becomes the center and relations are placed around.
– Cleared relationship: similar information is linked or placed adjacently.

These interfaces do not eliminate information, and we can see information as it is. In other words, this means no criteria described in this chapter are applied. When we use such systems, users should see the tendencies, nature, and unique points by themselves. Therefore, the outputs must be easy to see for supporting users, because they may want to know the reason why the tendencies, nature, and unique points appeared. The systems should prepare environments to ascertain the reasons; such systems can lead users to successful chance discoveries.

Tendencies, nature, and unique points differ from other outputs, and these appear if information holds various biases. After we give those biases to labels and interpret them, the meanings of the labels are implicitly inferred. Users may discover chances by discovering new labels, or by investigating backgrounds of labels. As captured biases and given labels differ from person to person for the same outputs, the person who discovers chances may have some luck. However, the possibility for discovery will increase if the person improves the information related to outputs. Therefore, it is necessary that the systems be able to answer questions that users may pose after seeing simple outputs.

17.4.2 Interface 2: Two-Dimensional Interface for Supplying Keywords

To support search engines, the system [17.7] supplies keywords related to search keywords and those keywords are arranged in a two-dimensional interface. Figure 17.1 shows an example output of the system. Each key frame includes a search keyword, and relational keywords are placed around the frame.

In this interface, the two-dimensional features are as follows:

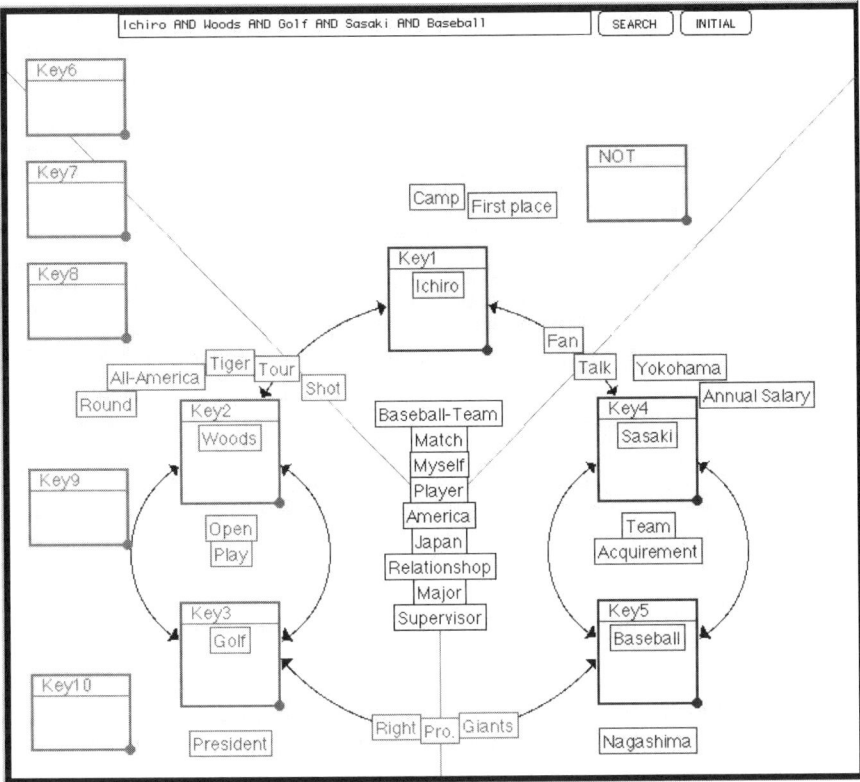

Fig. 17.1. Relational keywords for search keywords

- Order by directions: an initially input search keyword is placed at the top due to its importance.
- Use of the center: relational keywords related to all search keywords are placed at the center.
- Potential relationship:keywords related to search keywords are placed near each search keyword.
- Cleared relationship: search keywords are clustered and relations for search keywords are expressed by links. Relational keywords are distinguished by colors.

The criteria used for information selection are necessity, effectiveness, and quality. Information the user inputs is based on the user's needs. In this system, search keywords are input by users, so output information is limited. For information effectiveness, document frequency is applied to the system. That is, words appearing in many Web pages are treated as keywords. By using document frequency, we can know tendencies in the world because output keywords are extracted as words used by many people who create Web pages. For the sake of information quality, relational keywords are not extracted from all of the words in Web pages, but only from related search keywords. A word w is evaluated in (17.1), where S is the set of search keywords and $n(a)$ is the number of sentences appearing the word a [17.8].

$$\text{score}(w) = \prod_{s \in S} \frac{n(w \cap s)^2}{n(w)n(s)}. \tag{17.1}$$

Thus, the words related to many search keywords are evaluated in detail, and words appearing with the necessary sufficient conditions for search keywords in each sentence are divided by delimiters.

By using this system, a user may find a new word which is out of his/her vocabulary, and will want to ascertain the reason for its appearance. Such a user will probably discover unknown information using that new word as a new search keyword. This is the first step of chance discovery – that a user finds new information that already exists, and knows the backgrounds of the search keywords. Chances exist in places the user is not aware of; even if one piece of information was known to the world, the world can not know how that information will be interpreted by users. Therefore, we can say a system is effective if it can supply hitherto unknown information to a user.

17.4.3 Interface 3: Indication of Relational Topics and Examples

Figure 17.2 shows example sentences for relational topics for the search keyword, 'Document'. Relational topics are closely related to the search keyword, and greatly increase the number of hits. The most important sentences are extracted from each Web page, including both the search keyword and each topic keyword.

This list output is similar to search results by search engines, because it is difficult to adopt complex arrangements in a display if most outputs consist of texts or sentences. However, features of two-dimensional interfaces are also included in such an interface.

– Order by directions: topics are listed from the top by order of importance.
– Use of the center: since not all outputs are included in a display, a part that is wanted to be seen is moved onto the display.
– Potential relationship: a relational topic and its example sentences are arranged together.

Criteria used for information selection are necessity, effectiveness, adaption, and quality. As with interface 2, search keywords are input by users' inquiries. In addition to information effectiveness used in interface 2, information adaption is applied to those outputs. Increasing rate of a word w is evaluated as in (17.2), (17.3), and (17.4). That is, a word w is weighted by (17.3) with respect to its hit numbers for each month in the last year. More concretely, the words are evaluated if the hit numbers in the first half of the year were not so high but increased in the latter half.

$$\text{Ave}(w) = \sum_{k=-12}^{-1} \text{Hit}(w, k)/12, \tag{17.2}$$

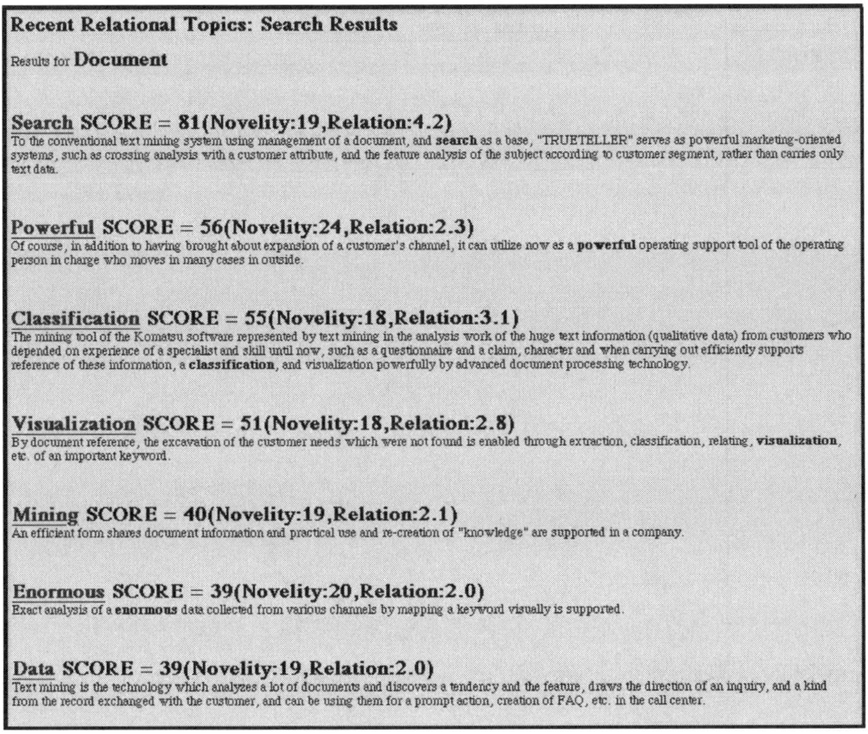

Recent Relational Topics: Search Results

Results for **Document**

Search SCORE = 81(Novelty:19,Relation:4.2)
To the conventional text mining system using management of a document, and **search** as a base, "TRUETELLER" serves as powerful marketing-oriented systems, such as crossing analysis with a customer attribute, and the feature analysis of the subject according to customer segment, rather than carries only text data.

Powerful SCORE = 56(Novelty:24,Relation:2.3)
Of course, in addition to having brought about expansion of a customer's channel, it can utilize now as a **powerful** operating support tool of the operating person in charge who moves in many cases in outside.

Classification SCORE = 55(Novelty:18,Relation:3.1)
The mining tool of the Komatsu software represented by text mining in the analysis work of the huge text information (qualitative data) from customers who depended on experience of a specialist and skill until now, such as a questionnaire and a claim, character and when carrying out efficiently supports reference of these information, a **classification**, and visualization powerfully by advanced document processing technology.

Visualization SCORE = 51(Novelty:18,Relation:2.8)
By document reference, the excavation of the customer needs which were not found is enabled through extraction, classification, relating, **visualization**, etc. of an important keyword.

Mining SCORE = 40(Novelty:19,Relation:2.1)
An efficient form shares document information and practical use and re-creation of "knowledge" are supported in a company.

Enormous SCORE = 39(Novelty:20,Relation:2.0)
Exact analysis of a **enormous** data collected from various channels by mapping a keyword visually is supported.

Data SCORE = 39(Novelty:19,Relation:2.0)
Text mining is the technology which analyzes a lot of documents and discovers a tendency and the feature, draws the direction of an inquiry, and a kind from the record exchanged with the customer, and can be using them for a prompt action, creation of FAQ, etc. in the call center.

Fig. 17.2. Search results for relational topics

$$\mathrm{Wei}(t) = \begin{cases} t + 6 \ (-12 \le t \le -7), \\ t + 7 \ (-6 \le t \le -1), \end{cases} \tag{17.3}$$

$$\mathrm{Time}(w) = \frac{\sum_{k=-12}^{-1} \mathrm{Hit}(w,k)\mathrm{Wei}(k)}{\mathrm{Ave}(w)}. \tag{17.4}$$

Example sentences for relational topics do not simply include relational words, but also those most closely related to search keywords. In other words, those topic sentences are summaries of Web pages where the search keyword is a viewpoint. The panoramic view system [17.8] is applied to this summarization.

By using this interface, users may discover new contexts within the keywords. Generally speaking, a word has various meanings, and a word is used in various contexts if the meanings are the same. Those contexts imply ideas, and if a user thinks of its applications, a discovery is expected to occur. In summary, it is important for us to store good-quality information to connect many possibilities to a single event, and it is also necessary to take concrete chances that may lead to success where users clarify their ideas within the scope of application to society.

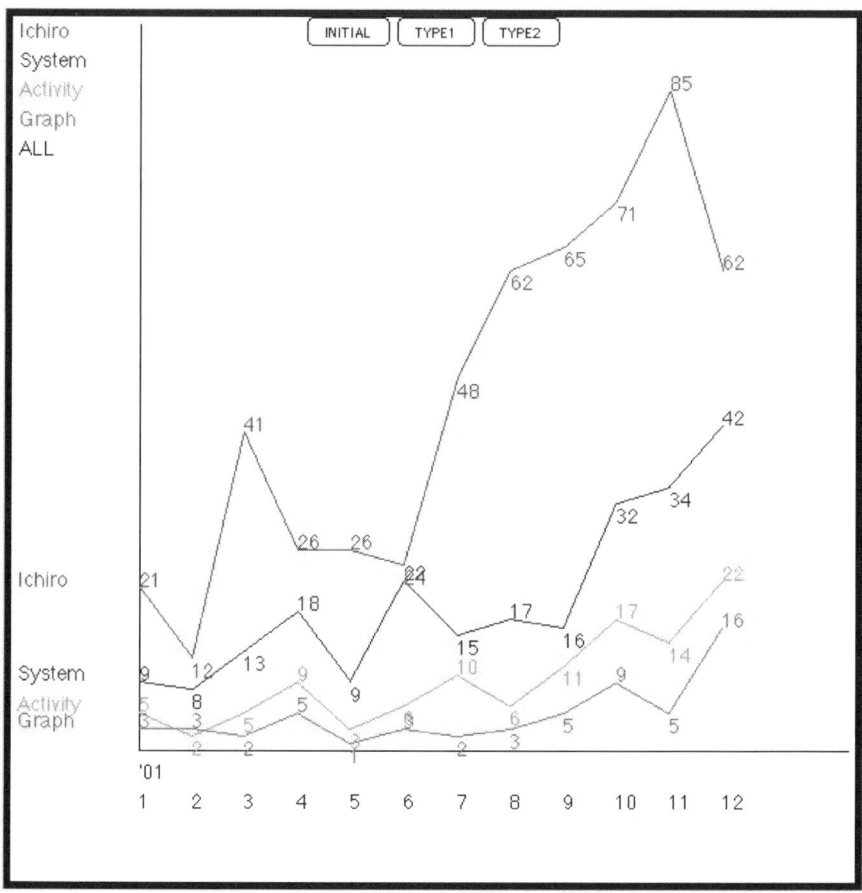

Fig. 17.3. Time-sequential viewer

17.4.4 Interface 4: Time Series of the Hit Numbers

Figure 17.3 shows line graphs expressing transitions of hit numbers for each month relating to words appointed by a user. In this interface, features of the two-dimensional plane are as follows:

– Order by directions: a high numbers on the graph approach the top. Time goes from left to right.
– Cleared relationship: graphs and search keywords are distinguished by colors.

This interface uses information necessity, adaption, and effectiveness. This graph shows time transitions of hit numbers. Inclination of the time transitions expresses information adaption, and also expresses effectiveness of its hit number on the vertical axis. These multiple graphs are useful for comparing to other keywords, as users can readily see relative adaption and effectiveness.

Users may be able to discover trends of the world by using this interface. When a season comes, information peculiar to the season will grow, and if an event occurs or a popular drama begins, information related to the event of the drama will begin to increase. Chance discovery is strongly related to being in and out of fashion. Users sensitive to the world would catch trends earlier than others. As information on the WWW exists at the time a user searches and goes network surfing, it is difficult to grasp the time transition of all the information simply by looking at Web pages. Therefore, we would like to take a chance for succeeding in the future by discovering biases based on time transitions from both modern and past information.

17.5 Conclusions

In this chapter, I described the criteria for information selection and two-dimensional interfaces for application. We require multi-sided analysis for shrinking enormous amounts of data, and we should understand visualized outputs by intuition for acquiring new ideas. Visualization methods will be heavily relied upon for aiding humans' comprehension by intuition, and promoting the urge for finding biased outputs. Such a system will provide clues for discovering chances.

For chance discovery, we should know some backgrounds and trends of the world via good information, and apply our potential and original ideas to the world as we see it. Thus, each aspect of chance discovery should be evaluated by effective critera to realize a chance discovery system.

References

17.1 http://www.google.com/

17.2 Khonen T (1990) The Self-Organizing Map, Proceedings of the IEEE, 78(9): 1464 – 1480

17.3 Lamping J, Rao R, Pirolli P(1995) A Focus+Context Technique Based on Hyperbolic Geometry for Visualizing Large Hierarchies, Proceedings of the ACM Conference on Human Factors in Computing Systems (CHI'95), pp.401–408

17.4 Ohsawa Y (2002) KeyGraph as Risk Explorer from Earthquake Sequence, Journal of Contingencies and Crisis Management, 10(3):119–128

17.5 Rocchio J (1971) Relevance Feedback in Information Retrieval, The SMART Retrieval System - Experiments in Automatic Document Processing, Prentice Hall Inc., Englewood Cliffs, NJ, pp.313 – 323

17.6 Skupin A (2002) A Cartographic Approach to Visualizing Conference Abstracts, IEEE Computer Graphics and Applications, 22(1):50 – 58

17.7 Sunayama W, Ohsawa Y, Yachida M (1999) Computer Aided Discovery of User's Hidden Interest for Query Restructuring, Lecture Notes in Artificial Intelligence 1721, Springer, Heidelberg, Germany, pp.68 – 79

17.8 Sunayama W, Yachida M (2002) Panoramic View System for Extracting Key Sentences based on Viewpoints and an Application to a Search Engine, Soft Computing Systems, 87: 863 – 870

18. *KeyGraph*: Visualized Structure Among Event Clusters

Yukio Ohsawa[1,2]

[1] PRESTO, Japan Science and Technology Corporation, 2-2-11 Tsutsujigaoka, Miyagino-ku, Sendai, Miyagi 983-0852, Japan
[2] Graduate School of Business Sciences, University of Tokyo, 3-29-1 Otsuka, Bunkyo-ku, Tokyo 112-0012, Japan
email: osawa@gssm.otsuka.tsukuba.ac.jp

Summary.

The most fundamental causes may be hidden and in severe cases unknown (not in the knowledge of a human nor a computer). These causal events might be occurring eternally, or be brought up from a sequence in the past and trigger events in the future. Here is presented *KeyGraph*, generalized from a document-indexing method to a method for extracting essential events and the causal structures among them from an event sequence.

18.1 *KeyGraph* for Abstracting Causalities in a Sequence

An event sequence in the real world may include causes and effects, and some events may be observable. Typically, the most fundamental causes are hidden (not observable) and in severe cases unknown (not in the knowledge of a human nor a computer). These causal events might be occurring eternally, or be brought up from a sequence in the past and trigger events in the future. I present *KeyGraph* [18.5] here, generalized from a document-indexing method, to a method for extracting essential events and the causal structures among them from an event sequence. Suppose text (string sequence) D is given, describing an event sequence sorted by time, with inserted periods ('.') corresponding to the moments of major changes. For example, let text D be:

$$D = 123\# \ 202\# \ 1\# \ 84\#. \ 76\#. \ 216\# \ 1\# \ 202\# \ 84\#. \ 249\# \ 84\#. \ 76\# \ 249\# \ ... \ (18.1)$$

Here, '$m\#$' means a known and observed event. Because it is a known event, it has an already defined ID number m. Here, 123# and 202# are events No. 123 and No. 202, which are the events occurring in the first and the second moments of observation. This example includes four '.'s because four major changes occurred. The definition of 'major changes' depends on the data domain. In the case of a document, periods are put on the end of each sentence. In the case of sales (position of sales: POS) data, periods are put in the end of each basket. Then *KeyGraph*, composed of the following steps *KeyGraph-step* 1 to *KeyGraph-step* 2, is applied to D. Let us introduce the outline of the algorithm of *KeyGraph* below; for details the reader is referred to [18.4].

[The algorithm summary of *KeyGraph*]

***KeyGraph-step* 1:** clusters of co-occurring frequent items (words in a document, or events in a sequence) are obtained as basic clusters, called *islands* . That is, items appearing many times in the data (e.g. the word '202#' in (18.1)) are extracted, and each pair of items occurring often in the same sequence unit (a *sentence* in a document, a bought set of items in each basket in sales data, etc.) is linked to each other, e.g. '202# - 1# - 84#' for (18.1). Each connected graph of those links and items forms one island, implying the existence of a common cause for the occurrence of belonging items.

***KeyGraph*-step 2**: items not so frequent as the ones in islands but co-occurring with multiple islands, e.g. '249#' in (18.1), are obtained as *hubs*. We can regard hubs as candidates of *chances*, i.e. items significant (assertions in a document, or latent demand in a supermarket dealing with POS data) with respect to the structure of relations among items.

18.2 The Extensible Semantics of *KeyGraph*

Table 18.1 shows the five major components of *KeyGraph*. We can apply *KeyGraph* to various data, where the components in Table 18.1 correspond to meaningful substances in the target world.

Table 18.1. Factors considered in *KeyGraph*

(1)	Event sequence D
(2)	Periods ('.'s): the moments of major changes
(3)	Islands: fundamental set of items, co-occurring frequently
(4)	Bridges: event–island co-occurrences representing causalities
(5)	Hubs: potentially significant events connecting multiple islands

Table 18.2. A document and a sequence of items bought in a store, each as a special case for *KeyGraph*: items for the same number correspond to each other, and to Table 18.1.

Case of a document:	
(1)	Actions of the author, i.e. writing words
(2)	Periods ('.'s): the ends of sentences
(3)	Basic concepts for the author
(4)	The flow of content, connecting basic concepts and assertions
(5)	The relations of asserted words to basic concepts
Case of POS data:	
(1)	The sequence of purchases of items by customers
(2)	The ends of baskets, i.e. of receipts
(3)	Fundamental set of items to be bought together
(4)	The combination of multiple fundamental sets of items, to make a certain life-style
(5)	New products attracting various types of customers

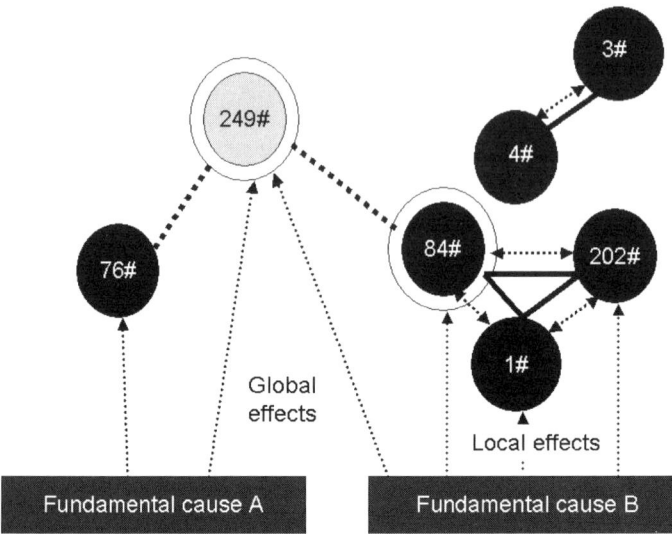

Fig. 18.1. An example of *KeyGraph*: circular nodes and thick (solid and dotted) lines show the output of *KeyGraph*, and other parts depict its causal semantics. Three islands are obtained from *D* in (18.1), each including event set 84#, 202#, 1#, 3#, 4#, and 76#. The double-circled nodes show the hub, connecting the two islands (one of #76 and the other of 84#, 202#, and 1# by thick dotted bridges)

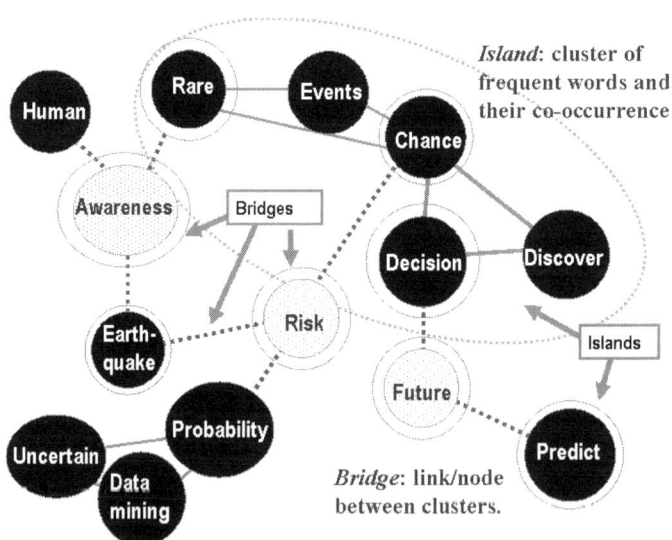

Fig. 18.2. An example output of *KeyGraph*. The black nodes representing frequent words are connected by solid lines and form clusters called islands, with underlying common contexts. Hubs in shadowed nodes, e.g. 'awareness', showing infrequent words are connected to multiple islands by dotted lines, showing rare but possibly essential words

18.3 *KeyGraph* Applied to a Document

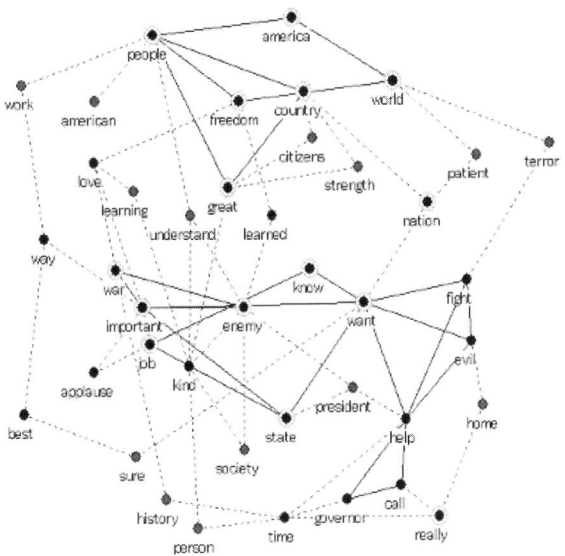

Fig. 18.3. *KeyGraph* applied to a document of speech

Given the definitions of nodes and links in *KeyGraph*, can the reader guess who made the speech which resulted in Fig. 18.3? The figure has two big islands, one including 'America', 'freedom', 'great country' representing the cheerful side of America. On the other hand, the figure also includes a large island having 'evil', 'enemy', 'war', etc. representing dark forces for the country.

Other small islands, 'best', 'really', etc. seem to be less relevant to the concept of a country than the words in the two large islands. 'love' is also separated from the large islands, but rather near to the upper island of cheerful concepts. Going between these islands, red lines connect things via the words in red nodes, e.g. 'terror', 'understand', 'learned' are connecting the cheerful and the dark sides. As the reader has already noticed, the source of this result is the speech of George Bush, in May, 2002. If the speech was made earlier than September 11, 2001, 'terror' would not have appeared. If the speech was made just after September 11, 2001, the words 'learning' and 'understand' would not have appeared, because the president was asserting that people on both sides learned and understood many things from the new occurrence of terrorism. 'Terrorism' is the keyword without saying, but it is hard to guess what 'understanding' and 'learning' meant in the speech if we are told just that these words are rare but significant. By looking at the structured graph, however, we can see the semantic position of these words in the overall document.

In the case of an interview, the document becomes more complex: many questions are asked and the interviewee answers in various contexts responding to the interviewers in one session. This tends to make it difficult to summarize the overall discourse. However, one interviewee should have a certain coherent motto (although it may have some self-inconsistencies) on which s/he makes all answers. For the interview session of George Bush on September 18, 2002, the result in Fig. 18.4 was obtained. According to the graph, the session had one large island made of 'nations', 'threat', 'fact', etc. surrounded by 'defiance', etc. meaning the looming threat of the next terrorism. On the other hand, the small islands on the left-hand side of the figure are connected, although weakly by dotted bridges, to make a loose cluster meaning the appreciations by Bush to his colleagues who have been fighting against terrorists. Going between the large island on the right and the cluster on the left, 'defying', 'ploy', 'latest', meaning the latest ploy and defiance of somebody is threatening the nation of the USA. In this way, *KeyGraph* aids the user in clarifying the semantic structure of discourse, even in complex cases such as discussions or conversations involving multiple participants.

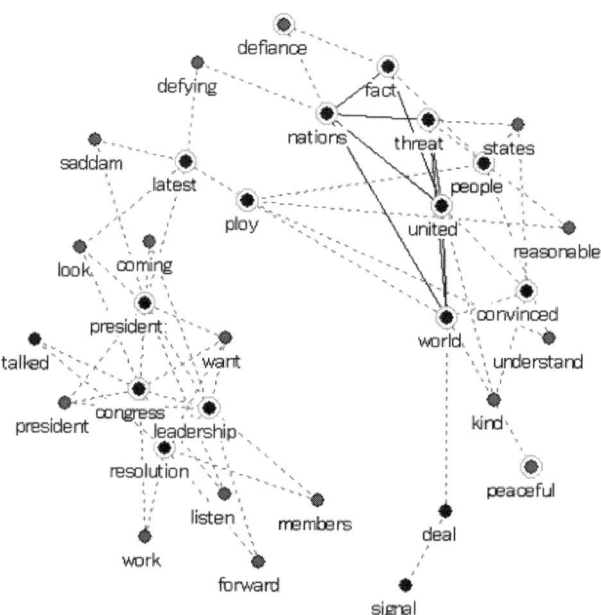

Fig. 18.4. The result of *KeyGraph* for the interview by George Bush on September 18, 2002

18.4 *KeyGraph* Applied to POS Data

This section deals with POS data, just as an exemplification of *KeyGraph* applied to a sequence of events. If the reader would like to read further about marketing applications, s/he is referred to Chap. 24 by Mizuno. Let us have such data as:

> **[POS data 1]** <2002.6.2> beer cigar magazine baby-diaper news-paper.
> <2002.6.9> cigar baby-diaper news-paper.
> <2002.6.22> food milk news-paper.
> <2002.6.23> beer food news-paper.
> <2002.7.7> magazine baby-diaper toy news-paper.
> <2002.7.21> cigar magazine beer baby-diaper news-paper.
> <2002.7.28> beer magazine food news-paper.
> <2002.8.25> food milk news-paper.
> <2002.9.1> food news-paper.
> <2002.9.8> food milk news-paper.
> <2002.9.22> food beer toy news-paper.
> <2002.9.23> cigar magazine beer news-paper.

Let us deal with this POS data of a single customer of a supermarket (by using a point-card, i.e. the card for each customer, a store can acquire this kind of sequence in their system). Figure 18.5 was obtained as the result of using *KeyGraph* on this data. You see 'toy' in a separated island from the large main island. Easily you can find that the dense cluster in the large island is meaningless for the marketing of the supermarket. For example, the links from 'food' seem meaningless because 'food' is linked to 'news-paper' (should be 'newspaper' but POS data usually has minor errors), 'milk', and 'beer' – commodities to be used in various contexts. From this,

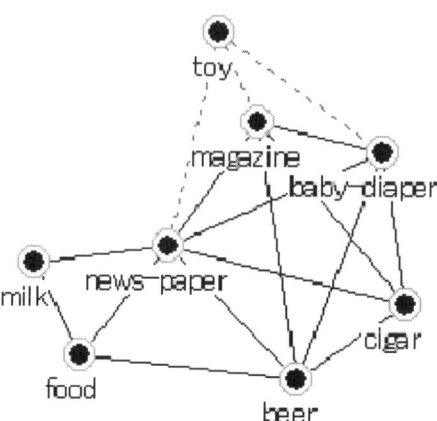

Fig. 18.5. Result for POS data 1

we should regard food as too large a category as it is involved in many situations of customers. It is clear that we should separate 'food' into subcategories.

In the next step we change the data as follows by classifying food into details as 'fish', 'meat', 'potato-chips', 'fruits', and 'vegetable' and apply *KeyGraph* to the following data.

[POS data 2]
<2002.6.2> beer cigar magazine baby-diaper news-paper.
<2002.6.9> cigar baby-diaper news-paper.
<2002.6.22> vegetable meat fruits milk fish news-paper.
<2002.6.23> beer potato-chips news-paper.
<2002.7.7> magazine baby-diaper toy news-paper.
<2002.7.21> cigar magazine beer baby-diaper news-paper.
<2002.7.28> beer magazine potato-chips fish news-paper.
<2002.8.25> vegetable meat milk fish news-paper.
<2002.9.1> vegetable fish news-paper.
<2002.9.8> fruits milk meat news-paper.
<2002.9.22> fruits beer toy news-paper.
<2002.9.23> cigar magazine beer news-paper.

In the new result in Fig. 18.6, you find 'news-paper' appears in the center, strongly relevant to everything in the data. This is apparently meaningless, even though newspaper really co-occurs with everything when the customer buys anything in the store. The fact, as guessed from the output, it should be that the customer often comes in to buy a newspaper. Even worse, such a common item bought constantly can not be a chance for making this customer a loyal one, i.e. one who spends much money in the shop, because it does not mean any change of his/her

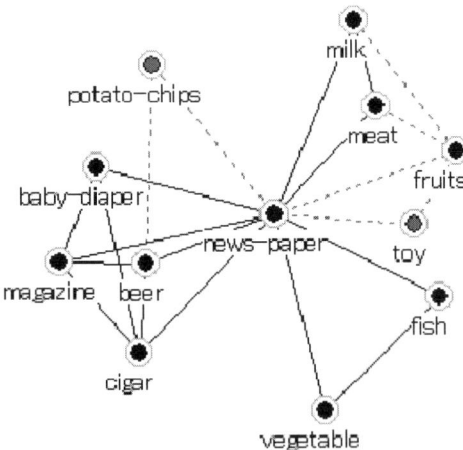

Fig. 18.6. 'Food' is now separated, but ... is newspaper so essential?

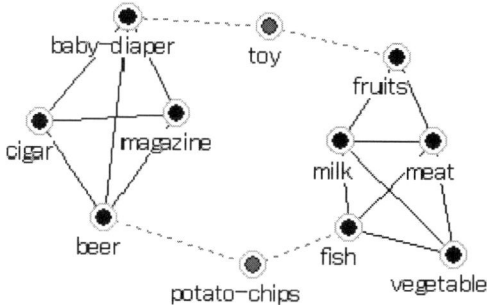

Fig. 18.7. The final result of *KeyGraph* for the case of POS data

behavior. Finally, we discarded 'news-paper' from the data and obtained *KeyGraph* in Fig. 18.7. The revision of results with processing of the data in hand, as in this example, can be done easily by replacing the items in the data reflecting the rising concerns of the user. As far as we know, *KeyGraph* has been dealing with megabyte data of sales, as well as a large document as a book and data of earthquakes for 10 years as in Chap. 22.

The correspondence analysis [18.2], on the other hand, has been used for data of the form as has been dealt with by *KeyGraph*. In the correspondence analysis, the co-occurrences among items are given as in the form of D in (18.1) and the results can be shown graphically as in Fig. 18.8. The links among nodes are not shown, but their positions in the $X - -Y$ field reflect the distances between each pair of nodes, where each distance is computed by the weakness of the co-occurrence of each pair of nodes. For the user, its differences from *KeyGraph* are:

1. The information to be expressed by the two dimensions is $2N$, where N is the number of nodes, where the links in *KeyGraph* represent $N \times N$ relations between nodes. As a result, some relations disappear from the output figure in the case of correspondence analysis.
2. Nodes going between multiple clusters are not clear in the correspondence analysis method, for the reason in 1.

For example, we can find that 'potato-chips' is going between two contexts, one of food consumption by a family and the other of a snack with beer and a magazine. It is not a rare event in a supermarket that a male customer comes in to buy a baby-diaper requested by his wife, with his beer and cigar. This is typical behavior of a young husband thinking about his family. That is, the left-hand side of Fig. 18.7 represents the feeling of amusement with light snacks of a young husband, and the right-hand side of the same figure shows things for a family as requested by his wife. In both contexts, the husband thinks about his baby and buys toys to have amusement with his child. That is, 'potato-chips' and 'toys' play a role of go-between of the left- and the right- hand sides of Fig. 18.7.

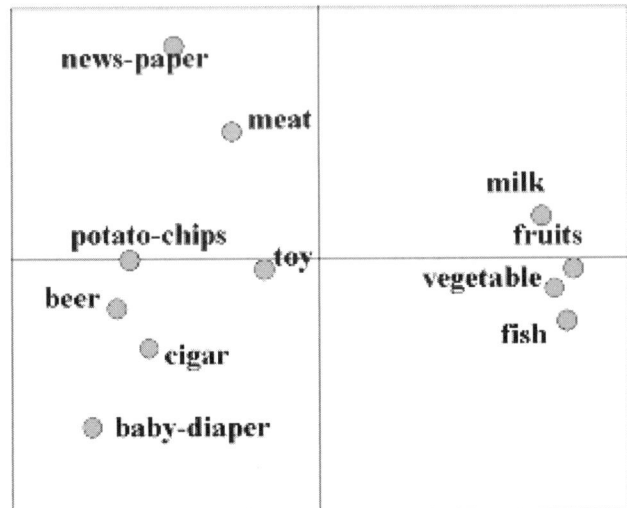

Fig. 18.8. The result of correspondence analysis for the same POS data as Fig. 18.7

On the other hand, the left half and the right half of Fig. 18.8 are separated without such a bridge as formed by 'toys' and 'potato-chips' and the figure rather looks like those of two customers, one a housewife and the other a single man. Even worse, the node 'potato-chips' is hidden in the left-hand cluster, not between the two clusters, because of the reason in 2. above. As a result, a user of correspondence analysis realizes only the existence of each cluster independently and hardly understands the relations between multiple clusters. Those users usually try to guess what the two (X and Y) axes mean, but this is often quite hard. In the case of Fig. 18.8, what do you readers think the axis X means?

In *KeyGraph*, as the tool we implemented in both Windows and Linux, we can set the values of various elements such as the *noise* items to be ignored from the data, parameters such as the number of nodes and links to appear in the window, and so on. These can be flexibly changed following a user's demand during the process of refining the output of *KeyGraph* as from Fig. 18.5 to Fig. 18.7. An important thing for a user is to use it flexibly using this interface. That is, a user should keep in mind the following know-how.

[What the user of *KeyGraph* should do]

1. First, look at the overall structure of *KeyGraph*. One should take care of all clusters and the connections between them, rather than looking at each island independently of other islands. Note: some users pay attention only to the hubs, i.e. the nodes between clusters. However, it is required to see the relation of hubs to neighboring islands, in order to understand the meaning of a hub in the overall data. For example, you can not understand what 'toys' means without looking at its connection to islands. You can not even guess what 'understand' means in Fig. 18.2 if you do not see the two large islands.

2. If the connection of multiple clusters means a possibility that you can merge the neighboring islands, imagine what it means to combine those islands. For example, in Fig. 18.7, the combination of the two large islands may imply the lifestyle of average husbands who can form a major market of a supermarket.

3. Note and record things in the figure that catch your attention.

4. If you find any newly interesting thing, take data relevant to the new interest (or just a weak concern). Go into the details of significant points you find by gathering necessary data.

   [What the user of *KeyGaph* should *not* do]

5. Pay too strong attention to a list of keywords/items, sorting by the significance. For example, the number of bridges connected from an item might seem significant. However, the information carried by *KeyGraph* is the binary (link of $N \times N$ pairs) relations of items rather than the one-dimensional (N at the largest) order in the list.

6. The extreme trust in automatic discovery of *KeyGraph*: some hubs might be really significant, but others are not. A user, instead of a machine, should finally judge the significance of each node by associating the output figure to the experiences s/he has.

For details of applying *KeyGraph* to marketing data, the reader is referred to [18.7]. In the remainder of this book, examples in natural science, marketing, social survey, etc., are given as applications of chance discovery. In some of these chapters, *KeyGraph* is used as an aiding tool, rather than an automatic tool of discovery. The reader is encouraged to see the interaction of the user and such data-mining tools.

18.5 Application to Web Links for Discovering Emerging Topics

By focusing on the analogy between a document and other textual data (data formed by readable letters), *KeyGraph* can be applied to a variety of topics. For example, *KeyGraph* has been adopted to

- finding areas with the highest risks of near-future earthquakes from data of observed past earthquakes [18.4],
- getting timely files from a visualized structure of one's working history [18.6],
- planning to guide concept understanding on the WWW [18.9]
- as well as to the data of event sequences. Here let us show a new application of *KeyGraph* – for discovering emerging social interests from the Web.

In a document D, high-frequency terms are used for expressing typical basic concepts, and term-pairs that frequently occur in the same sentences mean strong association throughout D. In this section, we extend the use of *KeyGraph* to another kind of data, i.e. a Web-page set (corresponding to D, a document) including Web pages (each corresponding to a sentence) having URL links, each corresponding to a word. That is, high-frequency links (which are the URLs pointing to other Web pages) in a collection W of Web pages show popular Web pages, and link-pairs which frequently occur in the same Web pages show strong relations, i.e. the co-citation, in W. Our fundamental hypothesis here is that the co-occurrence of terms in a document and the co-citation of Web pages are common in that both carry the underlying important shared context, and our strategy for applying *KeyGraph* is based on this analogy.

Speaking more formally, a Web page including a set of links to other pages (each URL put as $u1, u2, \ldots$) is translated to a sentence and a set of pages is given as a document as a whole in (18.2).

$W1$: $u1, u2, u3, \ldots$,

$W2$: $u1, u3, u5, \ldots$, (18.2)

$W3$: $u3, u7, u4, \ldots$.

A document is formed by combining (virtual) sentences as in (18.2), for each Web page in a collection. By this translation, we can obtain a 'document' reflecting the link structure of the WWW.

For example, Fig. 18.9 depicts the result of *KeyGraph*, where islands (solid lines and their touching black nodes), bridges (dotted lines), and hubs (double circles) are obtained. Some infrequent nodes can be obtained as a hub as depicted by p. 8. In this case, an island corresponds to an established Web community because each node shows a well-cited Web page and each link shows a strong tie among nodes in an island. Our aim is to detect significant emerging topics from Web pages (i.e. the node of p. 8 in Fig. 18.9) relevant to multiple established communities based on the assumption that Fig. 18.9 reflects the structure of the real world. We can expect that a graphical output of *KeyGraph* helps in understanding potential interests and the underlying relation between them, and leads us to the understanding of the structure of the interests of people in the real human society. This is a realization of looking at weak ties between strongly tied communities [18.1].

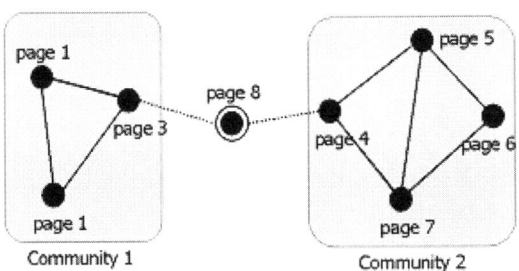

Fig. 18.9. An overview of *KeyGraph* for Web-page set

Experimental Examples and Discussion

We report our experiments where we applied *KeyGraph* to two sets of collections C_A and C_B, each containing 500 popular Web pages obtained by Google for the input query 'human genome', to follow the changes of the communities with time. The difference between the collections (i.e. documents) is the date: C_A

is obtained on November 26, 2000, and C_B is on March 11, 2001. After C_A and C_B were translated into two documents, the graphical outputs were obtained as in Fig. 18.10 and in Fig. 18.11 respectively. In the figures, the single-circle and double-circle nodes show *islands* and *hub* pages respectively, and links among nodes show *bridges*. For example, the hubs HGRI (http://www.nhgri.nih.gov), NCBI (http://www.ncbi.nlm.nih.gov), and Sanger (http://www.sanger.ac.ul) appear in both results. NHGRI (National Human Genome Research Institute) is one of 24 institutes, centers, or divisions that make up the National Institute of Health (NIH), the federal government's primary agency for the support of biomedical research. NCBI (National Center for Biotechnology Information) is also one of the institutes of NIH. Sanger (The Sanger Institute) is a research center to provide a major focus in the UK for mapping and sequencing the human genome, and genomes of other organisms.

These are the most contributing institutes for the Human Genome Project, an international scientific effort to map and sequence the 3 billion genetic codes, involving more than 1000 scientists from five countries (China, France, Japan, the UK, and the USA). On the other hand, the hubs such as GSC (http://genome.wustl.edu), TIGR (http://www.tigr.org), Celera (http://www.celera.com), and CNN (http://www.cnn.com) appear only in Fig. 18.11. GSC (The Genome Sequencing Center) is a leading contributor to the Human Genome Project and TIGR (The Institute for Genomic Research) is a center conducting large-scale human genome sequencing.

Let us focus on Celera (The Celera Genomics), an ambitious venture corporation sequencing the human genome from September, 1999. As you can see from Fig. 18.10 and Fig. 18.11, the situation around Celera changes dramatically from November, 2000, to March, 2001, i.e. Celera began to be supported by the cluster of the Human Genome Project and CNN, among the world's leaders in online news and information delivery. Looking back over the real events and situations, we can understand the leap of Celera.

In the field of the human genome, revolutionary events occurred in 2000 and 2001: The Human Genome Project team and Celera announced the completion of the draft sequence of the human genome in June, 2000, and subsequent articles were published in *Nature* [18.3] and *Science* [18.8] in February, 2001. Both are the most important milestones for the human genome analysis. In fact, J. Craig Venter, president and chief scientific officer of Celera, and Francis S. Collins, director of the Human Genome Project were celebrated by US President Bill Clinton and British Prime Minister Tony Blair for the progress of the human genome analysis at the White House on June 26, 2000.

Considering these real events and situations, the changes of the structures shown by Fig. 18.10 and Fig. 18.11 (e.g. Celera grew to be widely supported) are considered to reflect the real society.

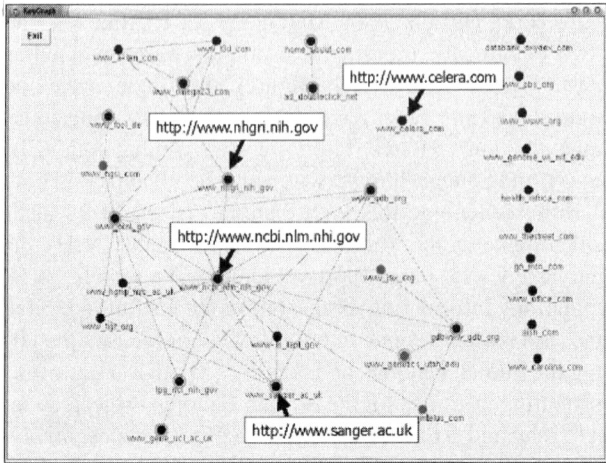

Fig. 18.10. The output of *KeyGraph* for the input query 'human genome' (November 26, 2000). We can recognize a large cluster, composed of the most contributed institutes for the Human Genome Project as NCBI (http://www.ncbi.nlm.nih.gov), NHGRI (http://www.nhgri.nih.gov), Sanger (http://sanger.ac.ul), etc. Celera (http://www.celera.com) is isolated from the big island

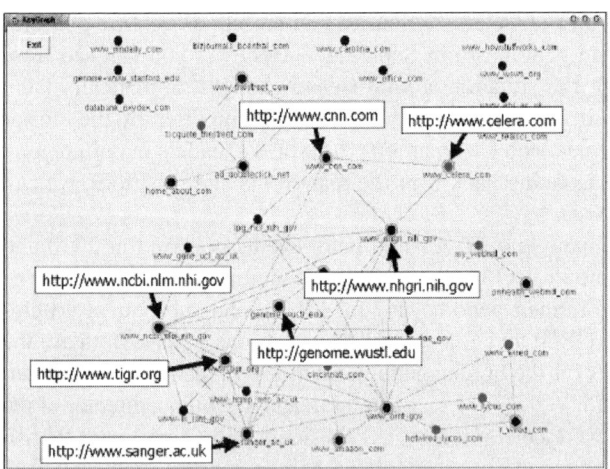

Fig. 18.11. The graphical output of *KeyGraph* for the input query 'human genome' (March 11, 2001). The major Web pages of the large cluster are almost the same as the cluster in Fig. 18.10, i.e. the Human Genome Project cluster. However, Celera (http://www.celera.com) began to be supported by multiple clusters, i.e. the Human Genome Project cluster and CNN (http://www.cnn.com), the mass-media cluster

18.6 Conclusions

In this chapter we introduced a method for aiding the understanding of significant and novel, i.e. emerging, events by visualizing their relation to people's familiar, i.e. frequent, sets of events. Here, the method *KeyGraph* is extended from a visual summarization method of a document to an event sequence. The applications shown here are to the speech of the US president, conversation, POS data, and Web pages linked to surrounding pages. In the current and future work, the interface and the algorithm of *KeyGraph* are being improved to be applicable to quick discoveries of essential chances.

References

18.1 Granovetter M (1973) *Strength of Weak Ties*, American Journal of Sociology, 8: 1360–1380

18.2 Greenacre MJ (1984) *Theory and applications of correspondence analysis.* Academic Press, London, UK

18.3 International Human Genome Sequencing Consortium (2001) *Initial sequencing and analysis of the human genome*, Nature, 409: 860–921

18.4 Ohsawa Y, Benson NE, Yachida M (1998) *KeyGraph: Automatic Indexing by Co-occurrence Graph Based on Building Construction Metaphor*, Proceedings of Advances in Digital Libraries Conference, pp. 12–18

18.5 Ohsawa Y, Yachida M (1999) *Discover Risky Active Faults by Indexing an Earthquake Sequence*, Proceedings of Discovery Science, Springer Verlag, Heidelberg, Germany, pp. 208–219

18.6 Ohsawa Y (1999) *Get Timely Files from Visualized Structure of Your Working History*, Proceedings of the 3rd Knowledge-Based Intelligent Engineering Systems & Allied Technologies, pp.546–549

18.7 Ohsawa Y, Fukuda H (2002) *Potential Motivations and Fountains of Chances*, Journal of Contingencies and Crisis Management, 10(3):129–138

18.8 Venter JC et al (2001) *The Sequence of the Human Genome*, Science, 291:1304–1351

18.9 Yamada S, Osawa Y (2000) *Navigation Planning to Guide Concept Understanding in the World Wide Web*, Proceedings of Autonomous Agents, pp. 114–115.

19. Discovering Deep Building Blocks for Competent Genetic Algorithms Using Chance Discovery via *KeyGraphs*

David E. Goldberg[1], Kumara Sastry[1], and Yukio Ohsawa[2]

[1] Illinois Genetic Algorithms Laboratory (IlliGAL),
Department of General Engineering,
University of Illinois at Urbana-Champaign, Urbana, IL 61 801, USA
email: {deg,ksastry}@uiuc.edu

[2] The Laboratory of Chance Discovery,
Graduate School of Business Sciences,
University of Tsukuba, Otsuka, Tokyo, Japan
email: osawa@gssm.otsuka.tsukuba.ac.jp

Summary.

In this chapter, we see whether chance discovery in the form of *KeyGraphs* can be used to reveal deep building blocks for competent genetic algorithms(GAs), thereby speeding up innovation in particularly difficult problems. On an intellectual level, showing the connection between *KeyGraphs* and genetic algorithms as related pieces of the innovation puzzle is both scientifically and computationally interesting. GAs represent that aspect of human innovation that tries to innovate through the exchange or *cross fertilization* of notions contained in different ideas; the *KeyGraph* procedure represents that portion of human innovation that pays special attention to and interprets *salient fortuitous events*. The chapter goes beyond mere conjecture and performs pilot studies that show how *KeyGraphs* and competent GAs can work together to solve the problem of deep building blocks; the work is promising and steps toward a practical computational combination of the two procedures are suggested.

19.1 Introduction

This volume brings together some of the latest work in *chance discovery*, and one of the things that makes chance discovery interesting relative to more common modes of machine learning and data mining is its emphasis on rare and possibly opportune events. Chance discovery is increasingly used [19.32, 19.34] to mine text and other data for the connection of rare events to more statistically regular events in the hopes of teasing out a meaningful trend. For example, in consumer marketing, the early behavior of market-leading innovative consumers can be unmasked with chance discovery techniques and exploited by savvy companies [19.32].

On the other hand, going back almost 20 years [19.10], genetic algorithms(GAs)–search procedures based on the mechanics of natural selection and natural genetics–have been likened to mechanistic versions of certain modes of human innovation , and a recent monograph [19.16] details the innovation connection, facet-wise theory, and algorithmic mechanism that make this connection tighter than ever.

Here, in this chapter, we bring these two seemingly disparate ideas together and wonder whether chance discovery in the form of *KeyGraphs* [19.35] can be used to reveal deep building blocks for competent genetic algorithms , thereby speeding up innovation in particularly difficult problems. On an intellectual level, showing the connection between *KeyGraphs* and genetic algorithms as related pieces of the innovation puzzle is both scientifically and computationally interesting. More will be said about the angle on innovation science as the paper progresses, but, briefly, GAs represent that aspect of human innovation that tries to innovate through the exchange or *cross fertilization* of notions contained in different ideas; the *KeyGraph* procedure represents that portion of human innovation that pays special attention to and interprets *salient fortuitous events*. This understanding of GAs and *KeyGraphs* is interesting on the face of it, but the linking of one sort of innovation to the other is a particular contribution of this work. Of course the paper goes beyond mere conjecture and performs pilot studies that should lead to an effective computational combination of the two procedures in the near future.

This chapter is structured as follows. We start with brief introductions to genetic algorithms, the innovation intuition, and competent GAs. The discussion of competent GAs leads to the identification of the *problem of deep building blocks*. This is followed by an introduction to *KeyGraphs*, and then the use of *KeyGraphs* as a possible solution to the problem of deep building blocks is discussed. Pilot studies of using *KeyGraphs* to discover deep building blocks in GAs are outlined, and preliminary experiments are presented and discussed. We conclude with a discussion of a number of future research directions and some discussion on the more immediate utility of these results.

19.2 GAs: Innovation, Competence, and Deep Building Blocks

This chapter[1] briefly reviews what genetic algorithms are, how they may be connected to human innovation, the recent program for making GAs more *competent* or scalable, and the problem of deep building blocks.

19.2.1 The One-Minute Genetic Algorithmist

Genetic algorithms are search procedures based on the mechanics of natural selection and genetics. In this section we briefly review what GAs are, their mechanics, and their power of effect.

Suppose we are seeking a *solution* to some *problem*. The first thing we must do to apply a genetic algorithm to that problem is *encode* it as an artificial *chromosome* or chromosomes. These artificial chromosomes can be strings of 1s and 0s,

---

[1] Portions of this section are excerpted from the first author's new book, *The Design of Innovation: Lessons from and for Competent Genetic Algorithms* (Kluwer Academic Publishers, Boston, MA 2002) and are copied with the permission of Kluwer Academic Publishers.

parameter lists, permutation codes, or even complex computer codes, and one of the surprises found over the last decade is how well selectionist schemes do when faced with widely different codings.

Another thing we must do in solving a problem is to have some means or procedure for discriminating good solutions from bad solutions. This can involve the usual elaborate computer simulation or mathematical model that helps determine what good is (the standard notion of an *objective* function), or it can be as simple as having a human intuitively choose better solutions over worse ones (what we might call a *subjective* function). It can even be an ecology-like process where different digital species *co-evolve* through an intricate mix of competition and cooperation. However it is done, the idea is that *something* must determine a solution's relative *fitness to purpose* (or *context*), and whatever that is will be used by the genetic algorithm to guide the evolution of future generations.

Having encoded the problem in a chromosomal manner and having devised a means of discriminating good solutions from bad ones, we prepare to *evolve* solutions to our problem by creating an initial *population* of encoded solutions. The population can be created randomly or by using prior knowledge of possibly good solutions, but either way a key idea is that the GA will search from a population, not a single point.

With a population in place, *selection* and *genetic operators* can process the population iteratively to create a sequence of populations that hopefully will contain more and more good solutions to our problem as time goes on. There is much variety in the types of operators that are used in GAs, but quite often (1) *selection*, (2) *recombination*, and (3) *mutation* are used.

Simply stated, selection allocates more offspring to better individuals; this is the survival-of-the-fittest mechanism we impose on our solutions. It can be accomplished in a variety of ways. Weighted roulette wheels can be spun, local tournaments can be held, various ranking schemes can be invoked; but, however we do it, the main idea is to *prefer better solutions to worse ones*. Of course, if we were only to choose better solutions repeatedly from the original database of initial solutions, we would expect to do little more than fill the population with the best of the first generation. Thus, simply selecting the best is not enough, and some means of creating new, possibly better individuals must be found; this is where the genetic mechanisms recombination and mutation come into play.

Recombination is a genetic operator that *combines bits and pieces of parental solutions* to form new, possibly better offspring. Again, there are many ways of accomplishing this, and achieving competent performance does depend on getting the recombination mechanism designed properly; but the primary idea to keep in mind is that the offspring under recombination will not be identical to any particular parent and will instead *combine parental traits in a novel manner*. By itself, recombination is not all that interesting an operator, because a population of individuals processed under repeated recombination alone will undergo what amounts to a random shuffling of extant traits.

Where recombination creates a new individual by recombining the traits of two or more parents, mutation acts by simply modifying a single individual. There are many variations of mutation, but the main idea is that the offspring be identical to the parental individual except that *one or more changes are made to an individual's trait or traits* by the operator. By itself mutation represents a 'random walk' in the neighborhood of a particular solution. If done repeatedly over a population of individuals, we might expect the resulting population to be indistinguishable from one created at random.

19.2.2 An Innovation Intuition for GAs

The previous section briefly described the mechanics of a genetic algorithm, but it gives us little idea of why these operators might promote a useful search. To the contrary, individually we saw how the operators acting alone were ineffectual, and it is something of an intellectual mystery to explain why such individually uninteresting mechanisms acting in concert might together do something useful. Starting in 1983 [19.10], the first author developed what has been called the *innovation intuition* to help explain this apparent mystery. Specifically, the innovation intuition likens the processing of selection and mutation together and that of selection and recombination taken together to *different facets of human innovation*, what are here called the *improvement* and *cross-fertilizing* types of innovation. We start first with the combination of selection and mutation and continue with the selection recombination pair.

Selection + mutation = continual improvement. When taken together, selection and mutation are a form of hill-climbing mechanism, where mutation creates variants in the neighborhood of the current solution and selection accepts those changes with high probability, thus climbing toward better and better solutions. Human beings do this quite naturally; in the literature of total quality management, this sort of thing is called *continual improvement* or, as the Japanese call it, *kaizen*. Others have had similar thoughts, for example the British author and politician Bulwer-Lytton[19.1, p. 118]:

> "Invention is nothing more than a fine deviation from, or enlargement on a fine model. ... Imitation, if noble and general, insures the best hope of originality."

Although this qualitative description is distant from an algorithmic one, we can hear the echo of mutation and selection within these words. Certainly, continuing to experiment in a local neighborhood is a powerful means of improvement, although it will have a tendency to be fairly local in scope, unless a means can be found for intelligently jumping elsewhere when a locally optimal solution is found.

Selection + recombination = innovation. One way of promoting this kind of intelligent jumping is through the combined effect of selection and recombination, and we can start to understand this if we liken their effect to that of the processes of human cross-fertilizing innovation. What is it that people do when they are being innovative in a cross-fertilizing sense? Usually they are grasping at a notion–a set of good- solution features–in one context, and a notion in another context and juxtaposing them, thereby speculating that the combination might be better than either notion taken individually. Again, the first author's thoughts on the subject were introspective ones, but others have written along similar veins, for example, the French mathematician Hadamard [19.23, p. 29]:

> "We shall see a little later that the possibility of imputing discovery to pure chance is already excluded. ... Indeed, it is obvious that invention or discovery, be it in mathematics or anywhere else, takes place by combining ideas."

Likewise, the French poet–philosopher Valéry had a similar observation, according to Hadamard [19.23, p. 30].

> "It takes two to invent anything. The one makes up combinations; the other chooses, recognizes what he wishes and what is important to him in the mass of the things which the former has imparted to him."

Once again, verbal descriptions are far from our more modern computational kind, but something like the innovation intuition has been clearly articulated by others.

With a basic understanding of the mechanics of genetic algorithms and an intuitive understanding of the innovation–GA connection, we now consider key aspects of the theory and practice of competent GAs.

19.2.3 Competent Genetic Algorithms

The last few decades have witnessed great strides toward the development of so-called *competent* genetic algorithms – GAs that solve hard problems, quickly, reliably, and accurately [19.15]. Competent GAs are a two-edged sword. From a computational standpoint, the existence of competent GAs suggests that many difficult problems can be solved in a scalable fashion. Moreover, it suggests that the usual casting about for a good coding or a good genetic operator that accompanies many GA applications can be minimized; if the GA can adapt to the problem, there is less reason for the user to have to adapt the problem, coding, or operators to the GA. From the standpoint of human innovation, the existence of competent GAs is tantamount to the creation of a computational theory of cross-fertilizing innovation. This has scientific ramifications that have yet to be explored.

In this section we briefly review some of the key lessons of competent GA design. Specifically, we restrict the discussion to selectorecombinative GAs and focus on the cross-fertilization type of innovation and briefly discuss key facets of competent GA design.

Using Holland's notion of a building block (BB) [19.27], the first author proposed decomposing the problem of designing a competent selectorecombinative GA

[19.13, 19.14, 19.17, 19.20]. This design decomposition has been explained in detail elsewhere [19.16], but is briefly reviewed in what follows.

Know that GAs process building blocks (BBs): the primary idea of selectorecombinative GA theory is that genetic algorithms work through a mechanism of *decomposition* and *reassembly*. Holland [19.27] called well-adapted sets of features that were components of effective solutions *building blocks* (BBs). The basic idea is that GAs (1) implicitly identify building blocks or subassemblies of good solutions, and (2) recombine different subassemblies to form very high performance solutions.

Understand BB hard problems: from the standpoint of cross-fertilizing innovation, problems that are hard have BBs that are hard to acquire. This may be because the BBs are deep or complex, hard to find, or because different BBs are hard to separate, or because low-order BBs may be *misleading*, or in other words *deceptive* [19.11].

Understand BB growth and timing: another key idea is that BBs or notions exist in a kind of competitive *market economy of ideas*, and steps must be taken to ensure that the best ones (1) grow and take over a dominant market share of the population, and (2) the growth rate can neither be too fast, nor too slow.

Understand BB supply and decision making: one role of the population is to ensure adequate *supply* of the raw building blocks in a population [19.27, 19.12, 19.21]. Randomly generated populations of increasing size will, with higher probability, contain larger numbers of more complex BBs. Furthermore, decision making among different, competing notions (BBs) is *statistical* in nature, and, as we increase the population size, we increase the likelihood of making the best possible decisions.

Identify BBs and exchange them: perhaps the most important lesson of current research in GAs is that the *identification and exchange of BBs* is the critical path to innovative success. First-generation GAs usually fail in their ability to promote this exchange reliably. The primary design challenge to achieving competence is the need to identify and promote effective BB exchange.

Efforts in principled design of effective BB identification and exchange mechanisms have led to the development of competent genetic algorithms.

Interestingly, the mechanics of competent GAs vary widely, but the principles of innovative success are invariant. Competent GA design began with the development of the *messy genetic algorithm* [19.19], culminating in 1993 with the *fast messy GA* [19.18]. Since those early scalable results, a number of competent GAs have been constructed using different mechanism styles:

Perturbation techniques include the fast messy GA (fmGA) [19.18], gene-expression messy GA (GEMGA) [19.28], linkage identification by non-linearity check GA (LINC GA), and linkage identification by monotonicity detection GA (LIMD GA) [19.31].

Linkage adaptation techniques such as linkage learning GA (LLGA)   [19.26, 19.25, 19.5].

Probabilistic model-building techniques such as the extended compact GA (ECGA) [19.24], iterated distribution estimation algorithm (IDEA) [19.4], Bayesian optimization algorithm (BOA), and hierarchical Bayesian optimization algorithm (hBOA) [19.38]. For further details on probabilistic model building GAs, the interested reader should look elsewhere [19.39, 19.37, 19.30].

The breadth of capability of competent GAs is difficult to fathom. The usual idea of optimization is restricted to fairly straightforward notions such as hill climbing and enumeration, but competent GAs zero in on global or near-global solutions in problems that are massively multi-modal, are noisy, and have difficult hierarchical and misleading non-linearities, because they have been constructed to face adversarially designed problems. This gives them striking capability to penetrate difficult problems with astonishing rapidity. Nonetheless, a more careful reading of the decomposition exposes an important Achilles heel of the method, one shared by many data-mining procedures and meta-heuristics. This difficulty is explored in the next section.

19.2.4 The Problem of Deep Building Blocks

Despite the progress made in competent GA design, certain types of problems can require solution times that thwart the best of known competent GAs. To better understand this we consider some details of competent GA population sizing.

The decomposition of competent GA design of the last section examined population sizing in GAs as largely an issue of building block supply and decision making. Regardless of the viewpoint one takes, the simple fact is that high-order building blocks require exponentially larger population sizes. That is, regardless of whether the population size is governed on supply grounds or decision grounds, population sizes n must grow as an exponential function of the building block order k required:

$$n \propto 2^k. \tag{19.1}$$

For small k, this requirement is no problem, but for larger values it is a deal breaker. Without delving into too many specifics, the usual population-sizing equation multiplies the BB-wise exponential term by a function that goes up as a polynomial function of the number of building blocks involved. The end result is a prohibitively expensive, yet theoretically tractable computation, for large-k or what here we shall term *deep* building blocks.

To put some flesh on the bones of our argument, consider a population sizing that goes as $n = cm2^k$, where m is the number of building blocks, k is the order of the deepest building block that must be revealed whole, and c a constant that varies with other problem parameters. Here we take $c = 1$ without loss of generality. We assume that we are solving a modestly sized problem with $m = 100$ building blocks, and compare population sizes required to solve problems with building blocks of order $k = 1, 5, 30$. The resulting population sizes are 100, 3200, and 1.07×10^{11} (or about 107 billion!). Clearly, the need to discover even modestly deep building blocks whole is a difficult task, and ways need to be found to expedite the solution of such problems.

In a later section, we engage chance discovery through *KeyGraphs* to try to do just that. In the next section, we briefly review the *KeyGraph* calculation to better understand how it might help us in this endeavor.

19.3 The *KeyGraph* Procedure: Overview and Intuition

The *KeyGraph* procedure is a graphical method for data mining originally developed for indexing a document [19.35, 19.33], and recently utilized for chance discovery [19.32, 19.34] including discovery of risky active faults of earthquakes [19.36]. Like other text- and data-mining algorithms, *KeyGraph* identifies relationships between terms and term clusters in a document. In particular, *KeyGraph* focuses on co-occurrence relationships, but one thing that sets *KeyGraph* apart is its emphasis on both high-probability and low probability events. The feature that attracts us to *KeyGraph* in this paper is its identification of low-probability terms and their linkage to high-probability terms and clusters of terms. The idea here is to see if *KeyGraph* can identify deep building blocks, and if so, under what conditions.

Here, we review the basic *KeyGraph* algorithm. We start by putting forth an innovation intuition for *KeyGraphs*, and we continue by reviewing the steps of *KeyGraph* processing. The next section follows up this discussion by suggesting how we might use the *KeyGraph* procedure to discover deep building blocks.

19.3.1 An Innovation Intuition for the *KeyGraph* Procedure

Previously, we likened the processing of genetic algorithms to different facets of human innovation, facets we called cross fertilizing and continual improvement. Here we make an analogous connection for *KeyGraphs*. In particular, we suggest that the *KeyGraph* procedure is a computational analog of the human tendency to mine and interpret *salient fortuitous events*.

The literature of the history of science and technology has lavished considerable attention on the role of chance events in the discovery of new ideas [19.3, 19.42, 19.8]. The Archimedes' bathtub moment, the discovery of penicillin, and the invention of post-it notes were all accompanied by a Eureka moment in which the scientist or inventor paid attention to a chance occurrence that seemed odd, but interesting. We will call such events *salient fortuitous events*, and we suggest that one of the reasons that *KeyGraphs* (and other chance discovery computations) are interesting is because they focus on the *unexpected* and how it is linked to that which is regular and well anticipated.

We will examine the details of the *KeyGraph* procedure in a moment, and the first part of the procedure has much in common with other data- and text-mining techniques. The interesting part of *KeyGraph* for our purposes (and for most chance discovery applications) is its emphasis of *rare events*. In the usual chance discovery application of *KeyGraphs*, everyone usually agrees upon the statistically significant data. The value of the procedure comes from interpretation of the low-probability events that link clusters of high-probability events. In other words, *KeyGraph* users usually find that the statistically regular terms (so-called *high-frequency terms*) need no discussion or interpretation. Instead, practitioners come to understand the need to focus on the *key terms* and try to understand how they may be exploited in the particular application.

Clearly we are using this innovation intuition for *KeyGraphs* in our study. In particular, we are attempting to use *KeyGraphs* to point out deep building blocks as salient fortuitous events. If we are successful, we will be able to solve problems with deeper building blocks more easily.

19.3.2 The *KeyGraph* Procedure: an Overview

Here we assume that a document, D, is composed of sentences and each sentence is composed of words. The main steps of the *KeyGraph* algorithm can be outlined as follows.

Document preprocessing, which consists of two tasks:
1. **Document compaction:** insignificant words from the document are removed using a user-supplied list of words and word stems. Word stems are used to reduce related words to the same root. For example, words like 'innovate', 'innovates', and 'innovating' are reduced to 'innovate' using a method proposed by Porter [19.41].
2. **Phrase construction:** here preference is given to longer phrases with higher frequency. A subset of ℓ_{phrase} words are chosen from the document and all possible phrases out of those words are constructed. A phrase that occurs with the highest frequency in the document is retained.

It should be noted that in this study we don't use any preprocessing. After preprocessing, if any, the document D is reduced to D', which consists of unique terms $w_1, w_2, \ldots, w_\ell$, where a term w_i refers to either a word or a phrase.

Extracting high-frequency terms: terms in D' are sorted by their frequencies in the document D and top n_{nodes} (high-frequency terms) are retained. These high-frequency terms are represented as nodes in a graph G. A set of the high-frequency terms is denoted by N_{hf}.

Extracting links: links represent *co-occurrence*–term-pairs that often occur in the same sentence. A measure for co-occurrence of terms w_i and w_j is defined as

$$\mathrm{assoc}\,(w_i, w_j) = \sum_{s \in D} \min\left(\left|w_i\right|_s, \left|w_j\right|_s\right), \tag{19.2}$$

where w_i and w_j are elements of the set N_{hf}, and $\left|w_i\right|_s$ is the number of times a term w_i occurs in a sentence s.

The assoc values are computed for all pairs of high-frequency terms in N_{hf}. The term-pairs are sorted according to their assoc values and top $N_{\mathrm{hf}} - 1$ tightly associated term-pairs are taken to be the links. The links between term-pairs are represented by the edges in G.

Extracting key terms: key terms are terms that connect clusters of high-frequency terms together. To measure the tightness with which a term w connects a cluster, the following function is defined:

$$\mathrm{key}(w) = 1 - \prod_{g \subset G} \left[1 - \frac{\mathrm{based}(w,g)}{\mathrm{neighbors}(g)} \right], \tag{19.3}$$

where g is a cluster, and

$$\mathrm{based}(w,g) = \sum_{s \in D} |w|_s |g - w|_s, \tag{19.4}$$

$$\mathrm{neighbors}(w) = \sum_{s \in D} \sum_{w \in s} |w|_s |g - w|_s. \tag{19.5}$$

$$|g - w|_s = \begin{cases} |g|_s - |w|_s, & \text{if } w \in g, \\ |g|_s & \text{if } w \notin g, \end{cases} \tag{19.6}$$

where $|g|_s$ is the number of times a cluster g occurs in a sentence s.

Qualitatively, $\mathrm{key}(w)$ gives a measure of how often a term w occurs near a cluster of high-frequency terms.

The *key* values are computed for all the terms in D and n_{key} top *key* terms are taken as *high-key terms*. These high-key terms are added as nodes–if they are not already present–in G and are elements of a set K_{hk}.

Extracting key links: For each high-frequency term $w_i \in N_{\mathrm{hf}}$ and each high-key term $w_j \in K_{\mathrm{hk}}$, $\mathrm{assoc}\,(w_i, w_j)$ is calculated. Links touching w_j are sorted by their assoc values for each high-key term $w_j \in K_{\mathrm{hk}}$. A link with highest assoc values connecting w_j to two or more clusters is chosen as a key link. Key links are represented by edges–if they are not already present–in G.

Extracting Keywords: nodes in G are sorted by the sum of assoc values associated with the key links touching them. Terms represented by nodes of higher values of these sums than a certain threshold are extracted as keywords for the document D.

More details are available in Chap. 18 of this book and in other references [19.35, 19.36, 19.9, 19.33]. In the next section, we examine how *KeyGraph* may be helpful in solving the problem of deep building blocks.

19.4 Can *KeyGraphs* Discover Deep Building Blocks?

We have likened selectorecombinative GAs to cross fertilizing innovation and we have likened the *KeyGraph* procedure to keeping an eye open for salient fortuitous events. Here we wonder whether we can use the *KeyGraph* procedure as a preprocessor to competent GAs to help identify deep building blocks in a population, where the deep BBs are not yet present in large numbers. This is an important question, because most GAs explicitly or implicitly assume that good building blocks are present in sufficiently large numbers that they constitute a statistically significant

sample. As discussed earlier, this assumption results in population-sizing requirements that grow exponentially with deep (high-order) BBs. From the standpoint of
a GA, the quest for less-costly methods of identifying potentially useful deep BBs
takes us from a point where searching for solutions to problems with deep building
blocks might be intractable to one where it may be practical.

In this section, we explain the design of a series of experiments to test the use
of the *KeyGraph* procedure in this manner. The results from those experiments are
also presented. As it turns out, the results are mixed, but they are interesting, and
they lead to a number of reasonable continuations of this work.

19.4.1 Design of Pilot Experiments

Our approach in designing pilot experiments for investigating the BB identification capability of *KeyGraph* is to *design* bounding *adversarial problems* that exploit
one or more dimensions of problem difficulty. Particularly, our pilot test problems
should possess the following properties:

- Building-block identification should be critical for successful innovation. That is,
  if the BBs of the problem are not identified and exchanged, it should be impossible to attain the global solution. We are interested in solving such a class of
  search problems where BB identification is essential for a cross-fertilization type
  of innovation.
- Building-block structure and interactions of the problem should be known to the
  researchers, but not to the problem solver (search method). Knowing the BB
  structure and their interactions a priori, makes it easy to analyze if a BB-identification technique–be it a competent GA, or *KeyGraph*–has successfully discovered individual building blocks. Furthermore, ensuring that the BB-identification
  methods work on such problems provides assurance that they would also identify
  BBs of real-world problems, where the BBs are not known a priori.
- The properties such as building-block size problem difficulty, should be tunable
  without significantly changing the function. For example, we should be easily
  able to change a building block of low order into a deep building block, by changing one or two parameters.

This adversarial and systematic design method contrasts sharply with the common
practices of using historical, randomly generated, or ad hoc test functions [19.16].

One such class of adversarial problems is the additively separable, *deceptive*
function [19.11]. Deceptive functions are designed to thwart the very mechanism
of selectorecombinative search by punishing any localized hill climbing and requiring mixing of whole building blocks at or above the order of deception. Using such
adversarially designed functions is a stiff test–in some sense the stiffest test–of algorithm performance. The idea is that if an algorithm can be an adversarially designed
test function, it can solve those problems and any easier than the adversary. Furthermore, if the building blocks of such deceptive functions are not identified and
respected by a selectorecombinative GA, then they almost always converge to the
local optimum.

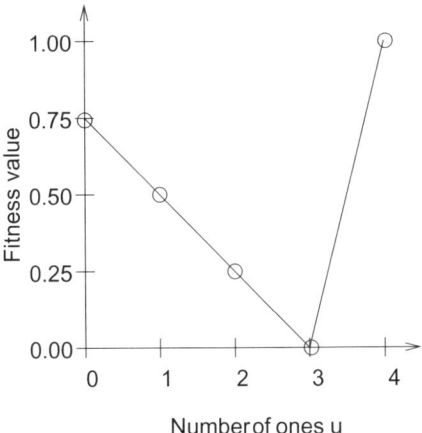

Fig. 19.1. A fully deceptive trap function with $k = 4$ and $\delta = 0.25$

In this study, we use deceptive *trap* functions [19.6, 19.7]. A fully deceptive k-bit trap function is defined as follows:

$$f\left(u(x_1, x_2, \ldots, x_k)\right) = \begin{cases} 1.0 & u = k, \\ (1.0 - \delta)\left(1 - \frac{u}{k-1}\right) & u < k, \end{cases} \tag{19.7}$$

where u is the unitation, or the number of ones in a portion of the bit string:

$$u\left(x_1, x_2, \ldots, x_k\right) = \sum_{i=1}^{k} x_i, \tag{19.8}$$

where x_i is the value of the ith bit and δ is the difference in the functional value between the good BB (all ones) and its deceptive attractor (all zeros). The difficulty of a trap function can be adjusted by modifying the values k and δ. The problem becomes more difficult as the value of k is increased and that of δ is decreased. A 4-bit deceptive trap function is illustrated in Fig. 19.1.

Given that the string length is ℓ, an additively separable test function is defined as

$$F\left(x\right) = f\left(u(x_1, x_2, \ldots, x_k)\right) + f\left(u(x_{k+1}, \ldots, x_{2k})\right) + \cdots$$
$$+ f\left(u(x_{(m-1)k+1}, \ldots, x_{mk})\right), \tag{19.9}$$

where F is the fitness function, x is the bit string (individual), and m is the number of BBs.

The important feature of additively separable trap functions is that if the good BB (all ones) in any particular partition is not identified, then the GA tends to converge to the deceptive attractor (all zeros) in that partition. Therefore, BB identification and mixing is critical to innovation success. Furthermore, notice that the problems are of bounded difficulty $k < \ell$. If the trap was of length ℓ, nothing could work better than enumeration or random search without replacement. However, given that the difficulty is bounded, a GA that identifies BBs and mixes them well has the opportunity to solve the problem in polynomial time [19.16].

In this study, we utilize two test problems:

1. **The 3×4 problem**: this problem consists of three uniformly scaled low-order building blocks (BB size, $k = 4$). That is, the 3×4 problem consists of three 4-bit fully deceptive trap functions ($m = 3$, $k = 4$):

$$F_{3 \times 4}(x) = f\left(u(x_1, x_2, x_3, x_4)\right) + f\left(u(x_5, x_6, x_7, x_8)\right)$$
$$+ f\left(u(x_9, x_{10}, x_{11}, x_{12})\right). \tag{19.10}$$

   Here, all the BBs are of low order and therefore they have statistically significant market share in a typically (subquadratically or linearly) sized population. In a randomly generated initial population, each building block will have a market share of approximately $(1/2^4)$th or 6.25% of the population.

2. **The $2 \times 4 + 12$ problem**: this problem consists of two uniformly scaled building blocks of low order (BB size, $k = 4$) and a deep building block (BB size, $k = 12$). That is, the $2 \times 4 + 12$ problem is composed of three deceptive trap functions, two of which have low-order BBs of size $k = 4$, and the other is a deep BB of size $k = 12$:

$$F_{2 \times 4 + 12}(x) = f\left(u(x_1, x_2, x_3, x_4)\right) + f\left(u(x_5, x_6, x_7, x_8)\right)$$
$$+ f\left(u(x_9, x_{10}, \ldots, x_{20})\right). \tag{19.11}$$

   Even though the regular BBs have a market share of 6.25% in the population, the deep BB has a market share of only $(1/2^{12})$th, or 0.024% in the population. Therefore, the probability of a deep BB occurring is insignificantly small in a typically sized population. Moreover, if a deep BB does not occur in a population, we have almost no chance of creating it from smaller pieces, because the very nature of a deceptive problem is to encourage the emphasis of the *wrong* smaller pieces. Yet, when a deep BB does occur in very few numbers (one or two individuals in the population of a thousand individuals have the deep building block), we need to identify it.

Here, for the sake of better visualization and understanding, we consider relatively short string lengths (12–20 bits) and modest BB counts (three BBs). By way of comparison, modern competent GAs can solve problems of bounded difficulty with hundreds and even thousands of decision variables to global optimality with high reliability and accuracy. Our use of modest string lengths and BB counts is justified by our immediate goal of understanding the possible role of *KeyGraphs* in solving the problem of deep BBs for competent GAs. Additional work will be necessary to more fully integrate *KeyGraphs* and working competent GAs if we are successful, but the pilot studies here are important first steps along the way to a working hybrid.

In the subsequent sections, we discuss the following pilot studies. First, we test the capability of competent GAs in identifying deep BBs. We consider two state-of-the-art competent GAs, the Bayesian optimization algorithm (BOA), and the extended compact genetic algorithm (ECGA). Next, we investigate whether *Key-Graphs* can identify low-order building blocks. Finally, we analyze the potential of *KeyGraph* in deep BB identification.

19.4.2 Competent GAs Fail to Identify Deep Building Blocks

Previous studies on competent GAs have demonstrated that they are highly capable of identifying and exchanging BBs of low order, and often do so in polynomial (usually subquadratic) time [19.18, 19.37, 19.24, 19.2, 19.29, 19.43]. However, their capability in identifying deep building blocks, to the best of our knowledge, has not been investigated. Therefore, it was logical for us to test if the competent GAs could identify deep BBs. Specifically, we investigated the deep-BB-identification capability of two state-of-the-art competent GAs–the Bayesian optimization algorithm (BOA) [19.39], and the extended compact genetic algorithm (ECGA) [19.24].

We used both BOA and ECGA to solve the $2 \times 4 + 12$ problem. The BOA and ECGA procedures employed can be broadly outlined as follows:

1. Initialize the population randomly.
2. Evaluate the fitness values of individuals in the population.
3. Perform selection to make more copies of better individuals. Selection biases the population towards highly fit individuals and therefore facilitates the identification of BBs.
4. Construct the probabilistic model of the selected individuals (Bayesian networks in the case of BOA and marginal product models (MPMs) in the case of ECGA). The probabilistic model refers to the building-block structure and interaction of the search problem.
5. Sample new individuals based on the probabilistic model.
6. Repeat steps 2–5 till one or more convergence criteria is satisfied.

The population size in each case was chosen in accordance with appropriate population-sizing models [19.40, 19.43]. However, in determining the population size, we assumed a problem with low-order BBs (5×4 problem). The population sizes resulting from the above assumption are less than that is required to ensure the presence of a significant number of raw deep BBs in the population. However, they are large enough to ensure a statistically significant presence of low-order BBs. Therefore, the competent GAs should be able to identify building blocks of low order without much difficulty.

If the deep BB of the $2 \times 4 + 12$ problem is correctly identified, then the individuals in the population should converge to the global optimum and the average fitness of the population should be 3.0 (fitness value of the global solution). On the other hand, if the deep BB is not identified then the population converges to the local optimum with a fitness value of 2.75.

First, we used BOA and ECGA to solve the $2 \times 4 + 12$ problem on a randomly initialized population, where the chances of the deep BB occurring in the population are slim to none. The performance of BOA and ECGA on the $2 \times 4 + 12$ problem is shown in Fig. 19.2. The figure plots average fitness of the population as a function

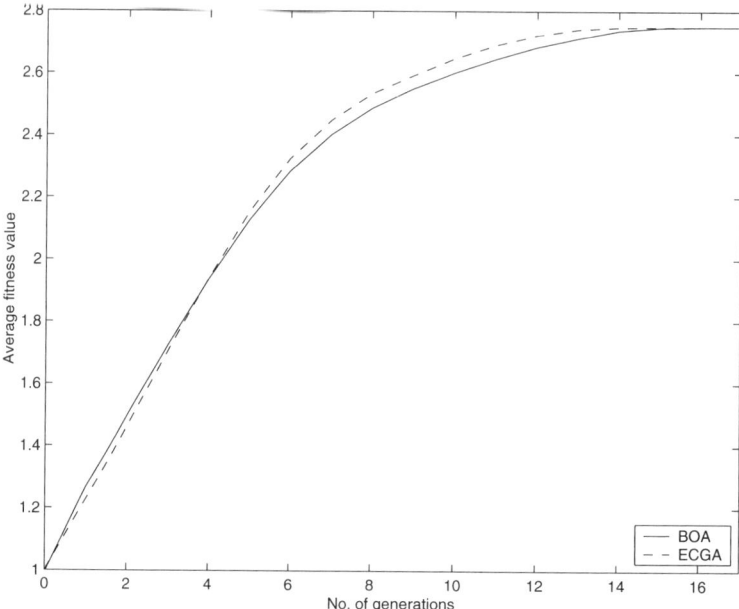

Fig. 19.2. Convergence of BOA and ECGA on the $2 \times 4 + 12$ problem. The average fitness of the population is plotted as a function of number of generations. The results are averaged over 50 independent runs. The results show that both BOA and ECGA fail to identify the deep BB and therefore converge to the local optimum (fitness value 2.75) in all 50 runs. Similar results are obtained even when the population is injected with a few copies of the deep BB

of the number of generations. The results are averaged over 50 independent runs. Results indicate that both BOA and ECGA, while successful in identifying low-order BBs, fail to discover deep BBs.

Usually in a typically sized (non-exponential) random population, there are no individuals with the deep BB, and, due to the deceptive nature of the $2 \times 4 + 12$ problem, we have no chance of creating the deep BB from smaller pieces. This makes it impossible for competent GAs to identify the deep BB. Therefore, a more interesting (and perhaps a more fair) question is to ask ourselves whether the competent GAs can identify the deep BB, if it is present in the population in small numbers (one or two individuals in a population of one or two thousand).

Therefore, we inject a couple of individuals into the selected population (after step 3 of the experimental procedure) in the first generation. We randomly choose two individuals from the selected population and for each individual replace the 12 bits, x_9–x_{20}, with the best BB (all ones). The remaining 8 bits (x_1–x_8) of both individuals are not modified during the injection. Note that the deep BBs are injected only once–after the first selection–in a GA run. The doping procedure ensures that the deep BB is at least present–albeit in very small numbers–in the population during the BB identification and exchange phase of the competent GAs.

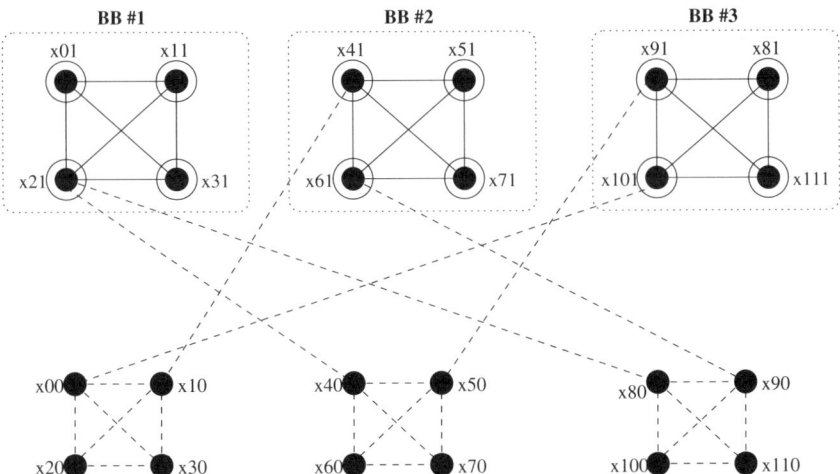

Fig. 19.3. *KeyGraph* results of three 4-bit trap functions. *KeyGraph* clearly identifies BBs as a cluster of high-frequency terms. The deceptive attractors are identified as weak clusters

Similar to the randomly initialized population, the competent GAs discovered the low-order BBs but failed to identify the deep BB. The injected deep BBs were lost in the subsequent–usually one or two–generations. Similar results were obtained even when as high as 5% of the population was doped with the deep BB.

The results presented in this section clearly demonstrate that both BOA and ECGA fail to identify the deep BB and that they lose the deep BB even when the population is doped with a very few number of individuals with the deep BB. Therefore, we need to utilize other techniques to identify deep BBs when they occur in very few numbers in the population. One possible technique for so doing is chance discovery through *KeyGraph*, which is discussed in the following sections.

19.4.3 *KeyGraph* and Identification of Low-Order Building Blocks

The previous section demonstrated that the competent GAs are unable to identify deep BBs when they are present in very few numbers in the population. One of the primary objectives of this study is to investigate the potential of employing *Key-Graph* for the identification and discovery of deep BBs. Though *KeyGraph* has been successful in solving a wide range of problems related to chance discovery and data mining, to the best of our knowledge, it has not been applied to identify BBs of a search problem. Therefore, we need to test whether *KeyGraph* can first identify low-order BBs. If *KeyGraph* is unable to identify BBs of low order, which is a much easier task than identifying deep BBs, then it would be less likely to succeed in identifying deep BBs. Therefore we tested the capability of *KeyGraph* in identifying building blocks of the 3×4 problem.

The algorithmic procedure employed in using *KeyGraph* for identifying the building blocks of the 3×4 problem is as follows:

1. Initialize the population randomly.
2. Evaluate the fitness values of individuals in the population.
3. Perform tournament selection [19.19] to make more copies of better individuals. Selection biases the population towards highly fit individuals and therefore facilitates the identification of BBs.
4. Convert the population of selected individuals into a document with a sentence referring to an individual and each word representing a bit value. For example, the following individual 10101100 is converted to 'x01 x10 x21 x30 x41 x51 x60 x70.'. This step is necessary because *KeyGraph* operates on a text document containing sentences constructed of words and phrases. A straightforward method of translating bit strings (individuals) is to convert each bit of an individual into a word, and each individual forms a sentence in the text document.
5. Run *KeyGraph* on the text document for building-block identification.

The above procedure refers to a single generation of a GA run. Therefore, the success of *KeyGraph* depends on whether it can identify BBs in the first generation. This requirement imposed on *KeyGraph* can be stringent, however: if the BBs are not identified in the early stages of a GA run, it becomes difficult, if not impossible, to identify BBs in the later stages of a GA run.

The results of *KeyGraph* on the 3×4 problem are shown in Fig. 19.3. In the 3×4 problem, each of the building blocks has a fair market share in the initial population (6.25% of the population). Selection further increases the market share of the BBs. Therefore, *KeyGraph* should identify each BB as a cluster of high-frequency terms. From Fig. 19.3, we can clearly see that *KeyGraph* indeed identifies individual BBs as a cluster of high-frequency terms (bits). Furthermore, attached to each BB is a cluster (with dashed lines) of the deceptive attractor of the other two BBs. The deceptive attractor is also identified by *KeyGraph* because, in the initial stages of the GA run, many individuals of the population get one BB right and the others are the deceptive attractors. Therefore, if an individual has the first BB, most probably the second and third partitions would have the deceptive attractor. Results shown in Fig. 19.3 indicate that *KeyGraph* is not only effective in identifying the BBs, but also can be used to visualize a GA population. It should be noted that results on larger problems (up to $m = 10$) were qualitatively similar to Fig. 19.3.

In this section, we demonstrated that *KeyGraph* is capable of identifying low-order building blocks using a simple 3×4 problem. Next, we discuss deep-BB identification using *KeyGraphs*.

19.4.4 *KeyGraph* and Identification of Deep Building Blocks: a Naïve Approach

The previous section showed that *KeyGraph* was indeed capable of identifying the building blocks of low order. This section investigates whether the *KeyGraph* procedure is capable of identifying deep BBs. In particular, we consider the $2 \times 4 + 12$ problem. As mentioned in Sect. 19.4.2, the chances of creating a deep BB from

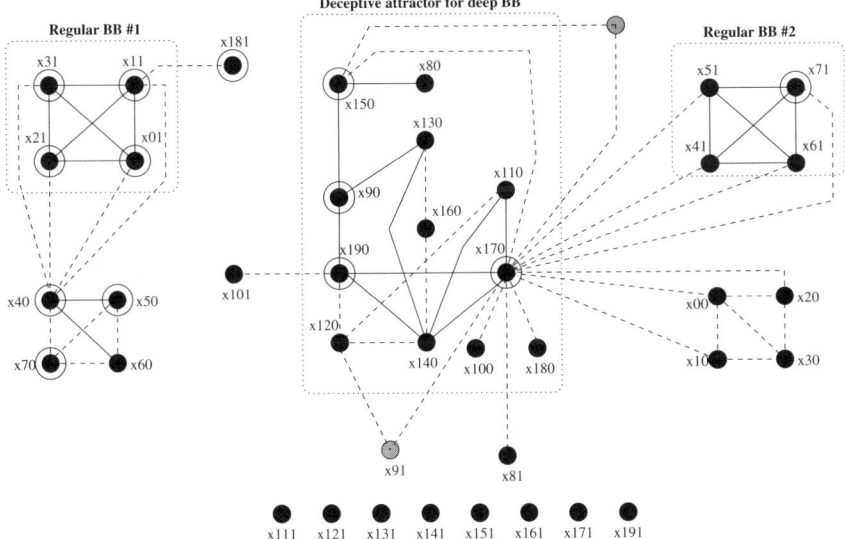

Fig. 19.4. Identification of the deep BB through *KeyGraph* using the naïve approach. Low-order building blocks are discovered as clusters of high-frequency terms. However, the *KeyGraph* fails to identify and discover the deep BB. The population was injected with two individuals and a population size of 2000 individuals was considered

smaller pieces is highly improbable and similarly its identification is also impossible if it has no market share in the population. Therefore, we inject one or two individuals with the deep BB into the selected population (after step 3 in the algorithm procedure mentioned in the previous section (Sect. 19.4.3)). Note that the injection procedure is similar to that of Sect. 19.4.2.

The *KeyGraph* on such a population is shown in Fig. 19.4. A population size of 2000 and tournament selection with a tournament size of 10 are used. Similar to the case of the 3×4 problem, the *KeyGraph* successfully identifies low-order BBs as clusters of high-frequency terms. The *KeyGraph* also discovers the deceptive attractor of the low-order building blocks. However, the *KeyGraph* does not identify the deep building block. Instead, it discovers the deceptive attractor of the deep building block as a cluster of high-frequency terms. Qualitatively similar results were obtained for other independent runs and for different population-size and tournament-size values.

A detailed examination of the internal mechanisms of the *KeyGraph* technique provides the reason for its failure in discovering the deep BB. We find that *KeyGraph* is effective in discovering rare events as a single-node cluster that are associated with clusters of high-frequency events. In other words, *KeyGraph* is effective if the key rare events occur in association with high-frequency events and if each of the rare events are independent of one another. However, in our case the deep BB is a rare cluster of events which may or may not be associated with other low-order building blocks. That is, individual events of the deep BB (genes belonging to the deep building block) are themselves not rare, but the cluster of genes with correct gene values is. Such a cluster of genes with correct gene values (ones) is the deep BB. Therefore, for the $2 \times 4 + 12$ problem the *KeyGraph* identifies parts of the deep BB as independent clusters of a single node, where each node represents the genes belonging to the deep BB.

Based on the above analysis, we reasoned that, in order to use *KeyGraph* in its current form, we should either convert the deep BB into a single node or convert the cluster representing the deep BB into a relatively higher-frequency one. In the following sections we propose two different techniques for circumventing the problem with the naïve *KeyGraph* approach.

19.4.5 Two Approaches for Identifying Deep Building Blocks

The two methods we propose to aid *KeyGraph* in identifying deep BBs are an *aggregation* method, where we change the methodology of converting a population of individuals to the text document, and a *doping* method, where the individuals with deep BBs are injected into the population before selection. Each of these methods are discussed in the following subsections.

Aggregation. In aggregation, instead of converting each bit into a word, we convert bits of the deep BBs into a phrase. Furthermore, the phrase indicates whether the individual has a deep BB or not. For example, the following individuals `1111 0000 010110101000`, `0000 1111 1010 0001 1100`, and `0000 1111 111111111111` are converted to 'x01 x11 x21 x31 x40 x50 x60 x70 xNotDeepBB.', 'x00 x10 x20 x30 x41 x51 x61 x71 xNotDeepBB.', and 'x00 x10 x20 x30 x41 x51 x61 x71, xDeepBB.' respectively. Qualitatively, aggregation transforms a deep BB from being a rare cluster of genes with correct values to a rare event 'xDeepBB'. Likewise all the other possible events (bit-value sequences) are converted to a single high-frequency event 'xNotDeepBB'.

The careful reader will recognize aggregation as cheating, but we do so to understand whether the *KeyGraph* procedure will work under the circumstances that better match its capability than those of the naïve tests. We consider how the aggregation might be accomplished after considering experimental results with the mechanism enabled.

The *KeyGraph* result on the population processed through aggregation is shown in Fig. 19.5. The figure shows that *KeyGraph* identifies the regular BBs as a cluster of high-frequency terms and the deep BB is discovered as a low-probability, high-key term (rare event). Figure 19.5 indicates that *KeyGraph* effectively discovers the deep BB even though it had only two copies in a population of 2000 individuals.

Aggregation is an effective method to discover deep BBs in a search problem; however, further investigation is required to develop efficient methods of aggregation. Specifically, we need to answer how one can aggregate bits belonging to a building block into a phrase indicating whether an individual has a deep building block or not. Perturbation techniques such as the gene-expression messy GA

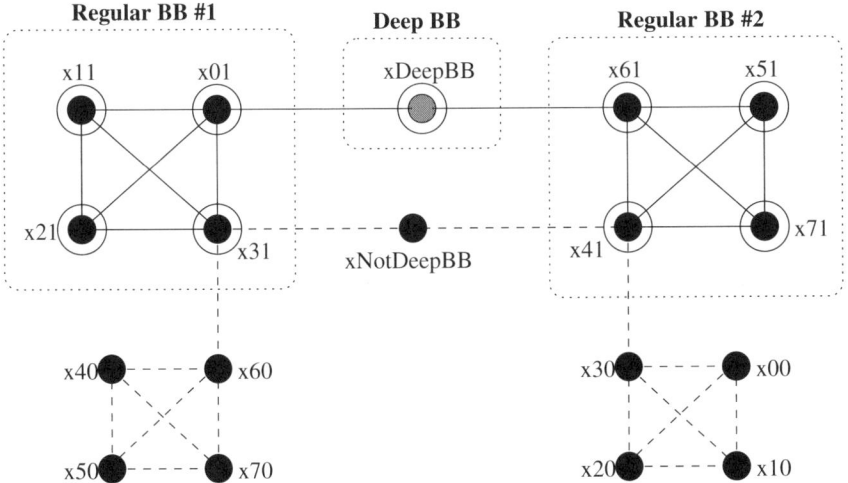

Fig. 19.5. Identification of the deep BB through *KeyGraph* using aggregation. The regular building blocks are represented by clusters of high-frequency terms and the deep BB is discovered as a low-probability, high-key event. Two individuals with the deep BBs were injected into a population of 2000 individuals

(GEMGA) [19.28] and LINC and LIMD GAs [19.31] can potentially be used to differentiate a bit-string partition as being either a building block or not. In other words, the perturbation techniques can provide candidates for the deep BB and thus perform the aggregation process. However, the results demonstrate that aggregation helps *KeyGraph* discover the deep building block.

Doping. The previous section investigated the utility of aggregation in aiding the discovery of the deep BB. This section discusses another technique, called *doping*, and its effect on discovering deep BBs.

In doping, a small number of individuals with deep BBs are injected into the population prior to selection. The injected population is then subjected to selection with high selection pressure (tournament size of 10–20), so that the deep BBs have relatively high market share as compared to the case of aggregation, but a low market share as compared to regular BBs.

The *KeyGraph* result on a doped population is shown in Fig. 19.6. About 0.1% of the population was injected with individuals with deep BBs. After selection about 3% of the population had individuals with deep building blocks. The figure shows that *KeyGraph* discovers regular building blocks as clusters of high-frequency terms and that it discovers the deep BB as a cluster of relatively low-frequency terms. Similar results were observed with different population sizes and tournament-size values, and for the deep BBs to be discovered about 2–5% of the population had to be doped with individuals containing deep BBs. It is interesting to note that competent GAs do not discover deep BBs even though up to 5% of the population is doped with the deep BB.

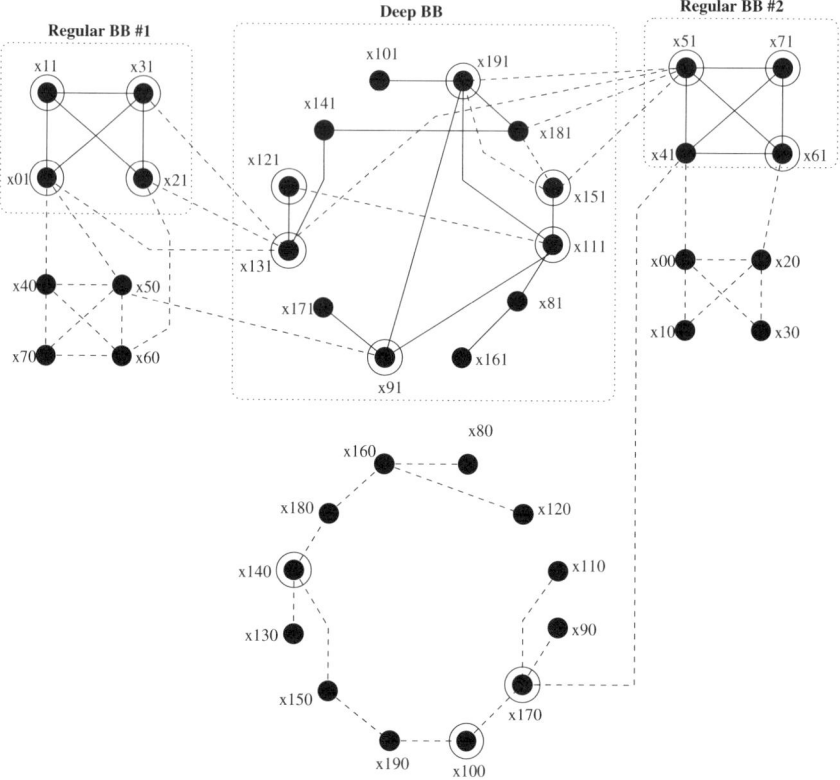

Fig. 19.6. Identification of the deep BB through *KeyGraph* using the doping method. The regular building blocks are represented by clusters of high-frequency terms and the deep BB is discovered as a cluster of relatively low-frequency terms. Approximately 0.1% of the population was injected with individuals containing deep BBs and after selection the market share of the deep BB grew to about 3% of the population size (2000 individuals)

Like aggregation, doping helps *KeyGraph* discover deep building blocks. As with aggregation, further investigation is required to design effective doping procedures. Specifically, we need to answer how to selectively dope deep BBs either by more injection of good BBs or by increasing the selection pressure as soon as deep BBs appear in the population. However, the objective of this study is to demonstrate the utility of *KeyGraph* in the discovery of deep BBs.

19.5 Future Work

This study demonstrates the utility of *KeyGraphs* in identification and visualization of deep building blocks in search and optimization problems. However, it has just scratched the surface, and much work is needed to develop practical competent GAs that capture both low-order BBs and deep BBs. Here we mention some of the key areas of future study which will enable us to do so.

Effective aggregation techniques: this work showed that aggregating the bits of the deep BB partition into a phrase is helpful. Perturbation techniques such as GEMGA, LINC GA, and LIMD GA can be used to preprocess the population and convert the deep partitions into a set of potential deep BB candidates.

Efficient doping methods: this study also showed that effective doping of the deep BB in the population can lead to its successful identification. Further studies are required to investigate how best to perform doping, especially in real-world problems, where the deep BBs are not known a priori.

Principled hybrids with existing competent GAs: existing competent GAs are very effective in identifying and exchanging regular building blocks. They do so in polynomial time (usually subquadratic). Therefore, an effective procedure to achieve innovation success for a search problem would be to combine *Key-Graphs* and existing competent GAs in a principled way [19.22, 19.44]. The result of such a hybrid should be the development of a solver which can identify both regular and rare notions.

These are not trivial challenges, but the pilot results of this study suggest that the investment is a potentially worthy one, in that it should permit deep BBs to be exploited sooner than would otherwise be possible in procedures that rely on statistically significant samples of the better substructures.

19.6 Conclusions

This paper has suggested how genetic algorithms and chance discovery may represent computational analogs of different facets of human innovation. In particular, selectorecombinative GAs, those that use selection and recombination mechanisms, have been discussed as being similar to that facet of human innovation that cross fertilizes different notions or subsolutions to create new, possibly better solutions to some problem at hand. Chance discovery, particularly chance discovery as exhibited by the *KeyGraph* procedure, has been considered as that aspect of human innovation where human beings pay more attention to salient fortuitous events, thereafter trying to understand or use the happy happenstance to effect useful change. The paper has reviewed GAs, *KeyGraphs*, basic mechanisms, and the innovation connection in some detail, and a combination of chance discovery and GAs has been suggested and partially tested.

In particular, *KeyGraphs* have been tried as a solution to the problem of deep building blocks. Large subsolutions of difficult problems are hard to evolve from lower-order building blocks, and sizing populations to include the statistically significant samples of high-order BBs is an exponentially growing nightmare. These facts place a premium on developing mechanisms to identify rare, salient fortuitous events when they occur, and to determine whether or not they may be used reliably as parts of a better overall solution to the problem at hand.

Pilot experiments were run on two test functions. A simple test function with uniformly sized and scaled BBs shows that *KeyGraphs* can identify and help visualize the usual low-order BBs that competent GAs regularly detect. A test function with two low-order BBs ($k = 4$) and one high-order BB ($k = 12$) was used to show that (1) ordinary competent GAs fail to exploit the deep BB, even when the population is seeded with a substantial number of copies of it and (2) a naïve implementation of *KeyGraphs* for deep BBs also fails. This led to two methods to get *KeyGraphs* to identify the deep BB. One technique, called aggregation, assumed that the deep BB could be identified whole as a unit. Under these circumstances, *KeyGraphs* are able to identify and visualize the deep BB, but assuming the existence of an aggregation procedure is tantamount to assuming local solutions to the linkage-learning problem. The other technique, called doping, assumes that the deep BB is present in significant numbers. Under this assumption, *KeyGraphs* are able to identify the deep BB without the aggregation assumption. Doping may be obtained in practice through significant selection pressures, and using high selection pressures may form a means of extracting chance occurrences of deep BBs from relatively small populations.

We think the results of the paper are interesting computationally, graphically, and scientifically. Computationally, the aggregation and doping pilot results are promising and deserve further inquiry to (1) understand the mechanisms involved theoretically and (2) develop practical deep-BB procedures integrating the best of competent GA practice and *KeyGraphs*. Graphically, one of the unintended benefits of this work has been the usefulness of graphical BB visualization using *KeyGraphs*. Highly dimensional optimization problems are notoriously difficult to visualize and the *KeyGraph* procedure has proved helpful in understanding graphically what is going on inside a run. Even if *KeyGraphs* were not useful for finding deep structure, this benefit would be worth exploring on its own. Scientifically, better understanding of different facets of human innovation, especially understanding that is computational and quantitative, is a useful thing in and of itself. The scientific contributions of this paper in this arena are largely conjectural, and we certainly do not claim to have found the fingerprint or smoking gun of human innovation in all its manifest complexity in our pilot experiments. But we believe that the connections we suggest are plausible, appeal to our own intuitions (and qualitative theories) about innovation, and deserve closer scrutiny from psychologists and other professionals who study these matters for a living. Regardless of whether these latter suggestions are widely accepted, we believe that the computational and graphical merits of the work deserve immediate scrutiny by those who are interested in chance discovery and genetic algorithms.

Acknowledgements. This work was sponsored by the Air Force Office of Scientific Research, Air Force Materiel Command, USAF, under grant F49620-00-0163, and the National Science Foundation under grant DMI-9908252. The US Government is authorized to reproduce and distribute reprints for government purposes notwithstanding any copyright notation thereon.

The views and conclusions contained herein are those of the authors and should not be interpreted as necessarily representing the official policies or endorsements, either expressed or implied, of the Air Force Office of Scientific Research, the National Science Foundation, or the US Government.

Kumara Sastry was also supported by a Computational Science and Engineering (CSE) fellowship, University of Illinois at Urbana-Champaign.

References

19.1 Asimov I, Shulman JA (eds) (1988) Isaac Asimov's book of science and nature quotations. Weidenfeld & Nicolson. New York, NY

19.2 Bandyopadhyay S, Kargupta H, Wang G (1998) Revisiting the GEMGA: Scalable evolutionary optimization through linkage learning. Proceedings of the IEEE International Conference on Evolutionary Computation, 603–608

19.3 Beveridge WIB (1957) The art of scientific investigation. W. W. Norton. New York, NY

19.4 Bosman P, Thierens D (1999) Linkage information processing in distribution estimation algorithms. Proceedings of the Genetic and Evolutionary Computation Conference, 60–67

19.5 Chen YP, Goldberg DE (2002) Introducing start expression genes to the linkage learning genetic algorithm. Parallel Problem Solving from Nature VII, 351–360 (Also IlliGAL Report No. 2002007)

19.6 Deb K, Goldberg DE (1993) Analyzing deception in trap functions. Foundations of Genetic Algorithms 2: 93–108 (Also IlliGAL Report No. 91009)

19.7 Deb K, Goldberg DE (1994) Sufficient conditions for deceptive and easy binary functions. Annals of Mathematics and Artificial Intelligence. 10: 385–408 (Also IlliGAL Report No. 92001)

19.8 Fine GA, Deegan JG (1996) Three principles of Serendip: insight, chance, and discovery in qualitative research. Qualitative Studies in Education, 9(4):434–447

19.9 Chance Discovery by Stimulated Group of People, Application to Understanding Consumption of Rare Food, *Journal of Contingencies and Crisis Management*, 10(3):129–138

19.10 Goldberg DE (1983) Computer-aided gas pipeline operation using genetic algorithms and rule learning. Doctoral dissertation, University of Michigan, Ann Arbor, MI. (University Microfilms No. 8402282)

19.11 Goldberg DE (1987) Simple genetic algorithms and the minimal, deceptive problem. In Davis, L. (ed), Genetic algorithms and simulated annealing. (Chapter 6) Morgan Kaufmann, Los Altos, CA, pp. 74–88

19.12 Goldberg DE. (1989) Sizing populations for serial and parallel genetic algorithms. Proceedings of the Third International Conference on Genetic Algorithms. pp.70–79 (Also IlliGAL Report No. 88004)

19.13 Goldberg DE (1991) Six steps to GA happiness. Paper presented at Oregon Graduate Institute, Beaverton, OR

19.14 Goldberg DE (1993) The Wright Brothers, genetic algorithms, and the design of complex systems. In Schaffer J.D. (ed), Proceedings of the Symposium on Neural-Networks: Alliances and Perspectives in Senri 1993. Senri International Information Institute, Osaka, Japan, pp. 1–7

19.15 Goldberg DE (1999) The race, the hurdle, and the sweet spot: Lessons from genetic algorithms for the automation of design innovation and creativity. In Bentley, P. (ed), Evolutionary Design by Computers (Chapter 4), Morgan Kaufmann, San Mateo, CA, pp.105–118

19.16 Goldberg DE (2002) Design of innovation: Lessons from and for competent genetic algorithms. Kluwer Academic Publishers, Boston, MA

19.17 Goldberg DE, Deb K, Clark JH (1992) Genetic algorithms, noise, and the sizing of populations. Complex Systems, 6: 333–362 (Also IlliGAL Report No. 91010)

19.18 Goldberg DE, Deb K, Kargupta H, Harik G (1993) Rapid, accurate optimization of difficult problems using fast messy genetic algorithms. Proceedings of the International Conference on Genetic Algorithms, pp.56–64 (Also IlliGAL Report No. 93004)

19.19 Goldberg DE, Korb B, Deb K (1989) Messy genetic algorithms: Motivation, analysis, and first results. Complex Systems, 3(5): 493–530 (Also IlliGAL Report No.89003)

19.20 Goldberg DE, Liepens G (1991) Theory tutorial. Tutorial presented at the 1991 International Conference on Genetic Algorithms, La Jolla, CA

19.21 Goldberg DE, Sastry K, Latoza T (2001) On the supply of building blocks. Proceedings of the Genetic and Evolutionary Computation Conference, pp.336–342 (Also IlliGAL Report No. 2001015)

19.22 Goldberg DE, Voessner S (1999) Optimizing global-local search hybrids, Proceedings of the Genetic and Evolutionary Computation Conference, pp. 220–228 (Also IlliGAL Report No. 99001)

19.23 Hadamard J (1945) The psychology of invention in the mathematical field. Princeton University Press, Princeton, NJ

19.24 Harik GR (1999) Linkage learning via probabilistic modeling in the ECGA. IlliGAL Report No. 99010. Department of General Engineering, University of Illinois at Urbana-Champaign, Urbana IL

19.25 Harik GR (1999) Learning gene linkage to efficiently solve problems of bounded difficulty using genetic algorithms. Doctoral dissertation, University of Michigan, Ann Arbor, MI. (Also IlliGAL Report No. 97005)

19.26 Harik GR, Goldberg DE (1997) Learning linkage. Foundations of Genetic Algorithms, 4:247–262 (Also IlliGAL Report No. 96006)

19.27 Holland JH (1975) Adaptation in natural and artificial systems. University of Michigan Press, Ann Arbor, MI

19.28 Kargupta H (1996) The gene expression messy genetic algorithm. Proceedings of the International Conference on Evolutionary Computation, pp.814–819

19.29 Kargupta H, Bandyopadhyay S (1998) Further experimentations on the scalability of the GEMGA. Lecture Notes in Computer Science, 1498: 315–324

19.30 Larrañaga P, Lozano JA (eds) (2002) Estimation of distribution algorithms. Kluwer Academic Publishers, Boston, MA

19.31 Munetomo M, Goldberg DE (1999) Linkage identification by non-monotonicity detection for overlapping functions, Evolutionary Computation, 7(4): 377–398

19.32 Ohsawa Y (2001) The scope of chance discovery. In Terano, T, et al. (eds), New Frontiers in Artificial Intelligence, Springer-Verlag, Berlin, Germany, p.413

19.33 Ohsawa Y (2002) KeyGraph as risk explorer from earthquake sequence. Journal of Contingencies and Crisis Management, 10(3): 119–128

19.34 Ohsawa Y (2002) Chance discoveries for making decisions in complex real world. New Generation Computing, 20: 143–163

19.35 Ohsawa Y, Benson N.E, Yachida M. (1998) KeyGraph: Automatic indexing by co-occurrence graph based on building construction metaphor. Proceedings of Advanced Digital Library Conference, pp.12–18

19.36 Ohsawa Y, Yachida M (1999) Discover risky active faults by indexing an earthquake sequence. Proceedings of International Conference on Discovery Science

19.37 Pelikan M (2002) Bayesian optimization algorithm: From single level to hierarchy. Doctoral dissertation, University of Illinois at Urbana-Champaign, Urbana, IL (Also IlliGAL Report No. 2002023)

19.38 Pelikan M, Goldberg DE (2001) Escaping hierarchical traps with competent genetic algorithms. Proceedings of the Genetic and Evolutionary Computation Conference. 511–518

19.39 Pelikan M, Goldberg DE, Lobo FG (2002) A survey of optimization by building and using probabilistic models. Computational Optimization and Applications, 21: 5–20 (Also IlliGAL Report No. 99018)

19.40 Pelikan M, Goldberg DE Cantú-Paz E (2000) Bayesian optimization algorithm, population sizing, and time to convergence. Proceedings of the Genetic and Evolutionary Computation Conference, pp.275–282 (Also IlliGAL Report No. 2000001)

19.41 Porter MF (1980) An algorithm for suffix stripping. Automated library and information systems, 14(3): 130–137

19.42 Roberts RM (1989) Serendipity: Accidental discoveries in science. John Wiley & Sons, New York, NY

19.43 Sastry K, Goldberg DE (2000) On extended compact genetic algorithm. Late Breaking Paper in the Genetic and Evolutionary Computation Conference, pp.352–359 (Also IlliGAL Report No. 2000026)

19.44 Sinha A, Goldberg DE (2001) Verification and extension of the theory of global-local hybrids. Proceedings of the Genetic and Evolutionary Computation Conference, pp.591–597 (Also IlliGAL Report No. 2001010)

Part V
Keys Combined to Applications

Finally, we look at chance discoveries realized so far on the methods above. Keys and methods shown above for chance discovery are combined in each example. Even when a data-mining method looks like the main actor of discovery, imagination or communication can not be missed to discover a chance. This part concludes the whole book with on-going challenges in the real world. The reader is always invited to propose new applications and new methods of chance discoveries.

20. Enhancing Daily Conversations

Yasuyuki Sumi and Kenji Mase

ATR Media Information Science Laboratories, 2-2-2 Hikaridai, Seika-cho, Kyoto 619-0288, Japan
email: sumi@atr.co.jp, mase@atr.co.jp

Summary.
This chapter presents a notion of enhancing our daily conversations for increasing opportunities to encounter new ideas and future partners for collaboration. We show two systems. One is AIDE, a system which facilitates online discussion with visualization of the discussion structure and a virtual discussant. The users of AIDE can mutually notice the similarity and difference among their viewpoints against common topics. The other is AgentSalon, a system which facilitates casual face-to-face chatting in the real-space setting such as museums and conference sites. AgentSalon has a big screen showing conversations among animated agents belonging to users. By observing a chat of the agents, the users can effectively obtain appropriate topics: that is, it tempts them to follow the chat.

20.1 Introduction

We spend a lot of time every day engaged in informal conversations in various situations such as daily chatting with neighbors and colleagues, informal talking after a business meeting or in a coffee break during a conference. Such informal conversations are in many instances very important in forming various kinds of communities because they establish the common ground among people necessary for the community. The informal exchange of ideas and thoughts can establish common ground, such as common interests and mutual understanding. While they may be initially immature, they are indispensable for further joint and collaborative activities. They are essential in starting and maintaining discussions in a precollaborative group that will later become a community.

This chapter presents two systems which enhance our daily conversations for increasing opportunities to encounter new ideas and future partners for collaboration. One is AIDE, a system which facilitates online discussion with visualization of the discussion structure and a virtual discussant. The users of AIDE can mutually notice the similarity and difference among their viewpoints against common topics. The other is AgentSalon, a system which facilitates casual face-to-face chatting in the real-space setting such as museums and conference sites. AgentSalon has a big screen showing conversations among animated agents belonging to users. Contents of the conversation include opinion exchange about the users' experiences so far, mutual recommendations of exhibits on behalf of them, etc. By observing a chat of the agents, the users can effectively obtain appropriate topics: that is, it tempts them to follow the chat.

20.2 AIDE: Augmented Informative Discussion Environment

This section presents a system called AIDE (augmented informative discussion environment) which facilitates our daily conversations [20.15, 20.9]. AIDE is an online chat system with conversation spaces to be shared by the users. The spaces are automatically visualized with a method that statistically structures conceptual spaces containing text objects and their keywords [20.18]. Also employed is a technique that extracts texts relevant to the conversation spaces from a text base [20.13].

Colab [20.14] is a pioneering system for electronic conferences. Its targets are brainstorming in electronic conversation environments, organizing fragments of ideas extracted there, and sharing information; these are similar to our targets. Colab, however, can not lead to a novel form of collaboration by making the best use of computers because this would only reproduce meetings using traditional tools such as a pen, paper, and chalkboard in some electronic form. Our purpose is to create a new form of collaboration with computer-augmented environments which actively offer such information that can not be offered by the traditional passive tools.

Some systems that help co-ordination in conversation have been proposed, e.g. [20.3, 20.20]. Their aim is to support information sharing among groups by processing the relationships among utterances and positions of participants during conversation in collaborative work. However, these systems force their users to converse following some conversation models prepared by their designers beforehand. That is, the users must attach their positions or relationships with others to all utterances. Our system does not require the users to specify any extra information during a conversation; in contrast, it offers them hints of relationships among utterances.

20.2.1 System Overview of AIDE

AIDE is a client–server–type chat system. This system can be both centralized/distributed and synchronous/asynchronous. Figure 20.1 illustrates the configuration of AIDE, and Fig. 20.2 depicts a scene in which two users are using AIDE in the face-to-face synchronous mode. Figure 20.3 is an example snapshot of the common screen image, which is also viewable on each user's client machine.

The main window of AIDE shown on the left of Fig. 20.3 includes a window with which a user can submit his/her utterances and a window that lists all collected utterances. AIDE is characterized by the following three subsystems.

Discussion viewer shows discussion spaces that visualize the structures of conversations. These spaces are information spaces shared among all participants in the conversations.

Conversationalist is a virtual participant who automatically extracts texts relevant to the conversation from an external text base and autonomously throws them into the discussion spaces.

Personal desktop is a desktop in which users can enter the phase of individual thought. The users can personalize shared information by duplicating and modifying the discussion spaces with it.

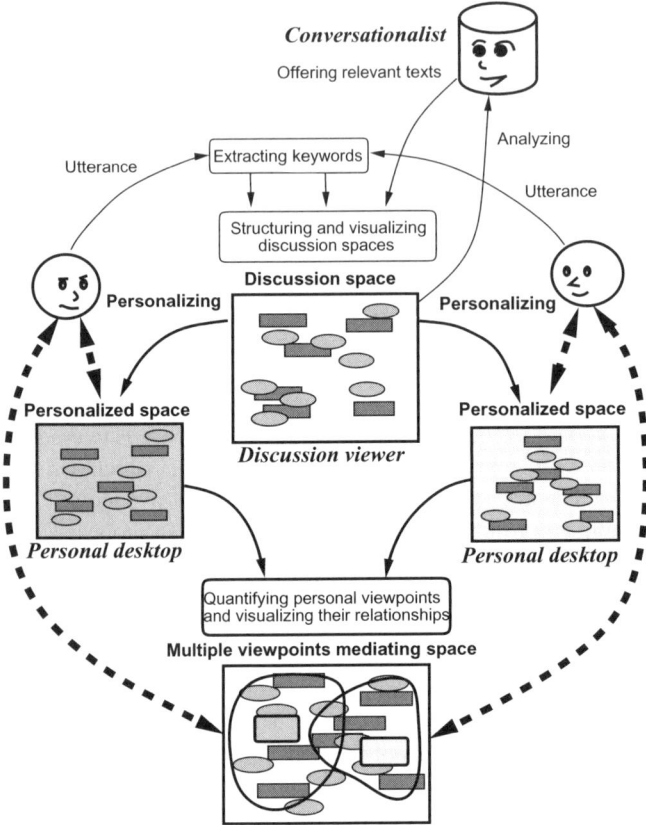

Fig. 20.1. Configuration of AIDE, and viewpoint sharing among participants using AIDE

Fig. 20.2. AIDE used in face-to-face synchronous mode

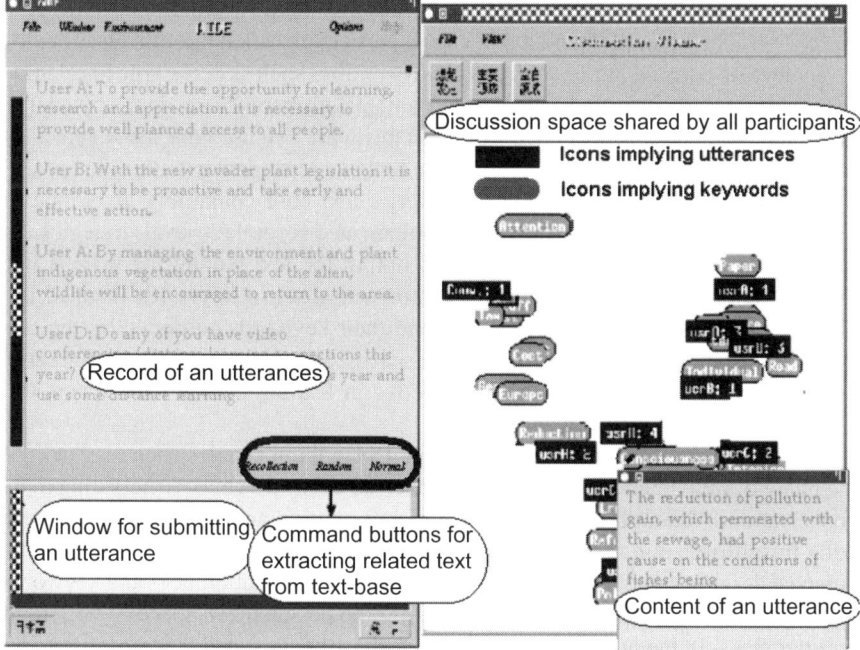

Fig. 20.3. Usage of AIDE

A user participates in conversation with the graphical user interface shown in Fig. 20.3 on a client machine. The server machine manages information on the users' utterances and discussion spaces which visualize the structures. When a user submits an utterance, the server automatically extracts keywords from the text along with their importance values, and, according to updated information, calculates and redisplays discussion spaces on all of the users' client machines.

In the discussion spaces, icons are used to indicate the utterances up to that point, and their keywords which are automatically extracted are mapped[1]. The discussion spaces are two-dimensional spaces which visualize the relationships between utterances and their keywords; a pair of utterances with more common keywords is located closer together and these common keywords are mapped around the pair [20.18]. All users can participate in conversation and understand the global structure and relationships among multiple topics (clusters of icons in the space) by viewing the shared discussion spaces. The discussion spaces visualize the relationships among the utterances based on such objective and simple information as the co-occurrences of keywords; this has the effect of making users notice new relationships instead of temporal relations. Hence, the discussion viewer and record of utterances on the main window are complementary.

---

[1] Each user can decide whether to show icons of utterances and keywords together or not.

Next, we explain the information-retrieval subsystem. As mentioned above, this subsystem is being implemented as a virtual participant called the conversationalist. To achieve its purpose, we have implemented abilities such as being able to calculate the timing of utterances and to judge the contents of utterances, perhaps by analyzing other utterances during conversation. In this work, we used the subsystem as an ordinary information-retrieval system, which works in response to user requests.

The information-retrieval subsystem has a text base containing texts indexed with keyword vectors beforehand[2]. We implemented several retrieval strategies, but this chapter explains only one method to output the text having the biggest normalized inner product of its keyword vector with a set of keywords mapped in a discussion space when requested. Texts and their keywords outputed by the subsystem are also thrown into the discussion spaces, and this causes a reconfiguration of the spaces. These results may be effective in leading human participants to a wider thought space and new ideas.

Lastly, we explain the personal desktop. Using them enables each user to enter the phase of individual thoughts whenever he/she wants to while participating in the conversation. Although the presentation of information and the method of visualizing this information are the same as those for the discussion viewer, users of the personal desktop can freely move icons, remove or modify utterances and keywords, and add new texts such as private memos into the personalized space as with regular utterances. In the next section, we will explain the personalization of discussion spaces in personal desktops and a method for mutually understanding participants' personal viewpoints during conversation using the results.

20.2.2 Mutual Understanding in Conversation

Personalizing Discussion Spaces. Since emerged clusters of utterances with many common keywords in the discussion spaces display the global structure and local information of the conversation simultaneously, not only the participants themselves but also an outsider can easily browse the conversation. While the discussion spaces visualize the structures of the conversation with an average viewpoint, they consequently may be unsuitable for any participant's viewpoint.

For that reason, we have prepared the personal desktop, where each participant can personalize information from a discussion space by duplicating the discussion space and doing the following operations;

– remove unattractive utterances and add private texts into the personalized space instead; and
– raise the importance values of attractive keywords and remove unattractive ones.

These data modifications are reflected in a restructuring of the space.

---

[2] For the presented version, this text base contains articles from a Japanese contemporary encyclopedia. The number of articles is about 10000 and the number of keywords extracted beforehand is about 40000.

Restructured spaces in personal desktops reveal each participant's individual viewpoints, namely, in the different personalized spaces, even the same pair of utterances from the same conversation can be mapped at relatively different positions. The sharing of such information by all participants can make all of them mutually understand each other. However, only preparing the environment for personalizing information is insufficient for explicitly utilizing the personal viewpoints and their relationships in collaborative work. Accordingly, in the next part, we propose a method that facilitates the mutual understanding of personal viewpoints by quantifying the personal viewpoints revealed in personalized spaces and visualizing their relationships.

Visualization of Individual Viewpoints and Their Relationships. Here, we describe how to quantify the personal viewpoints revealed in the personal desktops and to newly visualize their relationships.

We propose the following procedure, which does not postulate any special operations except the personalization of the discussion space of each user, and, accordingly, quantifies the users' viewpoints and visualizes their relationships as a by-product of the personalization (refer to the lower part of Fig.20.1).

1. Each user freely builds his/her own personalized space using a personal desktop as mentioned in the previous part.
2. The system newly creates a *viewpoint-object* that quantifies each user's viewpoint from information in the personal desktop. This is an object that has all of the keywords existing in the user's personalized space. These keywords have importance values, which are the mean values of those in the personalized space.
3. The system forms a *multiple-viewpoint mediating space*, which is a mediated space from multiple personalized spaces and visualizes the relationships between the viewpoints. This space is constructed from the sum of sets of utterances (including private texts given by each user) and keywords in the personalized spaces, and the viewpoint objects generated by the previous process. This space is structured by the same mechanism used with the discussion viewer and personal desktops.

The multiple-viewpoint mediating space has utterances and keywords commonly inherited from the discussion space. The space visually mediates the multiple-users' viewpoints and leads them to mutual understanding. Moreover, the space including private texts given in the personalized spaces encourages the users to mutually exchange and share private knowledge and ideas.

20.2.3 Experiments and Evaluation

The Effects of Discussion Spaces and the Information-Retrieval Function. We have preliminarily experimented on AIDE with sets of articles posted in online news, records of discussions by a group of close researchers using email, and so on. Here, we describe one example of the experimental usage of AIDE in detail to explain the implementation of the proposed method. This experiment was done by a group of people in an organization, i.e. usrA, usrB, and usrC. The subject of the conversation was 'recycling used paper in our office'. This experiment was done in one day, and they participated in the conversation in their spare time using their own desktop machines. The number of submitted utterances of usrA, usrB, and usrC were four, three, and four, respectively. The final status of the discussion space is shown in Fig. 20.4[3].

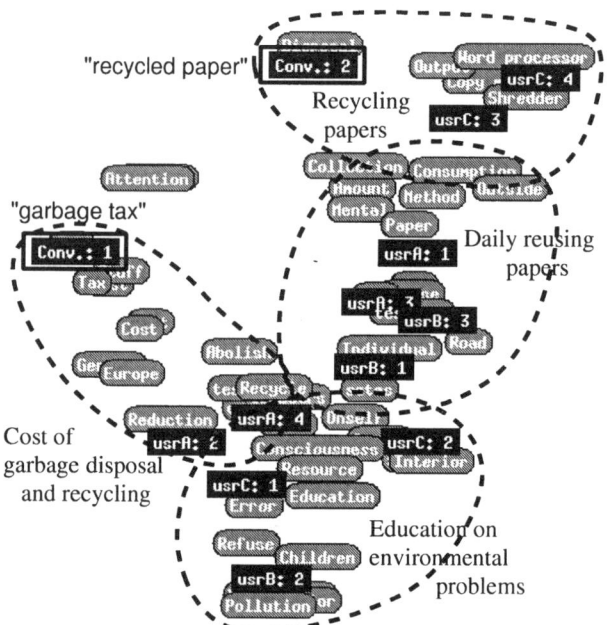

Fig. 20.4. An example of discussion space on a subject 'recycling paper'

Rectangular icons in the figure imply utterances, showing whose utterance and submission order. 'Conv.' seen in the figure shows an utterance by conversationalist. However, we handled this as an ordinary information-retrieval system instead of a virtual participant autonomously making utterances in this experiment. Oval icons imply keywords automatically extracted from the utterances, and the number of these were 208.

We can roughly understand the contents of the conversation by viewing clusters of utterance icons and keyword icons scattered around them, and we intuitively understand their topological relationships. For example, as noted in Fig. 20.4, we can understand that topics of the conversation were expanded from 'recycling used papers' to 'environmental problems related with garbage disposal and recycling' and 'educational issue'.

---

[3] This experiment was done in Japanese. The following examples are translated by the authors.

The information-retrieval function was used twice during the conversation. After each user input one utterance, this presented an utterance `Conv.:1` (an article on 'garbage tax') in response to a request of one of the users. An utterance given by `usrB` just after that did not mention this topic but an educational issue concerned with environmental problems, and a few utterances followed this topic. Since the focus of the conversation was a little stalemated, the information-retrieval function was used, and then `Conv.:2` (an article on 'recycled paper') was given. This made the focus go back to the original subject, i.e. effectively reusing papers. Moreover, a description about 'cost of garbage disposal and recycling' in `Conv.:1` gave stimuli to `usrA` and the following further discussion.

Personalized Spaces with Different Viewpoints. Here, we describe the building of personalized spaces by two users, i.e. `usrA` and `usrC`, derived from the discussion space shown in Fig. 20.4. Figures 20.5 and 20.6 show the respective results.

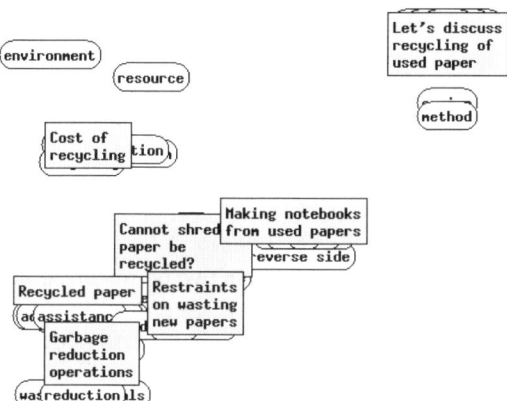

Fig. 20.5. UsrA's personalized space with the viewpoints of 'means of recycling'

The tags of the utterance icons are changed to phrases indicating the utterances by the authors with the function of a personal desktop. The same utterance appearing in both Fig. 20.5 and 20.6 is given the same tag.

In the case of `usrA`, the discussion space was personalized with the viewpoint of 'means of recycling'. As a result, utterances concerned with an educational issue were removed from `usrA`'s personalized space, and, in contrast, a text about 'ecological material' (mapped at the upper left of the space) was newly added. Here, this text was obtained as a related text to `usrA`'s personalized space by the information-retrieval function of AIDE. But this does not mean that texts added to personalized spaces are always obtained using this function. The number of keywords remaining in `usrA`'s space was 68. The keywords that have relatively high values of impor-

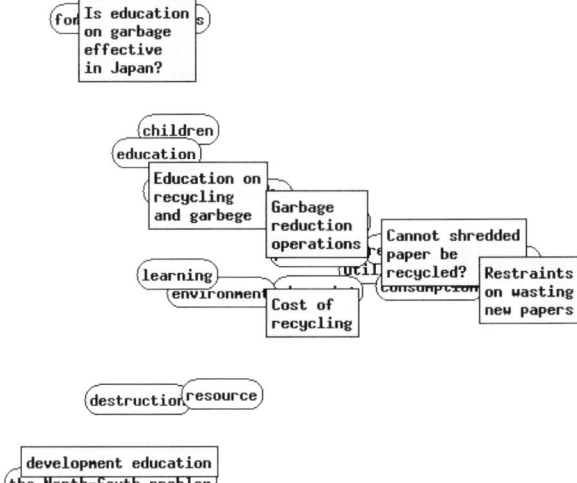

Fig. 20.6. UsrC's personalized space with the viewpoints of 'raising public spirit'

tance were {recycling, waste materials, nature, environment, cost}, which would be keywords of usrA's viewpoint object afterward.

In the case of usrC, his personalized space was built with the viewpoint of 'raising public spirit', and, consequently, many utterances were removed except for five utterances related to this. In contrast, he selected and added a new text about 'development education' (mapped on the lower left of the space) that was obtained using the information-retrieval function as usrA did. The number of keywords in his space was 69, and the prior keywords were {education, awareness, children, society, foreign countries}.

Note that even if a certain utterance is selected in two personalized spaces, each user has his/her own different interpretation of this. Specifically, four utterances ('garbage reduction operations', etc.) were selected in both personalized spaces, but sets of keywords regarded as important in the spaces severally differed: usrA gave higher values to keywords {paper, shredder, cost, collection} related with concrete means of recycling; however, usrC gave higher values to keywords {awareness, nature, protection} related with social consciousness. This difference was reflected in the difference of the structure of the personal spaces.

Mediating Multiple Viewpoints. Figure 20.7 shows a multiple-viewpoint mediating space (for short, MVM space) automatically created from the personalized spaces of usrA and usrC. Mapped icons of utterances and keywords are the sum of those in the two users' personalized spaces, and there are 11 and 111 of them, respectively. The MVM space also includes viewpoint-objects that imply the two users' individual viewpoints.

We can read several effects of MVM spaces by the example shown in Fig. 20.7. First, we notice that the MVM space is not a simple pile of the two personalized spaces, and the structure of that definitely differs from that of the initial discussion

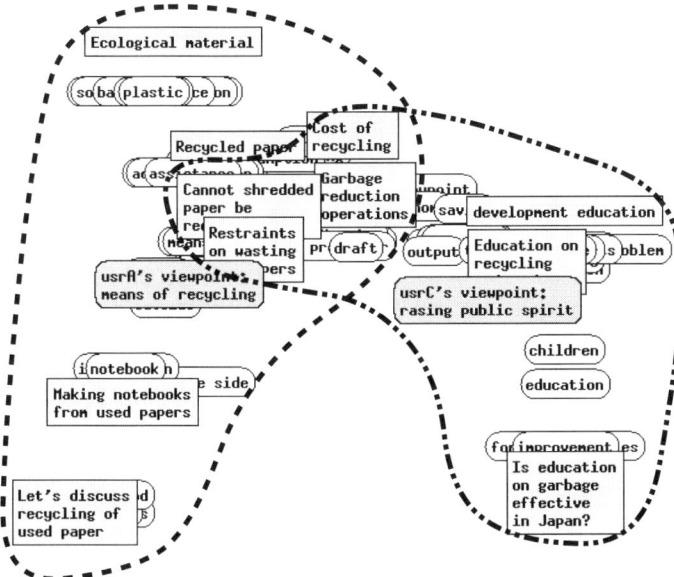

Fig. 20.7. An example of the space visualizing relationships between two participants' viewpoints

space shown in Fig. 20.4. This MVM space reveals different and shared parts between viewpoints of usrA and usrC. Moreover, while the initial discussion space also includes insignificant information for both of the two users, since it contains all information from the conversations, the MVM space can be regarded as a refined new common ground for the two users.

Second, MVM spaces include each user's private texts that show their interests and viewpoints; for example, the space shown in Fig. 20.7 has usrA's 'ecological material' (upper left icon in the space) and usrC's 'development education' (upper right icon). Such visual information facilitates users in intuitively catching their companions' intention and in sharing mutual personal knowledge.

Lastly, we point out the effect of reducing keywords, namely, worthless keywords are removed and the number of keywords is reduced in MVM spaces. This refines the structure of MVM spaces, which can be a new common ground for users. It is noteworthy that all we need to obtain an MVM space is each individual's operation of personalizing a shared discussion space; this method does not require any special operation for negotiation or co-ordination between users.

20.3 AgentSalon

This section presents another system, called AgentSalon [20.17], for facilitating face-to-face knowledge sharing and creative discussion among people having shared interests and experiences. AgentSalon has a big display for use by two to five users simultaneously. The display shows personal agents belonging to the users, and they chat to exchange their users' interests and experiences. Our intention is to encourage knowledge exchange and discussion among the users by tempting them to follow the chats via prompting by their personal agents.

AgentSalon was designed as a subsystem of our ongoing project to construct a personal guidance system for exhibition tours. The aims of the project are to build personal software agents that can navigate visitors through exhibition sites such as museums, trade shows, and conferences, and to facilitate new encounters and knowledge sharing between visitors and exhibitors with shared interests [20.19].

The user of our system carries PalmGuide, a hand-held guidance system, while touring an exhibition. A personal guide agent runs on PalmGuide and provides tour-navigation information, such as exhibit recommendations, according to the user's context, i.e. personal interests and spatio-temporal situation. The guide agent running on PalmGuide can migrate to and provide personalized guidance on individual exhibit displays or information kiosks that are located throughout the exhibition sites. This guide agent records its user's personal profile, touring records, and personal ratings and comments on individual exhibits he/she has visited, and this information is used for personalizing the presentation of individual exhibits, and matchmaking with other users having shared interests and visiting records [20.16].

We prototyped AgentSalon as a kind of information kiosk assumed to be located in a meeting place of an exhibition site, with a large touch-panel screen. AgentSalon is anticipated to be collaboratively used by two to five users as a place for chatting and exchanging experiences during touring. AgentSalon indirectly tempts users to follow *interesting* topics by the chat of their personal agents.

In order to achieve our goal to facilitate encounters and creative conversation among people, we adopt a method whereby agents participate in users' face-to-face conversation. Some researchers have already proposed agents participating in human conversations. Nagao and Takeuchi [20.10] focused on sociality and multi-modality of an anthropomorphic agent participating in two peoples' conversation. Isbister et al. [20.4] proposed a helper agent who provides conversation topics to first-meeting users in a virtual meeting space. The conversationalist of AIDE, shown in the previous section, provides users in brainstorming sessions with relevant and unexpected topics by monitoring the discussion contents and automatically searching texts from a database. These agents commonly participate in users' conversations as a third person, not as an agent belonging to an individual user. The aim of AgentSalon is to facilitate new encounters and collaborative knowledge sharing/creation by utilizing the information of individual users.

There have been works to support knowledge sharing and creation such as: Meme Tag [20.2], an electronic name tag for facilitating interactions between conference participants; HyperDialog [20.11], a personal agent arranging a meeting between its user and others; systems to help collaborative Web browsing (e.g. Silhouettell [20.7] and Let's Browse [20.8]); asynchronous knowledge sharing using alter-ego agents [20.12, 20.6]; and an embodied presentation team of conversational agents [20.1]. However, their knowledge resources are commonly static information such as previously prepared knowledge bases. On the other hand, AgentSalon uses personal information constantly accumulated by personal agents on PalmGuides carried by users. Such information is embedded in the real world; therefore, information presented by AgentSalon has potential to instantly influence users' ongoing (touring) behavior and accelerate collaborative knowledge sharing and creation among communities.

20.3.1 System Overview of AgentSalon

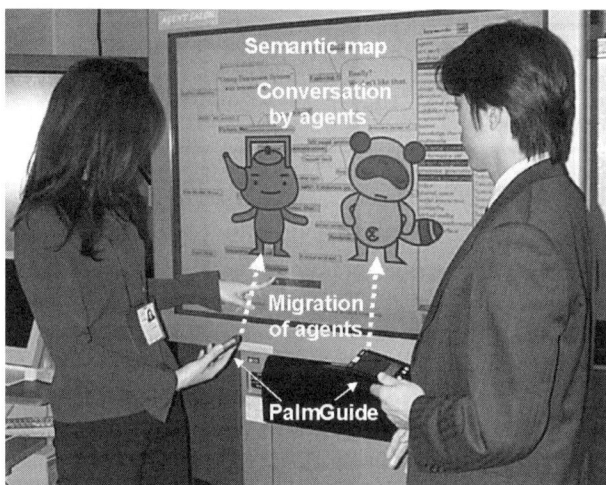

Fig. 20.8. AgentSalon in use

Figure 20.8 shows AgentSalon used by two users. The following is a scenario of using AgentSalon.

1. Personal guide agents on the PalmGuide of individual users migrate to AgentSalon with their users' personal information and are displayed as animated characters.
2. The migrating agents share their users' visiting records and interests and detect common as well as different parts in this information.
3. Based on the above results, the agents plan and begin conversations in front of the users. By observing the conversations, the users can efficiently and pleasantly exchange information related to an exhibit.
4. Because AgentSalon can access community information such as the information on each exhibit and other users' personal information via networks, users can browse detailed information about exhibits or users referred by agents.

As Fig. 20.9 illustrates, AgentSalon consists of the following three components.

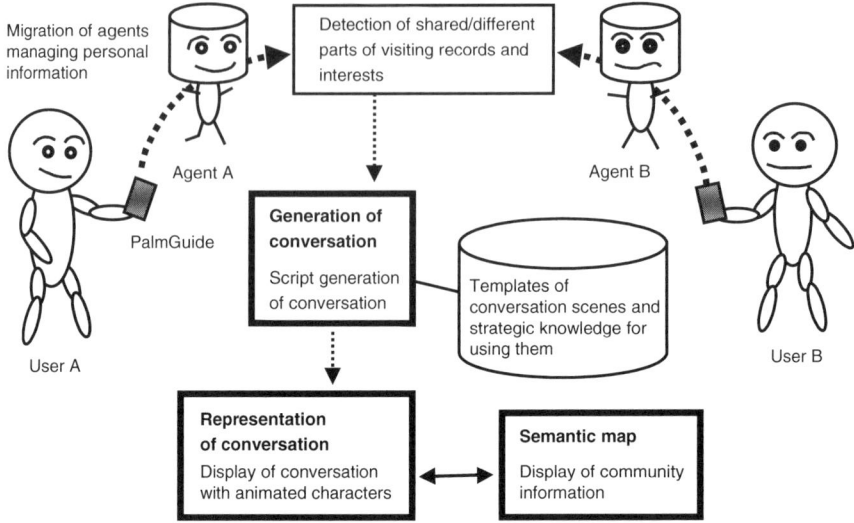

Fig. 20.9. System architecture of AgentSalon

Generation of conversation. Generates scripts of *interesting* conversations using personal information managed by agents. This is a knowledge-based system having utterance templates and strategic rules to tailor scripts depending on the context.

Representation of conversation. According to the generated scripts, this component controls and represents utterances and behaviors of animated agents by using Microsoft Agent. The streams of conversations, entrance and exit points of agents, and simple interaction with users are controlled by using JavaScript.

Semantic map [20.16]. A visual interface for browsing community information accumulated in the Web server. It shows semantic relationships between exhibits and the people involved with them and helps a user to associatively explore large information spaces according to his/her interests.

AgentSalon runs on Microsoft Internet Explorer. Semantic map is a Java applet running on the base of AgentSalon. The animated agents are displayed on the top of semantic map using Microsoft Agent. The display is a touch panel, so users can manipulate semantic map with their fingers and interact with their agents.

Because conversations of agents are controlled by JavaScript, animated agents on AgentSalon can use embodied gestures, e.g. manipulating icons of semantic map along with their utterances, achieved by executing a method of a semantic map applet by JavaScript.

In the current implementation, when a certain agent enters the salon, its user's icon appears in semantic map. At the same time, the icons of the exhibits that he/she has visited and evaluated as *interesting* appear and are linked with his/her icon. Therefore, it visualizes relationships (overlaps and differences) between touring experiences and the individual interests of users.

20.3.2 Generation of Conversation by Agents

This section describes a part of conversation generation, which is the main part of AgentSalon.

The goal of this part is to generate scripts of *interesting* conversations based on personal information managed by agents in the salon. In order to achieve this goal, we have been implementing a knowledge-based system to plan conversation by authoring reusable templates of conversations, and the strategic rules to use them.

We call comparatively independent and reusable sets of utterances a 'scene'. Conversation planning is triggered by entrance of a new agent, such that one or more scenes are displayed. Exit of agents from the salon is done on users' demand by touching the animated agents on the display.

Rules of script generation can be classified into the following two:

Object rules. Rules to make scenes by filling scene templates with personal information (e.g. user name, exhibit title, parametric numbers, etc.) on hand.

Meta rules. Strategic rules to select templates and object rules in order to make more effective conversation for stimulating users' meeting. They include editorial rules to smooth the stream of the whole conversation when combining several scenes.

Input data for the part of script generation are personal information, managed by the personal agent running on PalmGuide as follows:

− User name, affiliation, participating status (exhibitor or visitor), URL of homepage, and personal profile.
− Touring records, i.e. exhibit ID and visiting time.
− Personal evaluation (rating) of each exhibit.
− List of names of other PalmGuide users whom the user exchanged virtual business cards with, i.e. user ID and time of card exchange.
− Personal interests represented by keyword vector, updated when using semantic map.

The following provides examples of object rules that use personal information as above.

− One of the agents takes the initiative in a conversation and presents exhibits it (i.e. its user) has visited with a semantic map of the exhibition shown on the display. When another agent finds an exhibit among these which its user has visited, this agent discloses its user's evaluation and comments on the exhibit (by acting for its user). The agent can call the personal agents of the exhibitors and open the homepage of the exhibit for more detailed information.
− When two users' evaluations of a commonly visited exhibit are different (e.g. *user A* is interested in *exhibit 1*, but *user B* is not), their agents prompt a discussion about the exhibit. For example, the agent of *user A* says '*Exhibit 1* was interesting!', and then the agent of *user B* replies 'Really? We didn't like it.' By

observing the dialog, *user A* and *user B* can know that they have differing opinions about a shared experience (i.e. visiting *exhibit 1*), which efficiently leads them into a stimulating discussion.

– Suppose that *user A* has visited *exhibits 1, 2, 3,* and *4,* and *user B* has visited *exhibits 2, 4, 5,* and *6.* In this case, their agents will notice that the users have commonly visited *exhibits 2* and *4,* i.e. they share some interest in exhibits. Therefore, *user A*'s agent recommends *exhibits 1* and *3* to *user B,* and *user B*'s agent recommends *exhibits 5* and *6* to *user A.*

The following are examples of meta-rules, i.e. strategic rules to control agents' conversation.

– If the number of exhibits common to a certain pair of users is beyond a threshold, they are interpreted to have similar interests. Then, the scene for recommending the diverging parts of their visiting records is selected.
– If there is a user who has visited many more exhibits than other users, his/her agent takes the initiative in a conversation.
– If the difference of two users' age is large, the scene for revealing conflicts of evaluation ('interesting' and 'not interesting') is not used.
– If all users in the salon have not visited any exhibits, agents just introduce their users' profiles.

As seen above, object rules are reusable and applicable to various domains. On the other hand, soundness of meta-rules depends much on context and domains.

For example, while mutual recommendation of exhibits according to visiting history is useful in a museum application, recommending a presentation that has already been finished does not make sense. Exchange of personal evaluations on individual exhibits is an acceptable topic in amusement applications, such as theme parks. However, it may be provocative in academic conferences.

20.3.3 Implementation and Example

We prototyped the first version of AgentSalon as a service of the Digital Assistant Project to support participants in an academic conference held in July, 2000. However, data of the presentations and participants at the conference were in Japanese only, so we show an example, in this section, based on other data of our laboratories' open house in 1999.

Figure 20.10 shows an example display of PalmGuide. PalmGuide runs on PalmOS. A user can browse information for all exhibits and exhibitors (i.e. researchers of the laboratories), browse records of exhibit visiting and card exchanging, and get their agent's recommendation for their next visit according to the current time and his/her visiting records.

In the case of an exhibition site having exhibit booths, we can locate individual kiosk terminals to be infrared-connected by PalmGuides and to automatically collect users' electronic 'footprints' [20.16]. However, in the case of a conference

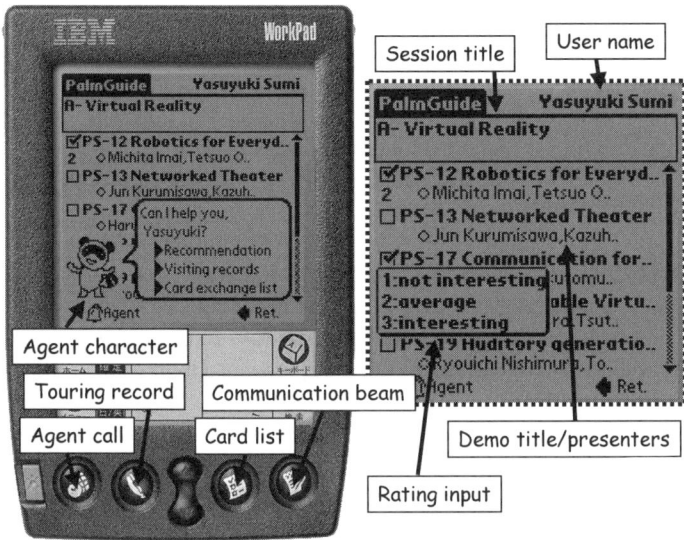

Fig. 20.10. An example of PalmGuide display

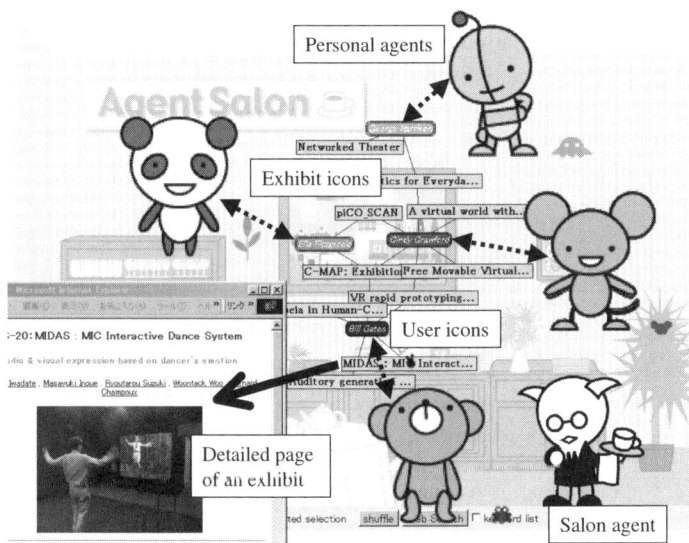

Fig. 20.11. An example of AgentSalon display

mainly having oral presentations, it is impractical for PalmGuide users to individually connect their PalmGuides to stationary kiosk terminals during oral sessions. Therefore, we provide users with check boxes on PalmGuide for checking presentations which they have attended. When the user checks a box of a certain exhibit, a dialog box prompts him/her to input evaluation (1: not interesting, 2: average, 3: interesting) of the exhibit. The evaluation data is utilized when calculating exhibit recommendations by PalmGuide.

Figure 20.11 shows an example display of AgentSalon. AgentSalon was designed using the metaphor of a salon for gathering and chatting together. On the display, there is always a master of the salon (the goat in the figure), which we call the 'salon agent'. Figure 20.11 shows the agent characters of four users. According to the display size, we limit the maximum number of personal agents in the salon to five at a time.

In the background of the animated agents, semantic map [20.16] is displayed for showing relationships between the agents' users and exhibits. Semantic map shows icons corresponding to users and exhibits they have visited as well as shows their interest by selecting 3 (interesting) on the evaluation dialog box shown in Fig. 20.10. Exhibit icons are linked with user icons based on their evaluations. Therefore, icons of users sharing interests on exhibits are indirectly connected and located nearby. Users can open detailed pages on exhibits and users by clicking their icons and associatively selecting keywords on semantic map. Semantic map is intended to be used for exploring related information and facilitating deeper discussion by users, which is triggered by conversation between agents.

We prepared eight kinds of agent characters, e.g. mouse and raccoon, that are selected by individual PalmGuide users upon user registration. Animation data includes about 40 kinds of actions such as greeting, moving, and pointing.

Table 20.1. An example of personal data of PalmGuide users

User name	Agent character	Visiting records (exhibit ID: evaluation)
Adam	Kettle	(PS-1:3), (PS-10:3), (PS-11:2), (PS-12:1), (PS-15:3), (PS-18:3), (PS-21:3)
Bill	Bear	(PS-1:3), (PS-2:3), (PS-10:3), (PS-11:1), (PS-19:3), (PS-20:3)
Cindy	Mouse	(PS-1:3), (PS-10:3), (PS-12:3), (PS-15:3), (PS-18:3), (PS-21:3)

Evaluation values 1: Not interesting, 2: Average, 3: Interesting

Now, we will describe an example of conversations performed by agents. Table 20.1 shows individual data of three users, Adam, Bill, and Cindy. These data are usually accumulated on individual PalmGuides, and used to generate conversations on AgentSalon.

Figure 20.12 shows sequential snapshots of AgentSalon where personal agents of the three users entered.

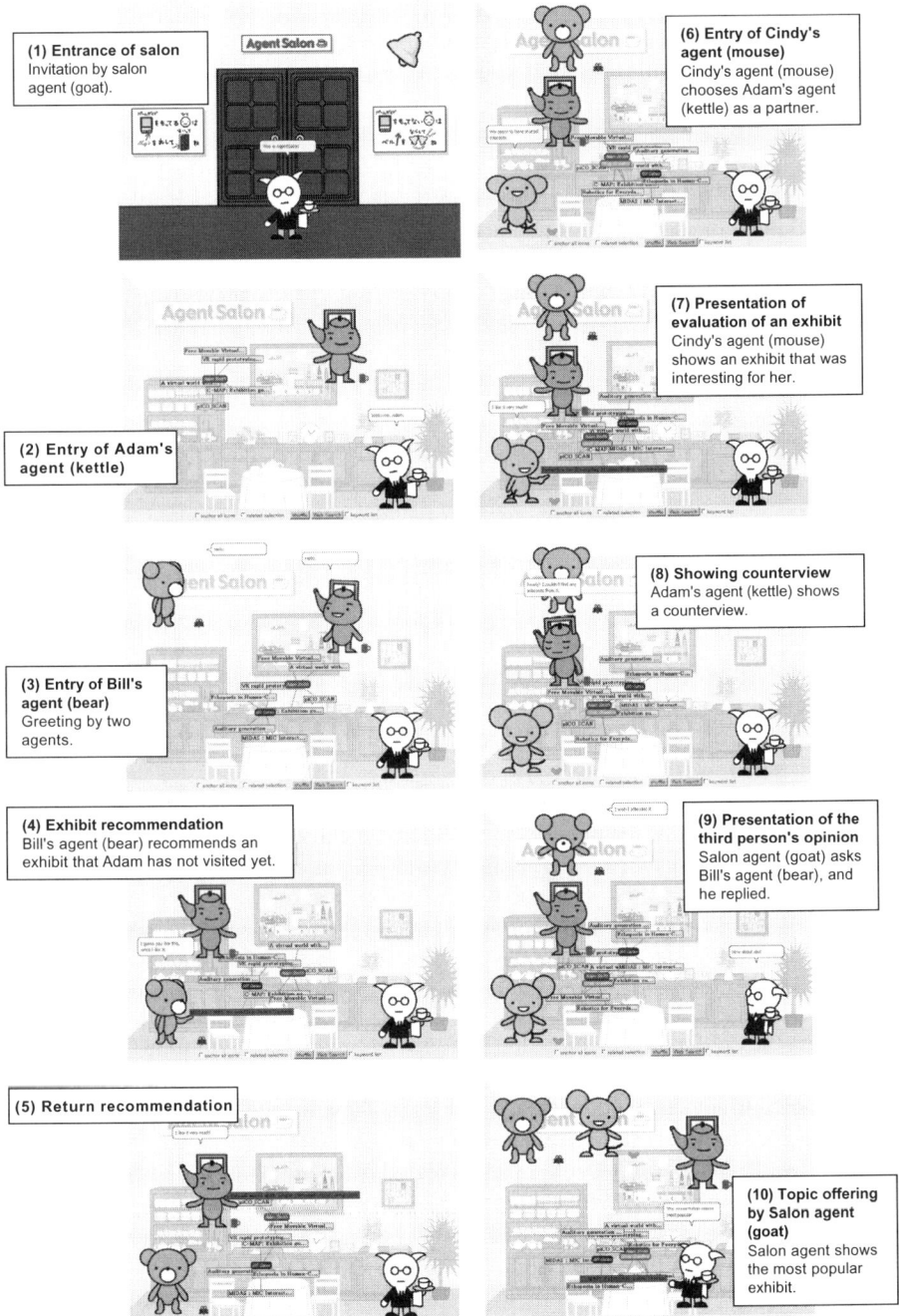

(1) Entrance of salon Invitation by salon agent (goat).

(2) Entry of Adam's agent (kettle)

(3) Entry of Bill's agent (bear) Greeting by two agents.

(4) Exhibit recommendation Bill's agent (bear) recommends an exhibit that Adam has not visited yet.

(5) Return recommendation

(6) Entry of Cindy's agent (mouse) Cindy's agent (mouse) chooses Adam's agent (kettle) as a partner.

(7) Presentation of evaluation of an exhibit Cindy's agent (mouse) shows an exhibit that was interesting for her.

(8) Showing counterview Adam's agent (kettle) shows a counterview.

(9) Presentation of the third person's opinion Salon agent (goat) asks Bill's agent (bear), and he replied.

(10) Topic offering by Salon agent (goat) Salon agent shows the most popular exhibit.

Fig. 20.12. An example of agents' conversations

(1) When there are no PalmGuide users, the entrance of the salon is displayed and the salon agent (goat) invites potential users.

(2) The agent (kettle) of the first user, Adam, enters the salon. At the same time, icons of Adam himself and five exhibits he had interests in are displayed in the background.

(3) The agent (bear) of the second user, Bill, enters the salon. The two agents greet. Icons of Bill and exhibits he was interested in are also displayed.

(4) Behind the scene, these two agents find that Adam and Bill have two exhibits in common among their visiting histories, and then select the scene for mutual recommendations of exhibits. In this example, Bill's agent (bear) points to the icon of PS-20, that Bill was interested in and that Adam has not visited yet, and says 'I guess you like this, since I like it.'

(5) Adam's agent (kettle) returns a recommendation by pointing to the icon of PS-15 and saying 'I like it very much!'

(6) The agent (mouse) of the third user, Cindy, enters the salon. After greeting, matchmaking between Cindy and the pre-existing users is calculated behind the scene. Concretely, the number of exhibits commonly visited by both Adam and Cindy is six, while for both Bill and Cindy it is only two. Therefore, Cindy's agent (mouse) selects Adam's agent (kettle) to speak to, by saying 'We seem to have shared interests.'

(7) The agents of Cindy and Adam select a discussion topic concerning an exhibit that both visited. In this example, they find an exhibit where the two users' evaluations were divided, and disclose this during a conversation. Concretely, they select PS-12 and Cindy's agent (mouse) says 'I visited this exhibit. I like it very much!'

(8) Adam's agent (kettle) replies 'Really? I couldn't find any interest in it.'

(9) In order to draw the third person, Bill, into the discussion, the salon agent (goat) asks Bill's agent (bear) 'How about you?' In this case, Bill's agent (bear) replies 'I wish I attended it' because Bill did not visit it.

(10) When no other events (e.g. entry of a new agent) happen for a while, the salon agent (goat) offers a topic. In this example, he points to the icon of the most popular exhibit among the three users, i.e. PS-1.

At the conference in July, 2000, AgentSalon was located in the lounge of the conference site to be freely used by participants. Though AgentSalon can be used by participants without PalmGuide as a browser of conference information, basically it is intended to be collaboratively used by PalmGuide users.

During the four-day conference, 40 of 65 PalmGuide users, that is, over 60%, accessed AgentSalon. While most of the users tried AgentSalon only two to five times, some users frequently (over ten times) used it. According to questionnaire results, most of the users recognized its effectiveness.

We could observe some 'regulars' frequently gathered in front of AgentSalon and having discussions by touching semantic map for over ten minutes. We could also observe some users exchanging their 'virtual cards' with their PalmGuides when meeting in front of AgentSalon: this implied that AgentSalon contributed towards supporting encounters among conference participants.

20.4 Conclusion

We have shown two systems, AIDE and AgentSalon, that enhance our daily conversations. For implementing these systems, we employed technologies of information visualization and interface agents providing topics related to the current context of users.

The philosopher Koestler [20.5] explains scientific discovery by the collision of two different planes of association ('contexts', in another word) in the scientist's mind. Our approach is intended to make the collision of different users' contexts visible, and then to facilitate the users to collaborate to create new knowledge.

References

20.1 Elisabeth André, Thomas Rist, Susanne van Mulken, Martin Klesen, and Stephan Baldes (2000) The automated design of believable dialogues for animated presentation teams. In Justine Cassell, Joseph Sullivan, Scott Prevost, and Elizabeth Churchill, editors, *Embodied Conversational Agents*, The MIT Press, London, UK, pp.220–255

20.2 Richard Borovoy, Fred Martin, Sunil Vemuri, Mitchel Resnick, Brian Silverman, and Chris Hancock (1998) Meme Tags and Community Mirrors: Moving from conferences to collaboration. In *Proceedings of CSCW'98*, ACM, New York, NY, pp.159–168

20.3 Jeff Conklin and Michael L. Begeman (1988) gIBIS: A hypertext tool for exploratory policy discussion. In *Proceedings of CSCW'88*, ACM, New York, NY pp. 140–152

20.4 Katherine Isbister, Hideyuki Nakanishi, Toru Ishida, and Cliff Nass (2000) Helper agent: Designing an assistant for human-human interaction in a virtual meeting space. In *Proceedings of CHI 2000*, ACM, New York, NY, pp.57–64

20.5 Arthur Koestler (1964) *The Act of Creation*. Peters, London, UK

20.6 Hidekazu Kubota, Toyoaki Nishida, and Tomoko Koda (2000) Exchanging tacit community knowledge by talking-virtualized-egos. In *Proceedings of Agents 2000*, ACM, New York, NY, pp.285–292

20.7 Masayuki Okamoto, Hideyuki Nakanishi, Toshikazu Nishimura, and Toru Ishida (1998) Silhouettell: Awareness support for real-world encounter. In Toru Ishida, editor, *Community Computing and Support Systems*, volume 1519 of *Lecture Notes in Computer Science*, Springer Verlag, Heidelberg, Germany, pp.316–329

20.8 Henry Lieberman, Neil W. Van Dyke, and Adrian S. Vivacqua (1999) Let's browse: A collaborative browsing agent. *Knowledge-Based Systems*, 12(8):427–431

20.9 Kenji Mase, Yasuyuki Sumi, and Kazushi Nishimoto (1998) Informal conversation environment for collaborative concept formation. In Toru Ishida, editor, *Community Computing: Collaboration over Global Information Networks*, chapter 6, John Wiley & Sons, Hoboken, NJ, pp.165–205

20.10 Katashi Nagao and Akikazu Takeuchi (1994) Social interaction: Multimodal conversation with social agents. In *AAAI-94*, pp.22–28

20.11 Katashi Nagao and Yasuharu Katsuno (1998) Agent augmented community: Human-to-human and human-to-environment interactions enhanced by situation-aware personalized mobile agents. In Toru Ishida, editor, *Community Computing and Support*

Systems, volume 1519 of *Lecture Notes in Computer Science*, Springer Verlag, Heidelberg, Germany, pp.342–358

20.12 Toyoaki Nishida, Takashi Hirata, and Harumi Maeda (1998) CoMeMo-Community: A system for supporting community knowledge evolution. In Toru Ishida, editor, *Community Computing and Support Systems*, volume 1519 of *Lecture Notes in Computer Science*, Springer Verlag, Heidelberg, Germany, pp.183–200

20.13 Kazushi Nishimoto, Shinji Abe, Tsutomu Miyasato, and Fumio Kishino (1995) A system supporting the human divergent thinking process by provision of relevant and heterogeneous pieces of information based on an outsider model. In *Proceedings of IEA/AIE-95*, pp.575–584

20.14 Mark Stefik, Gregg Foster, Daniel G. Bobrow, Kenneth Kahn, Stan Lanning, and Lucy Suchman (1987) Beyond the chalkboard: Computer support for collaboration and problem solving in meetings. *Communications of the ACM*, 30(1):32–47

20.15 Yasuyuki Sumi, Kazushi Nishimoto, and Kenji Mase (1997) Personalizing information in a conversation support environment for facilitating collaborative concept formation and information sharing. *Systems and Computers in Japan*, 28(10):1–8

20.16 Yasuyuki Sumi and Kenji Mase (2000) Communityware situated in real-world contexts: Knowledge media augmented by context-aware personal agents. In *Proceedings of the Fifth International Conference and Exhibition on the Practical Application of Intelligent Agents and Multi-Agent Technology (PAAM 2000)*, pp.311–326

20.17 Yasuyuki Sumi and Kenji Mase (2001) AgentSalon: Facilitating face-to-face knowledge exchange through conversations among personal agents. In *Proceedings of Agents 2001*, ACM, New York, NY pp.393–400

20.18 Yasuyuki Sumi, Ryuta Ogawa, Koichi Hori, Setsuo Ohsuga, and Kenji Mase (1996) Computer-aided communications by visualizing thought space structure. *Electronics and Communications in Japan, Part 3*, 79(10):11–22

20.19 Yasuyuki Sumi, Tameyuki Etani, Sidney Fels, Nicolas Simonet, Kaoru Kobayashi, and Kenji Mase (1998) C-MAP: Building a context-aware mobile assistant for exhibition tours. In Toru Ishida, editor, *Community Computing and Support Systems*, volume 1519 of *Lecture Notes in Computer Science*, Springer Verlag, Heidelberg, Germany, pp.137–154

20.20 Terry Winograd (1988) A language/action perspective on the design of cooperative work. *Human Computer Interaction*, 3(1):3–30

21. Chance Discoveries from the WWW

Naohiro Matsumura[1,2] and Yukio Ohsawa[1,3]

[1] PRESTO, Japan Science and Technology Corporation, 2-2-11 Tsutsujigaoka, Miyagino-ku, Sendai, Miyagi 983-0852, Japan
[2] Graduate School of Engineering, the University of Tokyo, 7-3-1 Hongo, Bunkyo-ku, Tokyo 113-8656, Japan
[3] Graduate School of Business Sciences, University of Tokyo, 3-29-1 Otsuka, Bunkyo-ku, Tokyo 112-0012, Japan
email: matumura@miv.t.u-tokyo.ac.jp, osawa@gssm.otsuka.tsukuba.ac.jp

Summary.

In this chapter, we introduce a method that can help understand significant and novel – i.e. emerging – topics. Here, *KeyGraph* is extended to be a method for the analysis and visualization of co-citations between Web pages. Communities, each having members (Web pages, their authors, and readers) with common interests are obtained as graph-based clusters, and an emerging topic is detected as a Web page relevant to multiple communities, corresponding to weak ties between strongly tied communities. An ultimate application of our method might be to understand the chances for governments and citizens, i.e. for discussing and deciding how we should deal with essential factors underlying emergent social events.

21.1 Introduction

We experience that new topics suddenly become popular. Such a topic, which might seem insignificant at first, can turn out to match our potential needs. *In The Tipping Point* [21.5], Gladwell describes this kind of phenomenon where a 'little' thing can make a big difference in the future. For example, how does a novel written by an unknown author become a bestseller? Why did the crime rate drop so dramatically in New York City? Gladwell calls these phenomena *social epidemics*, i.e. new topics sometimes behave just like outbreaks of infectious disease [21.5]. However, we can not detect the social epidemics (new topics) and their mechanisms in advance because the real world surrounding us is too complex to decode. Detecting a *tipping point*, in face of this obstacle, could be a big chance for one's activity, of which competitors are not aware. We interpret 'topics' in the broad sense that covers ideas, behavior, messages, products, and so on. Let us introduce some recent examples of new significant topics:

– **The mobile phone:** for the appearance of mobile phones, essentially two factors were present. First, mobile phones conquered the inconvenience of pagers whose users had to find a public phone when a pager rang. Second, mobile phones came to be equipped with the functions of the Internet and email services. Due to the synergy effects of these factors satisfying users' needs, mobile phones began to get popular.

- **Global warming:** the awareness of global warming realized the collaboration of automobile users and ecological preservation communities, and consequently brought about hybrid automobiles that have minimal exhaust emissions for preserving the Earth's ecology.
- **Human genome project:** many researchers in the field of artificial intelligence, biology, and medical science are collaborating on the human genome project to analyze the human genome and to reveal its effects. As we expect the conquest of fatal illnesses, the human genome project is in the limelight.

As we can easily realize from above descriptions, these topics are born when new collaborations of existing interests satisfy our potential needs or demands. Although the hidden factors might be 'submerged' in the human mind, we believe that a few signs can be mined from a database on human behavior reflecting the human mind. For this purpose, the Web is an attractive source of information because of its size and sensitivity to trends. The Web consists of an abundance of communities [21.10], each corresponding to a cluster of Web pages sharing common interests. Since a community means a chunk of shared interest, a Web page supported (or linked) by multiple communities is considered to satisfy their interests, and shows the movement direction of the wider human world, considering the synergy effects mentioned above. From this point of view, we are expecting the structure of the WWW to be a key to understand the real world. In this paper, we aim at revealing the structure of the WWW by using the *KeyGraph* algorithm [21.14], and then inspect whether the revealed structure of the WWW supports our detection of new significant topics.

21.2 Human Society on the WWW Structure

The WWW is a good source of information to detect the movement of human society, because it reflects the movement of the real world very quickly. On top of that, the WWW is a part of the human social network [21.1]. The creation of a hyperlink by the author of a Web page is an implicit type of 'endorsement' of the page being pointed to. By mining social interests contained in the set of such endorsements, we can obtain a better understanding of the movement of human society. To obtain Web pages reflecting human endorsements, let us overview related researches on Web communities by the link structure.

Discovery of related Web pages: Chakrabarti et al. have suggested using co-citation and other forms of connectivity to identify related Web pages [21.3]. Simply put, if page A points to both pages B and C, then B and C might be related. Terveen et al. used the connectivity structure of a Web page to find related Web pages [21.20]. Dean and Henzinger also found related pages only by the connectivity information where the input to the search process is the URL of a page [21.4]. The Netscape browser is equipped a 'what's related?' button that lists related pages to help us understand where to go next when we are surfing the Web or drilling for information [21.12]. Ohsawa et al. tried to discover Web pages that absorb attentions of people from multiple communities [21.18]. Topics in such pages can be triggers for personal or social progress of interests, beyond the bounds of existing communities. Kautz et al. made REFERRAL WEB, a social network graph designed to find an expert both reliable and likely to respond to the user [21.8].

Discovery of Web communities: the Web harbors a large number of communities – groups of content creators – each sharing a common interest that manifests itself as a set of Web pages. Although some communities have explicitly defined common interests (newsgroup, resource collections in portals, etc.), others are implicit. Kumar et al. defined a community on the Web as a dense *directed bipartite subgraph*, one whose nodes can be partitioned into two sets A and B such that every link in the subgraph is directed from a node in A to a node in B. They actually discovered over 100 000 communities from the entire Web [21.10]. The bipartite graph, however, comes to include pages of different interests if it is expanded to a wide area at the Web. As another use of links, Kleinberg [21.9] and Brin and Page [21.2] used link structures for ranking Web pages. Their main idea was based on mutual reinforcing, i.e. the more a Web page is referred to, the more authoritative the Web page becomes. The more authoritative a Web page becomes, the higher the Web page ranks. Thus, highly ranked Web pages tend to be the representative Web pages of communities.

21.3 Direct Relation and Co-citation

The Web can be viewed as a graph, where nodes represent Web pages and links represent the relation between Web pages. Two major relations of Web pages are at hand:

– **Direct relation:** a node represents a Web page and a link represents a hyper-link between two Web pages;
– **Co-citation:** a node represents a Web page and a link represents the relation of co-citation between Web pages.

We consider a community as a set of Web pages aiming at similar interests. Web pages in the same community do not frequently refer to one another. They may, for one reason, be in a competitive relation. In an extreme case, they are not aware of each others' presence because they keep secrets from each other. For example, a laboratory in the University of Tokyo (http://www.miv.t.u-tokyo.ac.jp), JST (http://www.jst.go.jp/EN/), and the University of Tsukuba (http://www.tsukuba.ac.jp) do not link to one another (although these are our affiliations); however our homepages have some hyper-links to theirs. In this sense, they are co-cited and have a relation. In the following, we clarify the difference between direct relation and co-citation.

The overview of a given area is obtained as follows: we first use query terms to collect a set of pages from the Google search engine[1]. We get a list of authoritative Web pages related to a given query. Then, we download the content of each Web page and extract hyper-links. Finally, we make a graph using one of the following strategies:

– **Algorithm 1:** obtain nodes representing the top authoritative Web pages and links representing the direct relation between them.
– **Algorithm 2:** obtain nodes representing the top authoritative Web pages and links representing the relation of co-citation.
– **Algorithm 3:** obtain nodes representing the most frequently cited Web pages and links representing the relation of co-citation.

In algorithm 1, a link has a direction in a natural sense; if page A points to page B, we make a link from page A to page B. In algorithms 2 and 3, assuming page $C \in C_{A,B}$ has hyper-links to both pages A and B, we calculate the strength of co-citation of A and B as:

$$\mathrm{rel}(A, B) = \sum_{C \in C_{A,B}} \frac{1}{\mathrm{OutDegree}(C)^2},$$

where $C_{A,B}$ represents a set of authoritative pages which points to both A and B, and $\mathrm{OutDegree}(C)$ represents the number of hyper-links in page C. This index is based on the random-surfer model [21.2], where a random surfer keeps clicking on successive links selected at random, and the probabilistic retrieval model [21.19]. A link by co-citation has no direction.

An example for a query 'abortion – pro life' for algorithms 1, 2, and 3 is respectively shown in Fig. 21.1, 21.2, and 21.3. For each figure, the number of nodes is 32 and the number of links is 31. In Fig. 21.1, we can see well-linked Web pages in the center of the figure. These Web pages include organizations (.org domain site) such as 'National Right to Life Committee', 'Priests for Life', and 'Republican National Coalition for Life'. On the other hand, in Fig. 21.2, we see a less centralized structure. The Web pages in the center are still organizations, but the left-hand-side Web pages are companies' Web pages (.com domain site) and right-hand-side Web pages are AOL and Amazon pages. In Fig. 21.3, we see more clearly the clusters of organizations, companies, and free Web sites such as AOL and Geocities. In this case, although the nodes are not necessarily authoritative Web pages, we can find some interesting Web pages. For example, one Web page in the middle of the figure, `http://www.afterabortion.org`, provides an important source of information on the aftereffects of the abortion, but this site is currently ranked very low (below 300) on the list by Google. This Web page is well cited and often co-occurs with `http://www.abortionfacts.com` and `http://www.prolifeinfo.org`.

We show another example for the query term 'Web mining'. Web mining is a relatively new topic and the authoritative Web pages relevant to this topic are not

---

[1] Google is a search engine to which Brin and Page's algorithm [21.2] is applied. Google is available at `http://www.google.com/`.

tightly connected, as shown in Fig. 21.4. However, we can see the relation more clearly for Algorithms 2 and 3 as Fig. 21.5 and Fig. 21.6. In other words, co-citation can detect more subtle relations between Web pages than direct relations can. If query terms are even newer and more rare, e.g. 'Ichiro' in major league[2], there is no way of finding the communities by direct links. In fact, if we make a graph by algorithm 1 for the query, neither of the two nodes are linked. By using the co-citation information, we can find communities even before participants realize that they have formed their own community.

In summary, the link structure of Web pages is a good source of information; however, if we look at direct links, we can not find emerging topics. Instead, we should focus on the co-citation information for detecting emerging topics. In the following, we focus not only on communities but also on Web pages implying a big change in the real world.

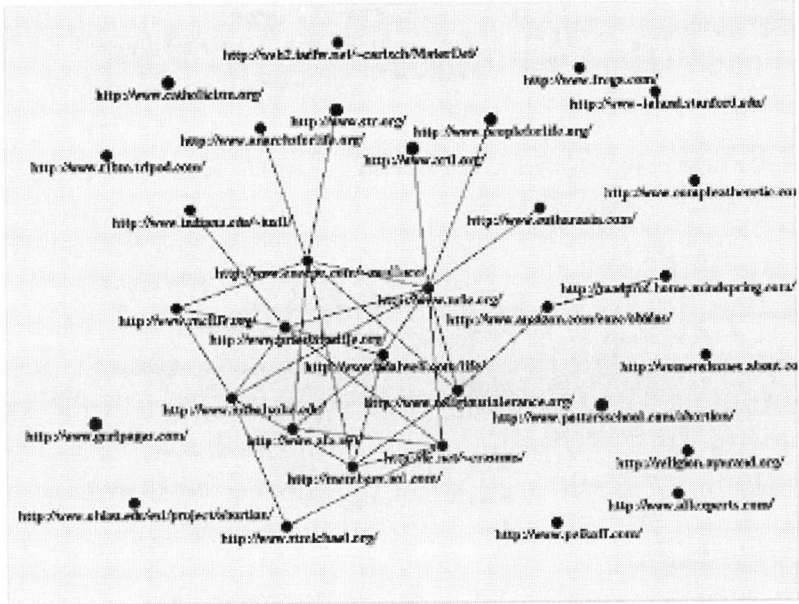

Fig. 21.1. The structure of Web pages related to 'abortion – pro life' by algorithm 1

---

[2] Ichiro is the first Japanese fielder in Major League Baseball and won the leading hitter title and MVP in the season 2001.

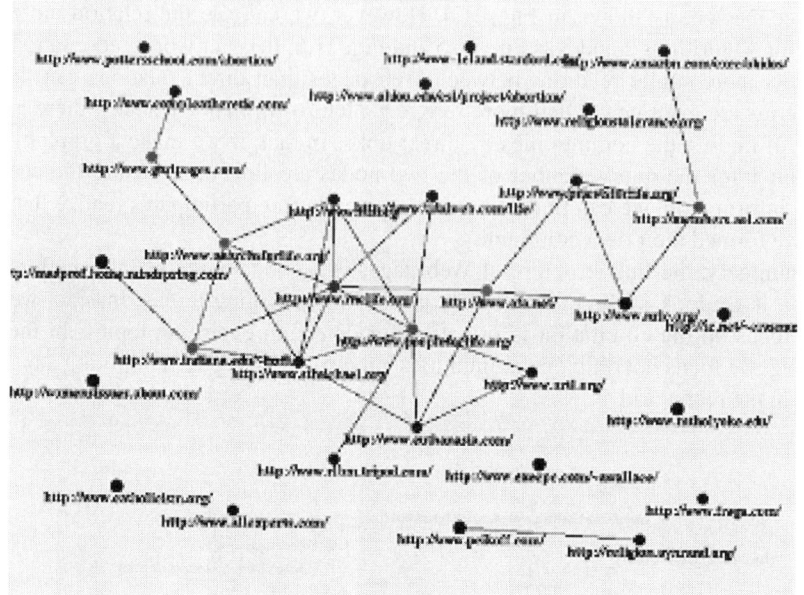

Fig. 21.2. The structure of Web pages related to 'abortion – pro life' by algorithm 2

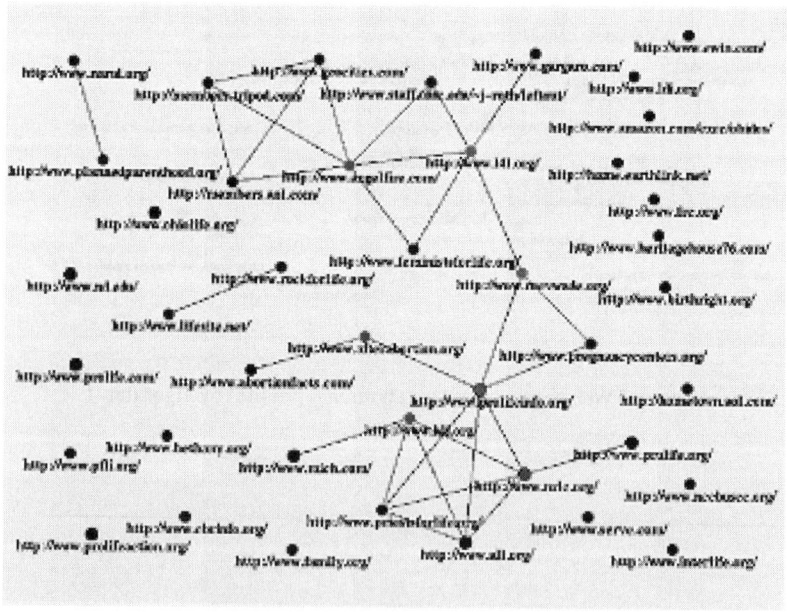

Fig. 21.3. The structure of Web pages related to 'abortion – pro life' by algorithm 3

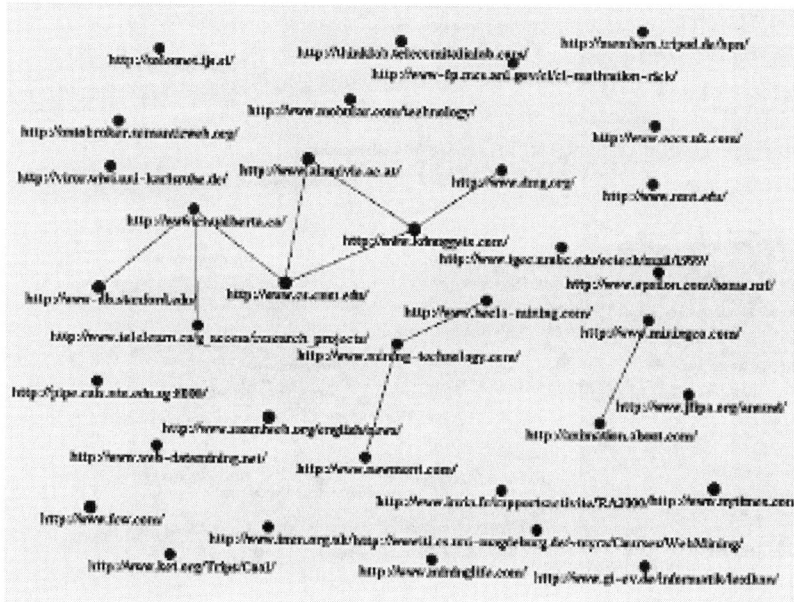

Fig. 21.4. The structure of Web pages related to 'Web mining' by algorithm 1

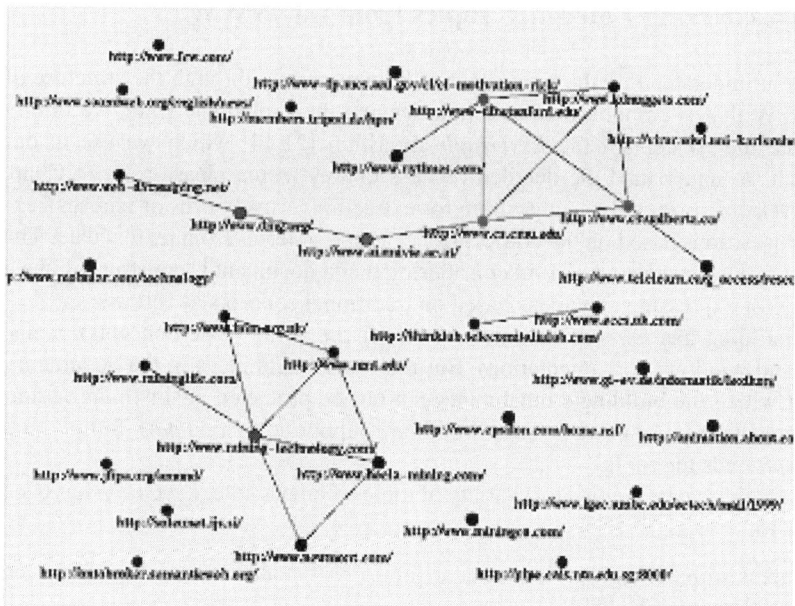

Fig. 21.5. The structure of Web pages related to 'Web mining' by algorithm 2

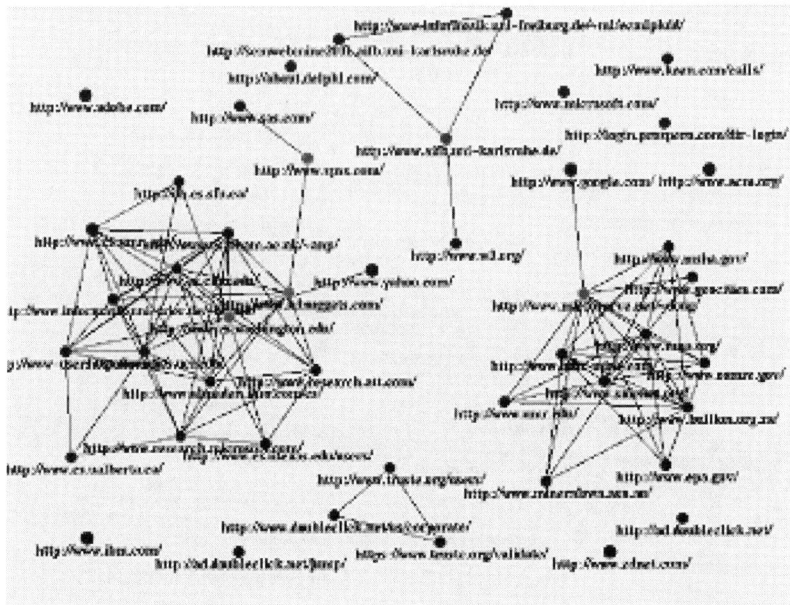

Fig. 21.6. The structure of Web pages related to 'Web mining' by algorithm 3

21.4 Discovering Emerging Topics from the WWW

We aim at understanding the movement of human society through the structure of the WWW that is composed of communities and their relations. Here, we briefly introduce the concept of the *KeyGraph* algorithm [21.14] which we use in our approach (to understand the detailed process of *KeyGraph,* please refer to Chap. 18). *KeyGraph*, originally an algorithm for extracting terms (words or phrases), expresses assertions based on the co-occurrence graph of terms from textual data. The strategy of *KeyGraph* comes from considering that a document is constructed like a building for expressing new ideas based on traditional concepts as follows:

'A building has *foundations* (statements for preparing basic concepts), walls, doors, and windows (ornamentation). But the *roofs* (main ideas in the document), without which the building's inhabitants can not be protected against rain or sunshine, are the most important. These roofs are supported by *columns*. Simply put, *KeyGraph* finds the roofs.'

KeyGraph can be applied to a variety of topics. For example, *KeyGraph* has been adopted to:

- find areas with the highest risks of near-future earthquakes from data of observed past earthquakes [21.15],
- get timely files from visualized structure of one's working history [21.16],
- computer plan to guide concept understanding in the WWW [21.22],
- make tools for shifting human context into disasters [21.12],

– discover potential motivations and fountains of chances [21.17].

Our approach is also focusing on the analogy between a document and other textual data (data formed by readable letters) to which *KeyGraph* can be applied. In a document D, high-frequency terms are used for expressing typical basic concepts, and term-pairs that frequently occur in the same sentences mean strong association throughout D. Here we extend the use of *KeyGraph* to another kind of data, i.e. a Web-page set (corresponding to D, a document) including Web pages (each corresponding to a sentence) having URL links, each corresponding to a word. That is, high-frequency links (which are the URLs pointing to other Web pages) in a collection W of Web pages show popular Web pages, and link-pairs which frequently occur in the same Web pages show strong relations, i.e. the co-citation, in W. Our fundamental hypothesis here is that the co-occurrence of terms in a document and the co-citation of Web pages are common in that both carry the underlying important shared context. Our strategy for applying *KeyGraph* is based on this analogy.

More formally, a Web page (whose URL is u_0) is translated to a 'sentence' as:

$$u_0 \quad u_1 \quad u_2 \quad u_3 \quad \cdots \quad u_i \quad \cdots \quad u_n, \quad (21.1)$$

where u_i $(i = 1, 2, 3, \ldots, n)$ are the URLs contained in the Web page. Combining the (virtual) sentences, shown in (21.1), for each Web page of a collection, forms a document. By this translation, we can obtain a 'document' reflecting the link structure of the WWW.

For example, Fig. 21.7 depicts the result of *KeyGraph*, where foundations (solid lines and their touching black nodes), columns (dotted lines), and roofs (double circles) are obtained. Some infrequent nodes as depicted by 'page 8' can be obtained as a roof. In this case, a foundation corresponds to an established Web community because each node shows a well-cited Web page and each link shows a strong tie among such nodes in a foundation. Our aim is to detect significant emerging topics from Web pages (i.e. the node of 'page 8' in Fig. 21.7) relevant to multiple established communities based on the assumption that Fig. 21.7 reflects the structure of the real world.

We expect that a graphical output of *KeyGraph* helps understand potential interests and the underlying relation between them, and leads us to the understanding of the structure of the interests of people in the real human society. This is a realization of looking at weak ties between strongly tied communities [21.6].

21.5 Experimental Examples and Discussion

We report our experiments where we applied *KeyGraph* to two sets of collections C_A and C_B, each containing 500 popular Web pages obtained by Google for the input query 'human genome', to follow the changes of the communities with time. The difference between the collections is the date: C_A is obtained on November 26, 2000, and C_B is on March 11, 2001.

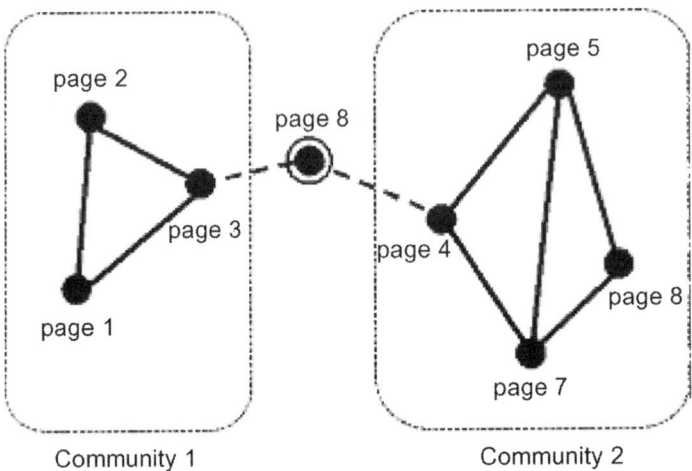

Fig. 21.7. An overview of *KeyGraph* for Web-page set

After C_A and C_B were translated into two documents, for each document *Key-Graph* outputs URLs as *roof* (asserted) keywords. The roofs for C_A and C_B are shown in Table 21.1 and Table 21.2, and the graphical outputs are shown in Fig. 21.8 and in Fig. 21.9 respectively. In the figures, the single-circle and double-circle nodes show *foundation* and *roof* pages respectively, and links among nodes show *columns*.

Comparing Table 21.1 with Table 21.2, we can recognize the movement among them. For example, the roofs: NHGRI (http://www.nhgri.nih.gov), NCBI (http://www.ncbi.nlm.nih.gov), and Sanger (http://www.sanger.ac.ul) appear in both tables. NHGRI (National Human Genome Research Institute) is one of 24 institutes, centers, or divisions that make up the National Institute of Health (NIH), the federal government's primary agency for the support of biomedical research. NCBI (National Center for Biotechnology Information) is also one of the institutes of NIH. Sanger (The Sanger Institute) is a research center that provides a major focus in the UK for mapping and sequencing the human genome, and genomes of other organisms. These are the most contributed institutes for the Human Genome Project, an international scientific effort to map and sequence the 3 billion genetic codes, involving more than 1000 scientists from five countries (China, France, Japan, the UK, and the USA). We can also understand this fact from Fig. 21.8 because these institutes are densely connected to each other.

On the other hand, the roofs GSC (http:// genome.wustl.edu), TIGR (http://www.tigr.org), Celera (http://www.celera.com), and CNN (http://www.cnn.com) appear only in Table 21.2. GSC (The Genome Sequencing Center) is a leading contributor to the Human Genome Project and TIGR (The Institute for Genomic Research) is one of the original centers conducting large-scale human genome sequencing. Note that TIGR already appeared in Fig. 21.8 as a node of foundation.

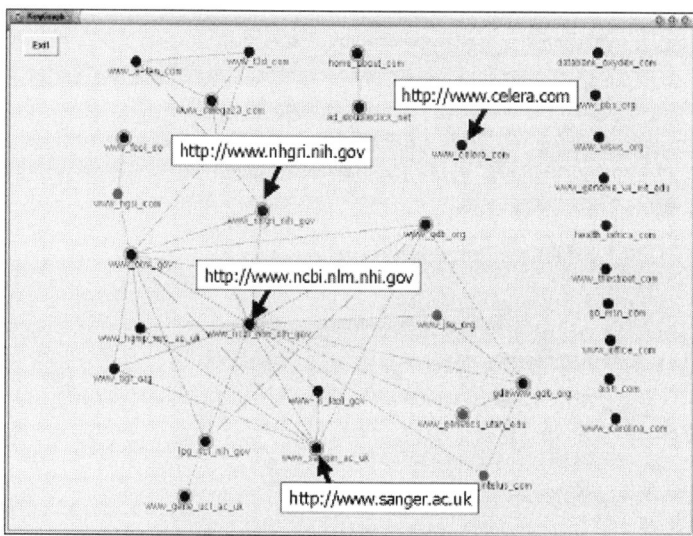

Fig. 21.8. The graphical output of *KeyGraph* for the input query 'human genome' (November 26, 2000). We can recognize a large cluster, which is composed of the institutes that contributed most to the Human Genome Project, such as NCBI (http://www.ncbi.nlm.nih.gov), NHGRI (http://www.nhgri.nih.gov), Sanger (http://sanger.ac.ul), etc. Note that Celera (http://www.celera.com) is isolated from the big cluster

Table 21.1. The textual output of *KeyGraph* for a collection of Web pages on 'human genome' (November 26, 2000)

URL	Affiliation
http://www.ncbi.nlm.nih.gov	National Center for Biotechnology Information
http://gdbwww.gdb.org	The Genome Database
http://www.ornl.gov	Oak Ridge National Laboratory
http://www.nhgri.nih.gov	The National Human Genome Research Institute
http://www.gene.ucl.ac.uk	The Galton Laboratory
http://www.ebi.ac.uk	European Bioinformatics Institute
http://www.gdb.org	The Genome Database
http://lpg.nci.nih.gov	CGAP Genetic Annotation Initiative
http://www.sanger.ac.uk	The Sanger Institute
http://www.genetics.utah.edu	Human Genetics Department in University of Utah

Here, let us focus on Celera (The Celera Genomics), an ambitious venture corporation sequencing the human genome from September, 1999. As you can see from Fig. 21.8 and Fig. 21.9, the situation around Celera changes dramatically from November, 2000 to March, 2001, i.e. Celera began to be supported by the cluster of the Human Genome Project and CNN, among the world's leaders in online news

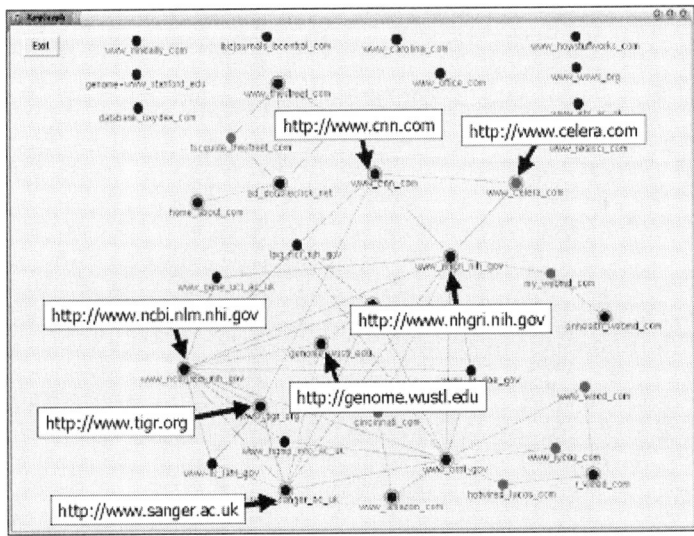

Fig. 21.9. The graphical output of *KeyGraph* for the input query 'human genome' (March 11, 2001). The major Web pages of the large cluster in this figure are almost the same as the cluster in Fig. 21.8, i.e. the Human Genome Project cluster. However, Celera (`http://www.celera.com`) began to be supported by multiple clusters, i.e. the Human Genome Project cluster and CNN (`http://www.cnn.com`), the mass-media cluster

Table 21.2. The textual output of *KeyGraph* for a collection of Web pages on 'human genome' (March 11, 2001)

URL	Affiliation
`http://www.ncbi.nlm.nih.gov`	National Center for Biotechnology Information
`http:.//www.nhgri.nih.gov`	The National Human Genome Research Institute
`http://www.ornl.gov`	Oak Ridge National Laboratory
`http://www.cnn.com`	CNN.com
`http://genome.wustl.edu`	Genome Sequence Center in Washington University
`http://onhealth.Webmd.com`	OnHealth Network Company
`http://www.gdb.org`	The Genome Database
`http://www.tigr.org`	The Institute for Genomic Research
`http://www.sanger.ac.uk`	The Sanger Institute
`http://www.celera.com`	Celera.com

and information delivery. Looking back on the real events and situations, we can understand the leap of Celera.

In the field of the human genome, revolutionary events occurred in 2000 and 2001. The Human Genome Project team and Celera announced the completion of the draft sequence of the human genome in June, 2000, and the subsequent articles were published in *Nature* [21.7] and *Science* [21.21] in February, 2001. Both are the most important milestones for the human genome analysis. In fact, J. Craig Venter, president and chief scientific officer of Celera and Francis S. Collins, director of the Human Genome Project, were celebrated by US President Bill Clinton and British Prime Minister Tony Blair for the progress of the human genome analysis at the White House on June 26, 2000.

Considering these real events and situations, the changes of the structures shown by Fig. 21.8 and Fig. 21.9 (e.g. Celera grew to be widely supported) are considered to reflect the real society.

21.6 Conclusions

In this chapter, we introduced a method that can help understand significant and novel – i.e. emerging – topics. Here, the algorithm of *KeyGraph* is extended to be a method for the analysis and visualization of co-citations between Web pages. Communities, each having members (Web pages, their authors, and readers) with common interests are obtained as graph-based clusters, and an emerging topic is detected as a Web page relevant to multiple communities, corresponding to weak ties between strongly tied communities. Experiments, an example of which is presented in this paper, show that the aimed effect of our method is realized.

The emergence of significant sites on the WWW as we seek to attract attention to various domains where social evolution occurs forms a key issue, because affairs in the real communities of man are reflected in the topics in the virtual communities on the WWW. This is true for the global communities among nations, as well as for local communities among humans. The CyPRG project [21.12], for example, seeks to understand the factors underlying the diffusion of the WWW into national, state, and local governments worldwide. In deepening their methods surveying each government in terms of organizational openness and internal effectiveness, it can be important to consider intergovernment interactions at global and local levels via real and virtual ties. An ultimate application of our methods in this article might be to understand the chances for governments and citizens, i.e. for discussing and deciding how we should deal with essential factors underlying emergent social events.

References

21.1 Adamic LA and Adar E (2001) Friends and Neighbors on the Web, Report of the Internet Ecologies Project, XEROX Palo Alto Research Center, Palo Alto, CA, electronically published: http://www.parc.xerox.com/istl/groups/iea/papers/web10/index.html

21.2 Brin S, Page L (1998) The Anatomy of a Large-Scale Hypertextual Web Search Engine, Proceedings of the 7th World Wide Web Conference, pp.107–117

21.3 Chakrabarti S, Dom B, Indyk P (1998) Enhanced Hypertext Categorization using Hyperlinks, Proceedings of ACM SIGMOD International Conference on Management of Data, pp.307-318

21.4 Dean J, Henzinger MR (1999) Finding Related Pages in the World Wide Web, Proceedings of the 8th World Wide Web Conference, pp.1467–1479

21.5 Gladwell M (2000) THE TIPPING POINT: How Little Things Can Make a Big Difference, Little Brown & Company, Boston, MA

21.6 Granovetter M (1973) Strength of Weak Ties, American Journal of Sociology, 8:1360-1380

21.7 International Human Genome Sequencing Consortium (2000) Initial sequencing and analysis of the human genome, Nature, 409:860-921

21.8 Kautz H, Selman B, Shah M (1997) The Hidden Web, AI magazine, 18(2):27-36

21.9 Kleinberg JM (1999) Authoritative Sources in a Hyperlinked Environment. Journal of the ACM, 46(5):604-632

21.10 Kumar R, Raghavan P, Rajagopalan S, Tomkins A (1999) Trawling the web for emerging cyber-communities, Proceedings of the 8th World Wide Web Conference, pp.1481–1493

21.11 La Porte TM, Demchak CC, Friis C (2001) Webbing Governance: Global Trends across National Level Public Agencies, Communications of the ACM, 44(1):63-67

21.12 Nara Y, Ohsawa Y (2000) Tools for Shifting Human Context into Disasters, Proceedings of the 4th Knowledge-Based Intelligent Engineering Systems & Allied Technologies, pp.655–658

21.13 Netscape Communication Corporation: 'What's Related' web page, (http://home.netscape.com/escapes/related/.)

21.14 Ohsawa Y, Benson NE, Yachida M (1998) KeyGraph: Automatic Indexing by Co-occurrence Graph Based on Building Construction Metaphor, Proceedings of Advances in Digital Libraries Conference pp.12-18

21.15 Ohsawa Y, Yachida M (1999) Discover Risky Active Faults by Indexing an Earthquake Sequence', Proceedings of Discovery Science, pp.208-219

21.16 Ohsawa Y (1999) Get Timely Files from Visualized Structure of Your Working History, Proceedings of the 3rd Knowledge-Based Intelligent Engineering Systems & Allied Technologies, pp.546–549

21.17 Ohsawa Y, Fukuda H (2002) Chance Discovery by Stimulated Group of People - An Application to Understanding Rare Consumption of Food, Journal of Contingencies and Crisis Management, 10(3):129–138

21.18 Ohsawa Y, Matsumura N, Ishizuka M (2001) Discovering Topics to Enhance Communities' Creation from Links to the Future, Poster Proceedings in the $10th$ World Wide Web Conference, pp.104–105

21.19 Rolleke T, Blomer M (1997) Probabilistic Logical Information Retrieval for Content, Hypertext and Database Querying, Hypertext - Information Retrieval - Multimedia 1997, pp.147-160

21.20 Terveen LG, Hill WC, Amento B (1999) Constructing, Organizing, and Visualizing Collections of Topically Related Web Resources, ACM Transactions on Computer-Human Interaction 6(1):67-94

21.21 Venter JC, et al (2001) The Sequence of the Human Genome, Science 291:1304-1351

21.22 Yamada S, Osawa Y (2000) Navigation Planning to Guide Concept Understanding in the World Wide Web, Proceedings of Autonomous Agents, pp.114-115

22. Detection of Earthquake Risks with *KeyGraph*

Yukio Ohsawa[1,2]

[1] PRESTO, Japan Science and Technology Corporation, 2-2-11 Tsutsujigaoka,
   Miyagino-ku, Sendai, Miyagi 983-0852, Japan
[2] Graduate School of Business Sciences, University of Tokyo, 3-29-1 Otsuka,
   Bunkyo-ku, Tokyo 112-0012, Japan
   email: osawa@gssm.otsuka.tsukuba.ac.jp

Summary.

KeyGraph, the document-indexing (keyword-extraction) algorithm, is applied
for another purpose: extracting active faults with risks of near-future big earthquakes from earthquake sequences. This presents an exemplification of *KeyGraph*
as a tool for abstracting causalities from an event sequence. From the results here,
we can validate *KeyGraph* as a tool for showing *which* active faults are risky, as
well as for showing which words abstract a document finely. The risky faults empirically obtained by *KeyGraph* correspond finely with the real occurrence of earthquakes and seismologists' estimation of risks. Even more important, the risks are
accountable. That is, the relation between the obtained risky faults and faults quaking frequently are visualized by the output of *KeyGraph*.

22.1 The Human Society as a Complex System

At this beginning of this paper, note that troughs (seismic faults deep in the sea) and
active faults in shallower land crust are both called active faults or *faults* simply in
the paper. An earthquake occurs at an *active fault*, a boundary of two areas of land
crust moving in different directions, stressed by those movements. There have been
various methods for earthquake prediction. *Trenching survey*, i.e. mining a fault
for evidence of past earthquakes [22.7], can approximately estimate the risk, i.e.
whether we have a long time before the next earthquake at the fault. Such a direct
land mining is inefficient and expensive, and more significantly can not distinguish
near-future (within a few decades) earthquake risks from far-future (hundreds of
years) ones.

Mining data, as apposed to, for finding risky active faults is inexpensive and efficient. Seismic gaps, i.e. areas recent earthquakes did not occur *in* but only *around*,
may be regarded as risky [22.6]. However, a seismic gap may have no cause of
earthquakes by nature. Earthquake risks at future moments of an active fault were
estimated from its recorded history of earthquakes [22.16, 22.9]. However, for high-
accuracy predictions, interactions among multiple faults should also be considered,
which is quite difficult because underground interactions among faults are hidden
and complex. Although statistical approaches look at long range both in time and
space, the underground interactions of faults across a wide area, e.g. the overall land
of Japan, are hardly counted [22.3, 22.12, 22.13, 22.14].

In this paper, we present a method to aid the exploration for risky active faults
where big earthquakes may occur in the near future, with detecting hidden causalities among land-crust movements and activities of faults. Hereafter, we call our

system *fatal fault finder* or F^3 in short. F^3 finds risky faults by applying *KeyGraph*, which has been originally presented as a document-indexing algorithm [22.5], not to a document but to a sequence of *focal faults*, i.e. the faults where past earthquakes occurred. This strategy comes from the idea that *KeyGraph* is essentially an algorithm for extracting causal structures from a sequence of events, and finding keywords carrying assertions of a document was just one application of *KeyGraph*. That is, the cause (global land-crust movements) and the effect (stresses on active faults, to cause near-future earthquakes) are extracted from an earthquake sequence by *KeyGraph*, as well as the cause (basic concepts of the author) and the effect (expressing the author's assertions by keywords) are extracted from a document.

In the remainder, We first show *KeyGraph* as a method to find events playing major roles in the causal structure from an event sequence. Applying this to earthquakes, the seismic semantics of *KeyGraph* when applied to an earthquake sequence are clarified. The experimental results show risky areas and their shifts after big earthquakes in Japan, with seismological evaluations.

22.2 *KeyGraph* Applied to Finding Risky Active Faults

The *KeyGraph*-based earthquake-data miner is called *fatal fault finder* (F^3), composed of the following three steps. The two steps (EQ2) and (EQ2) are embodied by the steps in Chap.18, i.e. *KeyGraph-step1 and KeyGraph-step2* respectively. That is, the risky faults are obtained as the hubs affected and stressed by the fundamental (invariant) movement of the large land crusts.

(EQ1) Get the following input data:

Data 1: land-surface locations of faults, i.e. $F = (a, b)$ for every fault F where a and b are the positions of the two ends of F defined two-dimensionally by the longitude and the latitude.

Data 2: a sequence of earthquakes, where each earthquake Ei is given as (*time*i, *longitude*i, *latitude*i, *depth*i, *magnitude*i), for $i = 1, 2, ..., N$ sorted by *time*i (the time the ith earthquake occurred) where N is the number of observed earthquakes. (*longitude*i, *latitude*i), *depth*i, and *magnitude*i denote the two-dimensional position, the depth, and the magnitude of the ith earthquake respectively.

(EQ2) Make D: the distance from the epicenter (the two-dimensional position of the earthquake focus measured on the ground surface) x of each earthquake in data 2 to a fault $F(= (a, b))$ is computed. Then, F is regarded as the focal fault if it is the nearest to x of all the faults in data 1 (the third dimension, i.e. the depth of each earthquake, is useless because it is unknown even by seismologists at what angle each fault lies underground). D is made as the sequence of these obtained focal faults. Each string in D implies an event, i.e. an earthquake at the focal fault. Then, '.' is inserted after each earthquake stronger than $M\theta$, a fixed magnitude. This is because energetic earthquakes often change the movement of land crust, by releasing crust-distortion energy [22.1].

(EQ3) obtain risky faults: obtain the events of the highest values of *key* in D, and regard their focal faults as risky, i.e. strongly stressed.

For example, from data 2 in Table 22.1, we obtain D as in (22.1) including N '#'s. In this case, m# means an earthquake at fault no. m in data 1. Here, 123# and 202# are faults no. 123 and no. 202 given in data 1, which are the closest ones to the epicenters of the first and the second earthquakes, i.e. (142.804E, 42.140N) and (139.523E, 37.441N) respectively. (22.1) includes four '.'s supposing that four big earthquakes over magnitude $M\theta$ occurred. 249# is obtained as risky, in **(EQ3)**.

$$D = 123\# \ 202\# \ 1\# \ 84\#. \ 76\#. \ 216\# \ 1\# \ 202\# \ 84\#. \ 249\# \ 84\#. \ 76\# \ 249\# \ ... \ (22.1)$$

Table 22.1. An example of data 2 (artificial data)

	*time*i	*longitude*i	*latitude*i	*depth*i	*magnitude*i
1	85 01 01 01:17:52.24	142.804E	42.140N	10.2	2.1
2	85 01 01 01:30:49.92	139.523E	37.441N	158.7	3.1
3	85 01 01 01:52.24.20	146.804E	46.140N	230.2	4.2
...	...	...	...	...	...
5	85 03 07 01:17:52.24	142.804E	42.140N	10.2	2.1
6	85 03 07 01:25:52.00	142.001E	41.000N	9.2	2.7
7	85 03 07 01:39:49.92	135.523E	38.441N	15.1	3.6
...	...	...	...	...	...
N	85 12 31 19:51:53.45	138.469E	36.992N	12.9	2.2

22.3 The Seismological Semantics of *KeyGraph*

We apply *KeyGraph* because the clusters, i.e. foundations, and roofs correspond to meaningful components of earthquakes. Here let us show these correspondences.

22.3.1 The Seismological Semantics of (KG1)

The major assumption introduced here is that the source of earthquakes at active faults is the underground movement of *fundamental faults*. Fundamental faults are large and mostly hidden active faults having branches as faults appearing on the land surface.

We can suppose that frequent earthquakes at a fault occur because the fault can quake for a weak but high-frequency influence from a small part of a fundamental fault or from neighboring faults. That is, the influence from a small part of a fundamental fault causes some of its branch faults to quake, which triggers the chain of pair-wise co-quaking (quaking between the same pair of nearest '.' s) of other branch faults. These co-quakings of a pair of faults are caused by two faults locally pushing each other by propagating local stress via shallow land crust. Such local interactions are detected by step (KG1) as pair-wise co-quaking of faults. As shown later in the experimental results, each cluster in (KG1) is formed by faults located around one or a few lines on the surface of land. Interpreting these lines as fundamental faults, this supports the seismological semantics above of (KG1).

22.3.2 The Seismological Semantics of (KG2)

Direct stress from fundamental faults is great, because they correspond to long boundary lines of large land-crust areas with enormous mass, i.e. moving with enormous energy, called *active fault provinces* [22.8]. In fact, long faults have been causing big earthquakes. Therefore, active faults affected directly by fundamental faults are stressed strongly and may cause a big earthquake soon.

As a result, a fault which co-quakes – quakes often at near moments – with fundamental faults can be regarded as risky, if the co-quaking reflects a causal relation of the quakes. Such a causation may not be observed directly, because fundamental faults are mostly hidden underground deeply. However, the co-quaking of faults can be observed and counted from data like data 1 and 2 above, by employing step (KG2) of *KeyGraph*, which shows the co-occurrence of each event and an event cluster corresponding to a group of faults near a fundamental fault. Note that this co-occurrence is not pair-wise (i.e. fault-to-fault), but global (i.e. fault-to-cluster).

There are theoretically established prediction methods on fitting established time-series models, e.g. a Markov model, to earthquake occurrences at an active fault or in a certain fixed area [22.9, 22.1]. However, the background principles of earthquakes are known too little to be modeled probabilistically [22.10]. As well as such an analysis of precise local effects, it is meaningful to consider global causalities behind earthquakes, because earthquakes occur due to the stress from global activities of land crust. Recently, simplified physical models of the process leading to the collapse in the land crust have been utilized for probabilistic estimation of big earthquakes [22.15]. Yet, the interaction of active faults scattering widely has been considered in these models much less than their real impact. Especially in Pacific-rim Asia including Japan, thousands of active faults are guessed to be pushing each other in complex unknown connections.

In short, *KeyGraph* works for grasping such interaction of faults in a wide area. In this sense, *KeyGraph* fits for earthquakes than considering local seismic effects [22.6]. As introduced in [22.4], *KeyGraph* was first presented as a method for extracting keywords representing assertions and essential concepts, based on basic concepts expressed as local (pair-wise) co-occurrence of words [22.5]. Comparing Fig. 18.1 in Chap.18, with $\#m$ representing each active fault, with Fig. 22.1 showing *KeyGraph* applied to a document [22.4], we see an earthquake sequence and a document similar in the sense of structure, i.e. both have fundamental parts and rare elements called roofs essentially related to foundations. Both can now be seen as two applications of *KeyGraph* to sequential data: substituting each line in Table 22.1 with corresponding lines in Table 22.2 clarifies the analogy.

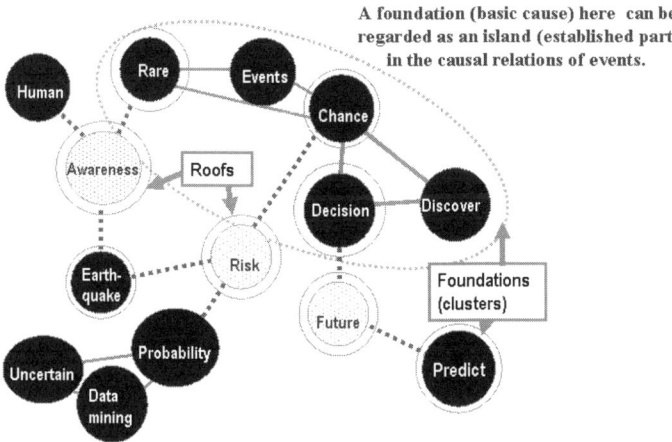

A foundation (basic cause) here can be
regarded as an island (established part)
in the causal relations of events.

Columns and a roof (concepts supported by foundations) form
a special type of bridge, i.e. a bridge activated by the islands.

Fig. 22.1. An example output of *KeyGraph*. The black nodes representing frequent words are connected by solid lines and form clusters called foundations, with underlying common contexts. The shadowed nodes, e.g. "awareness", showing infrequent words are connected with multiple foundations by dotted lines, showing rare but possibly essential words. Nodes linked to multiple strong columns are double-circled, meaning significant positions in the structure

Table 22.2. A document and an earthquake sequence as special cases for *KeyGraph*: items for the same number correspond to each other, and in Table 22.1 and this table

Case of a document:

(1)	Actions of the author, i.e. writing words
(2)	Periods ('.'s): the ends of sentences
(3)	Basic concepts for the author
(4)	The co-occurrence of words, for expressing a basic concept
(5)	The relations of asserted words to basic concepts

Case of an earth quake sequence:

(1)	The sequence of focal faults of past earthquakes
(2)	Periods ('.'s): moments of major changes in the movement of land crusts, i.e. moments of major earthquakes emitting large energy
(3)	Fundamental faults
(4)	Branch faults of one fundamental fault pushing each other locally
(5)	The stress to risky faults from the movement of fundamental faults

22.4 Experimental Support of the Semantics of *KeyGraph*

22.4.1 Risky Faults and Fundamental Faults

We dealt with 390 major active faults in Japan as data 1, and ten thousand earthquakes per year on average as data 2. These data were obtained by mixing data of JUNEC (Japanese University Network Earthquake Catalog) and data of JMA (Japan Meteorological Agency). The threshold value $M\theta$ used in step **(EQ2)** was set to 4.0.

First, let us show the performance of F^3 for the Kansai area of Japan. Here, data 2 was taken for earthquakes in 1985 in this area. Figure 22.2, whose upper half shows the active faults by the numbered lines, presents the output of F^3 in the lower half. Active fault no. 39 was obtained as risky. No. 39 is the Nojima fault, at which the south Hyogo earthquake of M7.2 occurred in 1995. This fault was selected to be the most risky from each earthquake sequence of each year from 1985 to 1992 in the Kansai area. We can say that *KeyGraph* detected the risk of the south Hyogo earthquake, because data 2 in this case was recorded before the big earthquake. The 51 (top 14%) risky faults obtained from the data of 1985 to 1992 all over Japan included all the real earthquake focuses after the time the data was recorded.

Another point in Fig. 22.2 is that we find fault no. 39 stressed by the cluster including two fundamental faults under areas (A) and (B) and faults no. 24 and no. 34 (each node forming one cluster). This corresponds to the seismological semantics presented above, and also to the prevalent seismological hypothesis that the southern area is moving to the west stressing (A) and the northern area (Eurasian plate) is relatively moving to the east stressing (B). Thus, fault no. 39 is located to be stressed by these movements of fundamental faults (A) and (B). These validate the seismological semantics of *KeyGraph* presented above.

22.4.2 Results of Japanese Risky Faults

(1) The snap-shot results from data 2 from 1985 to 1992 in JUNEC

From this range of data, the thick lines in Fig. 22.3 were obtained as risky faults by *KeyGraph*. Here, let us go into detailed evaluation of this result. The solid lined faults in each of the five frames in Fig. 22.3 were obtained by F^3 as the 13% riskiest faults in the frame from data 2 of earthquakes which occurred in the framed area. The output figures of *KeyGraph*, exemplified by Fig. 22.2, for these frames showed that local crust areas smaller than each frame stressed the risky faults. On the other hand, faults with the dotted lines in Fig. 22.3 were obtained as the 13% riskiest faults in overall Japan, from the same data set. The difference between the solid and the dotted lines comes from the activity localities, i.e. local (solid lines) and global (dotted lines) land-crust activities.

In Fig. 22.3, the thicker line (fault) was obtained as the riskier, i.e. we obtained the larger *key* value of *KeyGraph-step* 2. The circles in Fig. 22.3 depict the greatest 10 earthquakes after or near the years while data 2 was recorded. The nearest faults to all these epicenters were selected by F^3 as risky. We restricted the target data to

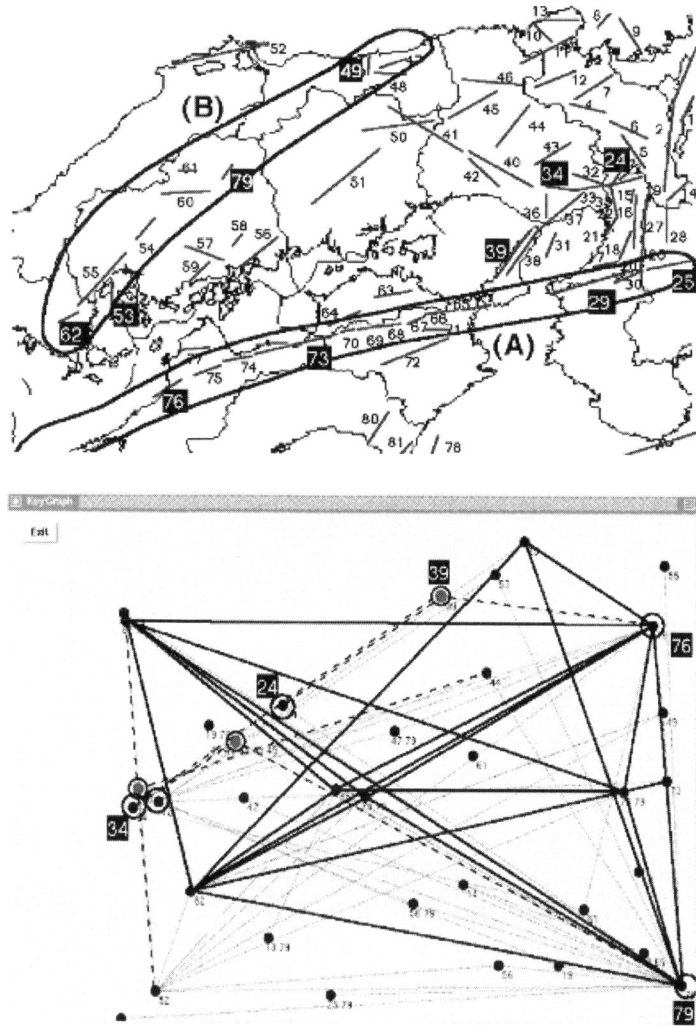

Fig. 22.2. The result of *KeyGraph* for the Kansai area. KeyGraph obtained the double-circled nodes in the lower figure as risky faults. The thick solid lines and black nodes form clusters, i.e. foundations, and the dotted lines show the stress from clusters to faults. Thick dotted lines represent the strongest stress

earthquakes before 1992 here, for showing that F^3 successfully detected future risks of earthquakes.

Faults in the output included the 12 riskiest faults (areas (A)–(L) in Fig. 22.3) to which seismologists pay attention, e.g. in [22.2, 22.8]. Further, believing in all the results of F^3 except only one probable error pointed out by staff in ERI (Earthquake Research Institute of University of Tokyo), the precision of risky faults obtained by F^3 was 98% (50/51). The 'only probable error' here was as shown in the thick

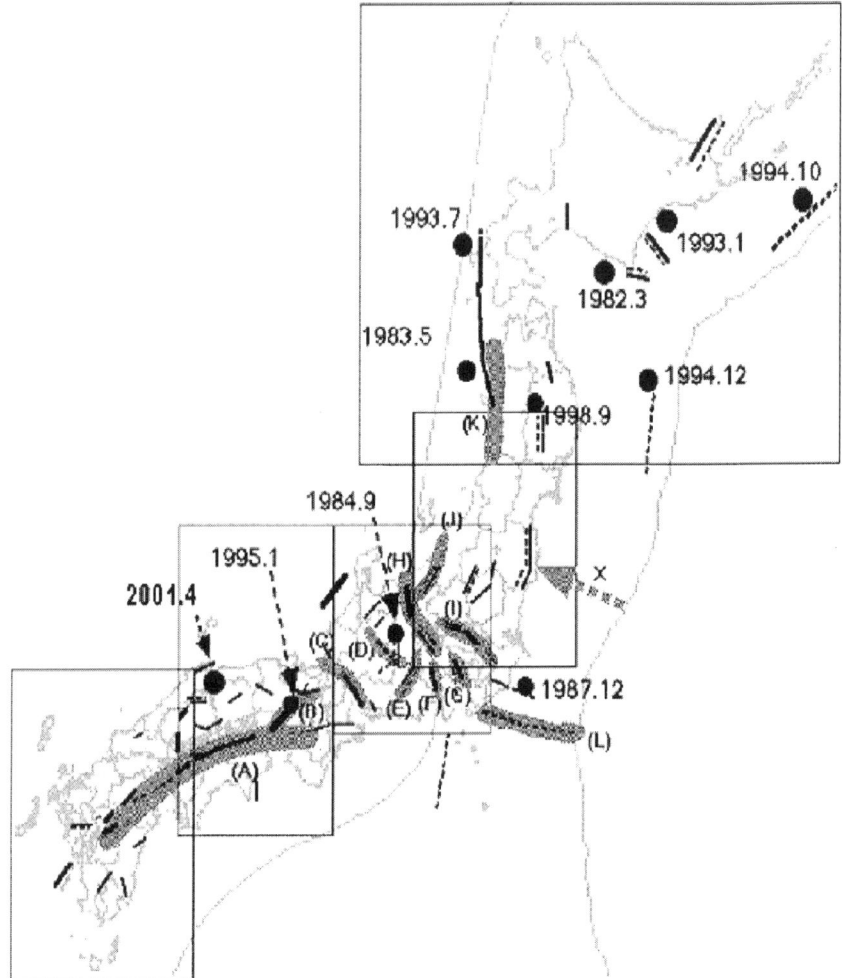

Fig. 22.3. Risky faults obtained by F3, from the data of earthquakes in each framed area (solid lines) and of all over Japan (dotted lines). The shadowed areas are estimated as risky according to seismologists

dotted arrow with X in Fig. 22.3. The sea-side fault in Fukushima prefecture, at the arrowhead, was taken as risky although seismologists regard this area as rather safer. If this is a failure, the reason is regarded as being that some earthquakes in the data, whose epicenters are near this fault but the real focus is far, i.e. on a deep extension of the *Pacific plate* (seen in the east of the trough at the tail of the arrow, if looked at from the surface of land), disturbed the risk detection.

It may be possible to reduce such failures by restricting our target to only shallow active faults to ignore deep earthquakes at troughs. However, in that approach, we have to give up some good results. For example, area (L) in Fig. 22.3 (south Kanto area) is said to be possible to quake greatly by seismologists, and the trough in area (L) was judged risky by F^3. If we ignore deep earthquakes, earthquakes in area (L) will be thrown away from data 2 and (L) becomes eliminated from the risky areas of F^3. For avoiding both errors, we have to consider the crust structures in the deep level under the ground. However, unfortunately, the deep underground structures of most faults are unknown.

(2) The shifts of risky areas

When a big earthquake occurs, fundamental faults having affected the quaked focal fault usually begin to stress other faults they can affect. This newly stressed fault tends to exist near the quaked fault, because of being affected by the same fundamental faults. For example, after the Tottori earthquake (M7.2) in 1943, the stress shifted to a near-by fault which caused a big (M7.1) earthquake in Fukui prefecture in 1948. This section shows that F^3 caught such phenomena in various parts of Japan. By this, we can validate the semantics of (KG2), i.e. a fundamental fault affects its branch faults with great energy and triggers the local co-quaking of branch faults.

The period from 1993 through 1995 was a term in which many big earthquakes occurred in Japan. Here, we compare the risky areas obtained from the earthquake history before 1993 and after January, 1995, for detecting stress shifts after big earthquakes. Figure 22.4 shows that the stress which had been concentrated in a narrow area including no. 39 faded with the south Hyogo earthquake in 1995 at no. 39, and shifted to the left end (west–south) of the map along the median tectonic line (area (A) in Fig. 22.2 and Fig. 22.3). This shows that the focal fault no. 39 of the south Hyogo earthquake had been stressed by the fundamental fault, i.e. the

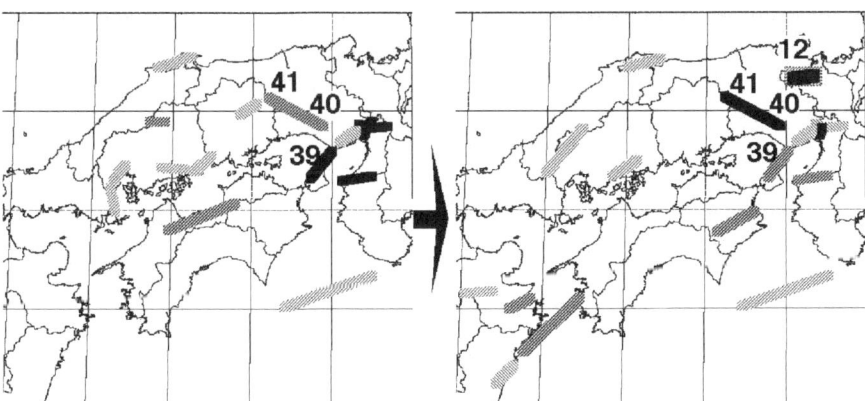

Fig. 22.4. The shift in the south of the stress in the Kansai area. The more densely black lines show faults with the stronger stress. The left-hand figure was obtained from data before 1993, and the right from data after January, 1995. No. 12 (the striped area) was of the highest key for every year after 1995 and quaking little now – a typical sign of near-future risk of a big earthquake. The illustrated faults are the riskiest 15 of 81 active faults in data 1 of the area

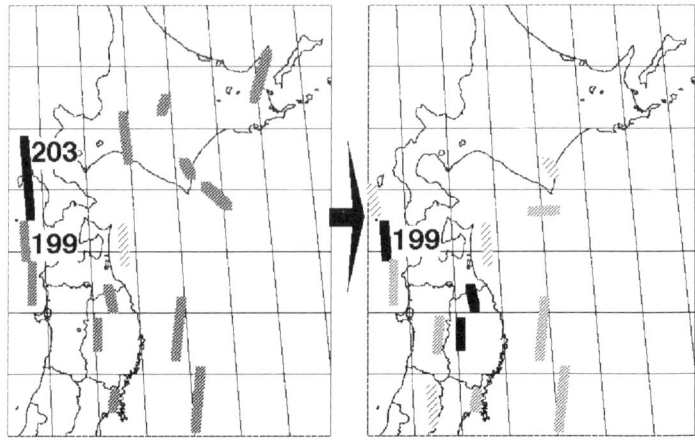

Fig. 22.5. The south shift of stress in the north of Japan, before and after the south-west Hokkaido ocean earthquake in 1992 which occurred near no. 233. The riskiest 14 faults of 70 in data 1 of this area are shown in each figure

median tectonic line. Further, because the movement (relative to other areas) of the area between areas (A) and (B) was speeded up after 1995 due to the big quake at fault no. 39, the stress at faults nos. 40, 41, and 12 came to be strengthened. This stress scattering along the median tectonic line is seen in Fig. 22.4.

Similarly, in the result for the northern part of Japan in Fig. 22.5, the stress in no. 203 shifted to no. 199 with the south-west Hokkaido ocean earthquake in 1992 near fault no. 203. A fundamental fault near both nos. 199 and 203 seems to affect both faults, which is supported by the fact that this area is near to the boundary of two large plates (Eurasian plate and North America plate). Also, the overall stress around Hokkaido shifted to the south, and currently the stress near volcano (e.g. Mt. Iwate) and the ocean areas of Tohoku and Kanto were obtained to be strong. In fact, the fault near Mt. Iwate had a severe earthquake in 1998. All in all, risky faults in the results of *KeyGraph* shifted in directions parallel to lines under which hidden large faults are supposed to exist (or boundaries of active fault provinces in [22.8]).

(3) Other risky areas

Risky faults according to *KeyGraph*, for areas other than those mentioned above, were as follows. The shifting of risky areas found are related to boundaries of active fault provinces or plates [22.8].

South area (Kyushu): the area around the median tectonic line and its south extensions are stressed (after the south-Hyogo quake, the stressed areas shifted to the south along this line). Southern extension of the line includes Taiwan, where M7.0 occurred in 1999.

Middle-west (Chubu and Kanto): the stress in the east of the Izu peninsula has been strengthened by the north shift of the Philippine plate. Faults in Gifu, Nagano, and Niigata prefectures, which are near the large fundamental fault Fossa Magna, are found to stay risky before 1993 and after 1995.

22.5 Conclusion

Applying an existing algorithm (*KeyGraph*) for the new purpose (earthquake-risk exploration), differing from the problem the algorithm previously aimed to solve (document indexing), has triple impacts.

1. Engineering impact: by finding events (e.g. an earthquake sequence) analogous to previously deal with data (e.g. document), we can extend the applicability of the previous algorithm to various problems.
2. Scientific impact: identifying the model, or the cause-and-effect structure of real events, is a major aim of science in general. When there is common structure in previous and new data, we can obtain a clue to modeling the new one, referring to the model for the previous one.
3. Social impact: some rare events are significant for human life, and earthquakes are the typical examples of negative significance. According to items 1 and 2 above, we can expect to extend the successful results in this paper to a new kind of events.

Among these, the social impact is being extended to techniques for *chance discovery*, i.e. discovery of significant events/situations for decisions, as opportunities and crises.

Acknowledgements. I express the highest gratitude to Prof. Kunihiko Shimazaki of the Earthquake Research Center in the University of Tokyo, Prof. Yoshihiko Ogata of the Institute of Statistical Mathematics, and other seismologists and statisticians who discussed with me the soundness of applying *KeyGraph* to the time-series data of earthquakes. They could not confirm my soundness, but mentioned the ways here of considering the structural mechanism of land crusts possibly reflect the underground activities, although this can not be verified by direct observation. This study has been conducted by the fund for Research Project on Discovery Science, as Scientific Research on Priority Areas of the Ministry of Education and Culture of Japan, from 1998 to 2000.

References

22.1 Hori T, Oike K (1993) Increase of intraplate seismicity in Southwest Japan before and after interplate earthquakes along Nankai trough, *Proc. Joint Conf. of Seismology in East Asia*, pp.103–106

22.2 Matsuda T (1997) *Active faults* 6th Edition, (in Japansese), Iwanami-Shinsho No 423, Iwanami, Tokyo, Japan

22.3 Ogata Y, Utsui T, Katsura, L (1995) Statistical Features of Foreshocks in Comparison with Other Earthquake Clusters, Geophysical Journal International, 121:233–254

22.4 Chance Discovery by Stimulated Group of People, Application to Understanding Consumption of Rare Food, *Journal of Contingencies and Crisis Management*, 10(3):129–138

22.5 Ohsawa Y, Benson NE, Yachida M (1998) *KeyGraph*: Automatic Indexing by Co-occurrence Graph based on Building Construction Metaphor, *Proc. Advanced Digital Library Conference* (IEEE ADL'98), 12—18.

22.6 Ohtake M (1994) Seismic gap and long-term prediction of large interplate earthquakes, *Proc. of International. Conf. on Earthquake, Prediction and Hazard Mitigation Technology}*, 61–69.

22.7 Okada A, Watanabe M, et al (1993) Active fault topography and trench survey at the central part of the Yangsan fault, southeast Korea, *Proc. of Joint Conf. of Seismology in East Asia*, 64–65

22.8 RGAFJ (The Research Group for Active Faults of Japan) edt., (1992) *Maps of Active Faults in Japan with an Explanatory Text*, Univ. of Tokyo Press.

22.9 Rikitake T (1976) Recurrence of great earthquakes at subduction zones, *Tectonophysics*, 35: 335—362

22.10 Savage JC (1992) The uncertainty in earthquake conditional probabilities, *Geophysis Res. Letter*, 19: 709–712

22.11 Suzuki Y, Matsuo M (1995) A probabilistic estimation of the expected accelerations of earthquake motion by inland active faults and its application to earthquake engineering, Lemaire M, Favre JL, Mebarki A (eds) *Applications of Statistics and Probability – Civil Engineering Reliability and Risk Analysis*, Ballema Press, Ramona, Canada, 635–641

22.12 Utsu T (1980) Spatial and temporal distribution of low-frequency earthquakes in Japan, *J. Phys. Earth*, 28:361-384

22.13 Utsu T (1983) Probabilities associated with earthquake prediction and their relationships., *Earthquake Prediction Research*, 2:105-114

22.14 Utsu T (1999) Representation and analysis of the earthquake size distribution: A historical review and some new approaches. *Pure Appl. Geophysics*, 155: 509-535

22.15 Vamvatsikos D, Cornell CA (2002) Incremental Dynamic Analysis, *Earthquake Engineering and Structural Dynamics*, 31(3):491-514

22.16 WGCEP (Working Group on California Earthquake Probabilities) (1995) Seismic hazards in Southern California: probable earthquakes, 1994 to 2024; *Bulletin of the Seismology Society of America*, 85: 379-439

23. Application to Questionnaire Analysis

Yumiko Nara[1] and Yukio Ohsawa[2,3]

[1] Osaka Kyoiku University, Osaka 582-8582, Japan
[2] PRESTO, Japan Science and Technology Corporation, 2-2-11 Tsutsujigaoka, Miyagino-ku, Sendai, Miyagi 983-0852, Japan
[3] Graduate School of Business Sciences, University of Tokyo, 3-29-1 Otsuka, Bunkyo-ku, Tokyo 112-0012, Japan
email: nara@cc.osaka-kyoiku.ac.jp, osawa@gssm.otsuka.tsukuba.ac.jp

Summary.

In this chapter we apply the *double helix* introduced in Chap. 1 to model the process of *chance discovery*, applied to the survey of peoples' behaviors on the information from the Internet and in the real world being motivated by unknown factors. This, at the end of this chapter, leads to the process model of chance discovery. Simply abstracting the double helix (in Chap. 1), we review what the process of chance discovery means and how it helps social scientists.

23.1 The Human Society as a Complex System

In previous social surveys based on questionnaires asking a certain number of questions to a number of interviewees, social scientists often applied statistical analysis (e.g. path analysis by tools such as AMOS) to the survey data. This corresponds to introducing the approximation of $m = n$ to the target sample set of people in Sect 1.1, where n is the number of all factors affecting the world being observed and m is the number of observable variables among the n variables. That is, this approximation means regarding all factors as observable and that each specific feature, beyond the ability of observable factors to present, of samples are ignorable or killing the influence of each other. However, with the dynamic changes in the social infrastructures, e.g. the rapid growth of the Internet, the easy publication of individual ideas and dynamic flows of peer-to-peer information are increasing the impact of specific and unpredictable appearance of personalities onto the whole society. Socialists, market researchers, politicians, and educators need to be aware of specific behaviors of a few questionnaire subjects making decisions and actions on emerging opportunities and risks [23.14].

In this chapter we apply the *double helix* introduced in Chap. 1 to model the process of *chance discovery*, applied to the survey of peoples' behaviors on the information from the Internet and in the real world motivated by unknown factors. This, at the end of this chapter, leads to the process model of chance discovery. Simply abstracting the double helix (in Fig. 1.4) as in Fig. 23.1, let us review what this means to social scientists. We will refer to Fig. 23.1 for the explanation of each step we made in the discovery process.

For inspecting the latent motivation of the behaviors of people, social scientists have been applying analysis methods developed in statistics in order to summarize the understandable part of data obtained from social surveys. However, hidden motivations of behaviors could be hardly identified because:

– In hypothesis-driven research, e.g. employing path analysis, we began from a concrete model about the relations among variables which had been already in our mind and ended with a validation of the model.
– In data-driven research, e.g. employing correspondence analysis, we began from given data and ended with a vague (sometimes misleading) interpretation of event-to-event relations.

The points of the double helix, having two helical subprocesses as in Fig .23.1, are at least three-fold for social scientists. First, the spiral process for humans progresses to deeper awareness of various behaviors of people. The mind of a human is the best sensor of the feeling of people, whereas a machine can not go beyond the data from chemical reactions in the brain. Second, the other helix, i.e. the process of a computer, receiving and mining data ('DM-a' and 'DM-b' in Fig. 23.1), has much capacity to deal with the various tools social scientists have in hand for statistical analysis: *KeyGraph* [23.7, 23.11] is one tool, and AMOS (see later) is another. Third, the interaction between these two helices is essential. That is, 'the subject data' monitoring the mind of the discoverer of chances has not been fed back to the process of research even though we have been paying attention to the major elements contributing to the mind and the social activities of humans.

23.2 A Chance Discovery for Understanding Chance Discovery as an Application of the Double Helix

Tracing the arrows in the process of chance discovery in Fig. 23.1, let us exemplify the human–computer interaction for understanding the influence of value criteria of individual people on their behaviors. The interaction includes data acquisition with the questionnaire led by Nara [23.5], supported by *KeyGraph* and the double helix model having been developed by Ohsawa [23.10]. The aim of this process is chance discovery, where the chances are the novel and significant behaviors of people living in the real world and Internet users, and the discovery aims at understanding the reasons why the new feelings came out and where they are going. This is significant for educators and marketers as well as for social scientists, as mentioned above. Furthermore, if we can understand how people discover new events/situations significant to take actions on, it will make a trustworthy model of the process of chance discovery.

23.3 A Double-Helix Process to See the Behaviors of People on the Web

In social sciences, the influence of the Internet on human behaviors is gathering attention. We briefly exemplify the double helical process for understanding the

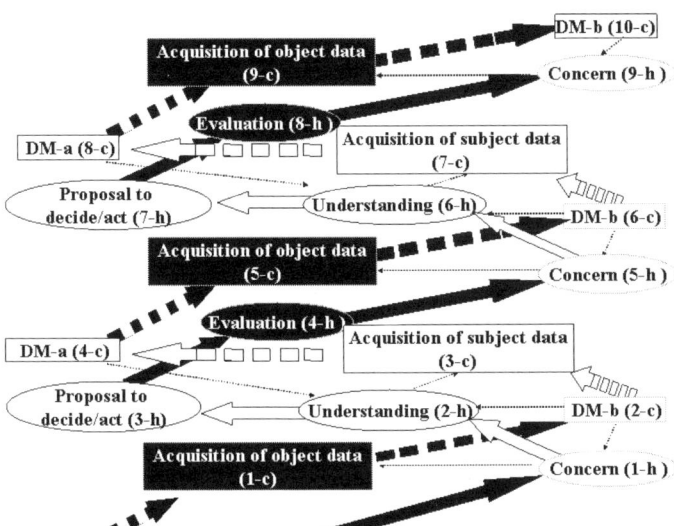

Fig. 23.1. The double-helical process of chance discovery (the numbers such as '1-h' correspond to each step below). The process continues from Chap. 1

influence of value criteria of individual people on their behaviors. The interaction includes data acquisition with the questionnaire, supported by *KeyGraph* (see [23.6] or Chap. 18) running with the double helical process. Hereafter, the number of each step with '-h' ('-c') appended, i.e. 'k-h' ('k-c') for integer k means the information process of a human (computer) in the increasing order of k. That is, the thick solid (dotted) arrows in Fig. 23.1 depict the steps on the side of human(s) (computer). Thin arrows show the interactions between a human and a computer.

(1-h) *Concerned* **with behavioral factors of people.** The first model by Nara was as in Fig. 23.2, composed of five nodes and five links each corresponding to her initial hypotheses for explaining the behaviors of people in the real and the virtual (Internet) world. In this figure, we find a sociological way of thought about behaviors of people, focusing attention on fundamental personalities.

(1-c) The object data: questionnaire results. Based on Fig. 23.2, Nara made questionnaires for university students. The students answered questions about their own behaviors and feelings about affairs in the real and the virtual worlds .

(2-c) DM-b: path analysis for the questionnaire data. The hypothesized relationships between behaviors and personalities, in Fig. 23.2, were mostly validated by path analysis applied to the questionnaire results. Here the usage of the Internet and subjects' empathy correlate with the ethical factors in the questions. In [23.5], Nara declared that four of the five essential hypotheses were validated.

(2-h) *Understanding* **of peoples' behaviors**
By clarifying why hypotheses became validated in (2-c), Nara mentioned 'external sanctions' as a factor distinguishing the real world from the virtual world for inhabitants. For example, 'sensation-seeking' people can not be satisfied in the real world as easily as in the virtual world, various behaviors are forbidden, and penalties are given apparently on violating the rules.

(3-c) Nara [23.5], the data on discoverer's thoughts
The subject data taken here is the document [23.5].

(4-c) DM-a: the result of *KeyGraph* **applied to [23.5]**
Data from [23.5] was processed by *KeyGraph* [23.6]. *KeyGraph* deals with data in the style such as:

$$Data = a1 \ a2 \ a3 \ldots . \ am.$$
$$b1 \ b2 \ldots .. \ bn.$$
$$c1 \ c2 \ c3 \ldots cp.$$
$$\ldots$$

Here, each line ending with a delimiter ('.') show the set of co-occurring items. If the data is a document, each line is a sentence composed of items corresponding to words in the sentence. If the data is questionnaire-result data, each line is the answer set of one examinee.

Briefly, *KeyGraph* obtains a set of item clusters called *foundations*, each composed of frequent items depicted by nodes and the links between items of the highest co-occurrences, i.e. occurring often within a predefined window of distance (one sentence, if the data is a document) in the data. Each foundation represents a basic concept underlying the data, for one who has some interest in the data domain. Then *KeyGraph* shows rare (not so frequent as in foundations) items co-occurring with multiple foundations. These can be regarded as candidates of 'chances', i.e. rare but significant items carrying emerging significance from/into the basic environments having been established as foundations. In the case of a document, a chance can be a word carrying new assertions significant for the author's decision to write the document, based on basic concepts Nara previously established in her mind (see Part IV for details on data-mining methods, including *KeyGraph*, for chance discovery).

Here, *KeyGraph* obtained Fig. 23.3 from paper [23.5] processed as a document—words such as 'sanction' and 'scales' appeared as the candidates of chances. As mentioned above, 'sanction' is a key newly emerging for her. 'Scales' means the questions in the questionnaire for measuring the consciousness of each examinee. This implics her invariant concern looking for deep motivations of people and the 'scales' for measuring the influence of those factors. This output made Nara aware of her own interests in these two items of which she was not conscious before.

(3-h) Proposal to realize ethical education in the Internet,
(4-h) evaluation of the results, and (5-h) new concerns
Because an external sanction is harder to be given in the virtual world than in the

real world, Nara proposed ethical education enhancing internal sanctions in people's minds. Yet, this proposal was too abstract to appeal to make a realistic decision. This is not an evaluation of the accuracy of a piece of knowledge discovered, but is an evaluation of the utility of a proposal based on a chance, i.e. the appearance of the concept 'sanction': she had to understand further what actions should be made for internal sanctions. Thus, her new concern led to the next step of seeing extraordinary (i.e. rare) behaviors and their causes, for finding effective actions to stop bad behaviors.

5-c) New object data: rare answers in the data
Due to the awareness of her own new interest in (4-h), Nara came to be interested in rare answers rather than frequent answers. In the framework of Fayyad et al.[23.1] (Fig. 1.1 in Chap. 1), this is a reflection to the 'selection' step. Here she had an alternative choice to make a new questionnaire for young people in the prison for focusing attention on rare bad behaviors, if she had no obstacle in doing it – the acquisition of data is a generalized step of selection.

(6-c) DM-b: *KeyGraph* applied to the set of a few answers and
(6-h) understanding the output figure
To the rare-answer set in the results of the questionnaire in (5-c), we applied *KeyGraph* again. Here, an individual examinee's set of answers was dealt with as a unit of co-occurrence, and Fig. 23.4 was obtained. Nara noticed the significance of the concept 'rational egoist tendency', i.e. personality to make crimes if no penalty can be given, matching with her rising interest in sanctions. The effect of visual data mining is the externalization of such unnoticed tendencies of subjects. Rare answers as 'I smoked before 20 years old" or "I drink carelessly till losing sense' were found to be a significant appearance of potential emotions of young people, relevant to deep-level personal tendencies affecting attitudes to IT tools. For example, we can apply established strategies of education in the real world for young smokers, for stopping various behaviors on the Internet coming from rational egoistic tendencies.

(7-h) New proposals Considering the fundamental factors in Fig. 23.2, i.e. (1) human relations in communities (2) information seeking, and (3) deep-level ethics relevant to rational egoistic tendency, feasible policies were proposed for the virtual society. For example, educators desire to teach young people to eliminate rational egoistic tendencies.

(8-h) evaluation and (9-h) new concerns Although the proposals in (7-h) might be reasonable, it is not concrete yet – this is another evaluation of the chance based on the utility of proposals. That is, the human tendency to seek information and their rationality in dealing with new information is above the reach of the purely ethical point of view. People may gather information for maximizing self-profit, and their rationality may differ on what context of business they are involved in. For example, extreme fans of sports players sometimes make extraordinary behaviors, even going beyond their own rational decisions. On the other hand, inside traders of stocks

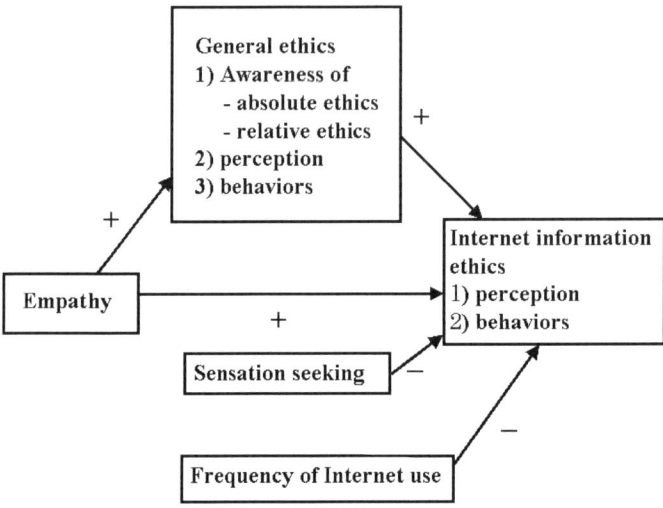

Fig. 23.2. The initial model of behaviors of people in/out of the Internet

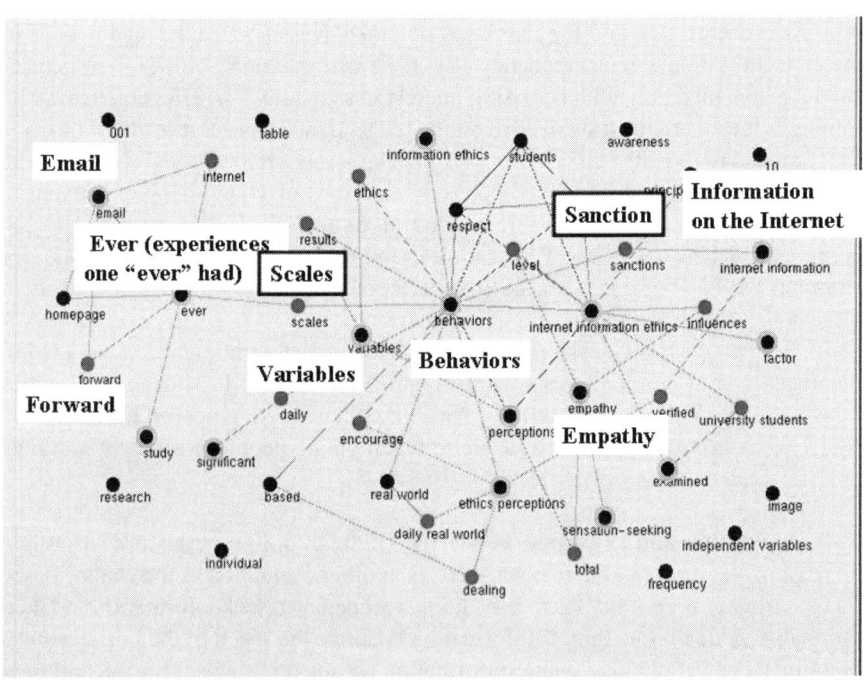

Fig. 23.3. The output of *KeyGraph*

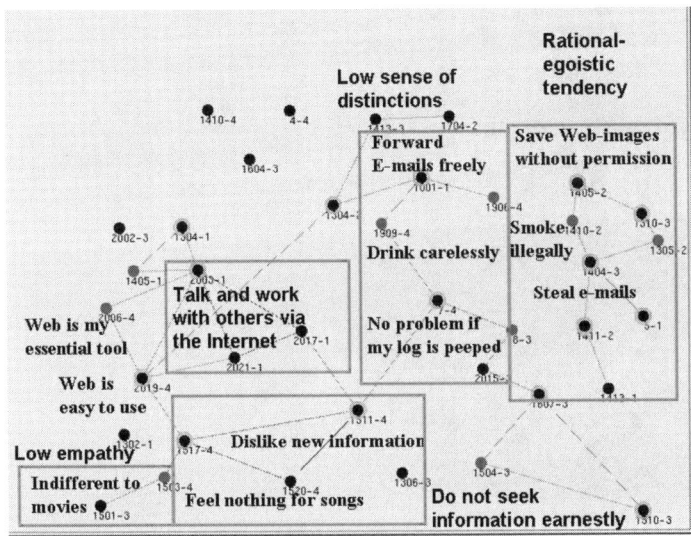

Fig. 23.4. The result of *KeyGraph* for the questionnaire result

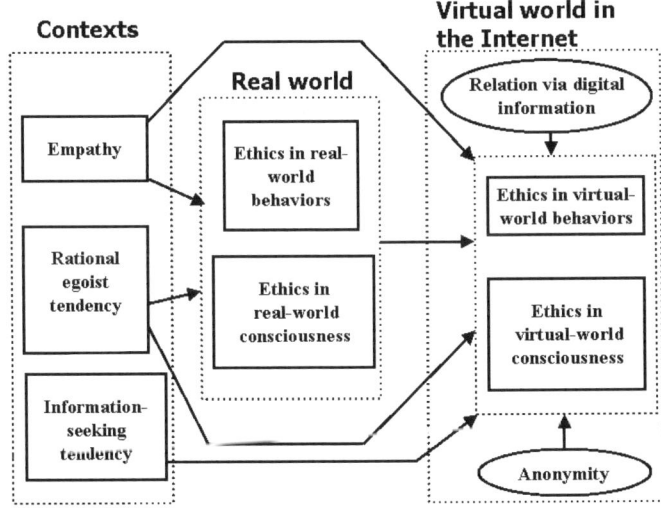

Fig. 23.5. A new model acquired from Fig. 23.4

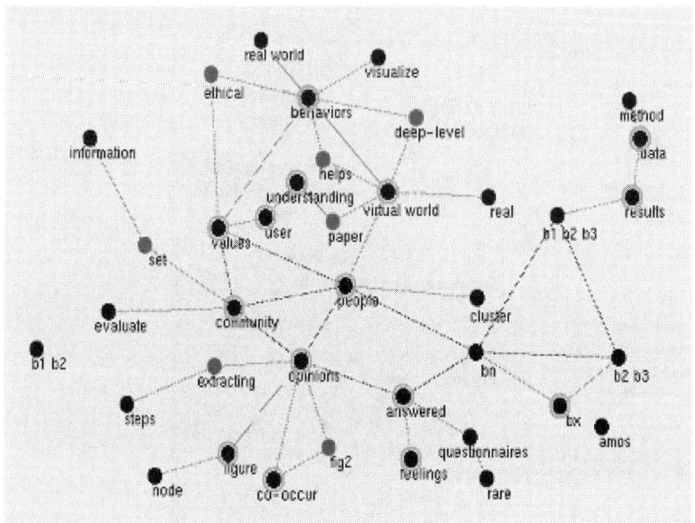

Fig. 23.6. The output of *KeyGraph*

are making illegal but rational decisions if they take the risk of imprisonment into account. Thus, we came to be concerned with business-sake information acquisitions and business-peoples' behaviors on the Internet and in the real world. Thus, we planned an inter disciplinary discussion about the decision-making process on the Internet, including us authors and other areas such as business management, as mentioned below.

(9-c) Merged data of subjects and objects: words in discussion and (10-c) the output of *KeyGraph* as DM-a/b We gathered a group of discussants to talk about the decision process with the information on the Internet, with looking at Fig. 23.4. This is a collection of object data in that we had discussants who are also objects in the environment we are studying. However, this is also a collection of subject data in that we joined the discussion. Based on the lesson learned in [23.8] on group-working discovery of chances, Ohsawa called eight discussants from a variety of areas. That is, we had three experts of information technologies, three of management science, and us two. In the discussion, information and its relevance to real actions seemed to be the central topic. We reached Fig. 23.7 as a model of decision making.

Yet, the excitement during the three hours rather disturbed the discussants from systematic understanding of where their arguments were going. We recorded the overall discussion and made a text file, as the data of subjects. Then, the text was processed by *KeyGraph*. Figure 23.8 was obtained as a result. As seen here (see the thick letters for the English translation), the personality factors made a significant bridge from the consciousness in the human mind to the embodied decisions and actions. That is, discussants paid strong attention to empathy, rational egoism, and

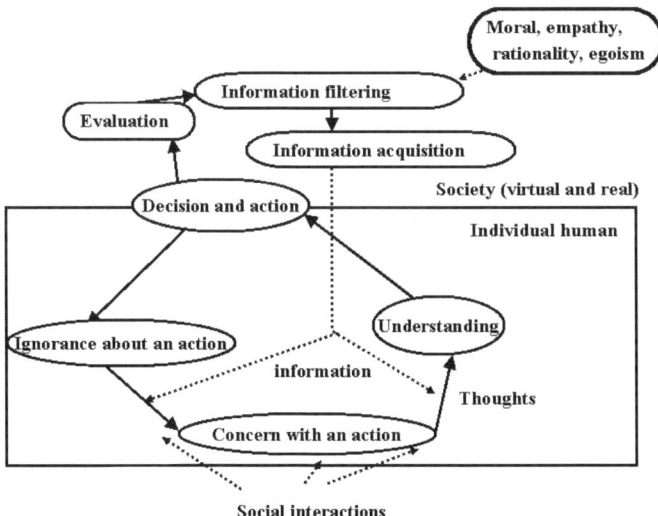

Fig. 23.7. The decision-making model in the real and the virtual worlds, drawn on the board in the discussion.

information-seeking tendency, as significant factors affecting human decisions. The change from Fig. 23.7 to Fig. 23.9 is clear evidence of the effect of showing Fig. 23.8.

(10-h) Understanding and (11-h) proposal

As a result, we obtained Fig. 23.9 as the new model of decisions on the Internet, affected by personalities. Two examples of the new proposals obtained were that:

1. According to the difference between the paths from 'information exchange in the real world' and 'information exchange in the virtual world' respectively to the process in the human individual's mind in Fig. 23.9, Web-based marketing to customers should be directed to customers who accept virtual (email) recommendations better than real-world (face-to-face) recommendations.
2. Multi-paths (via the Internet and face-to-face) to customers in business is required, for the same reason according to Fig. 23.9 as in (1).

Here we can also find that we may lose the point of a discussion during the time of exchanging complex opinions, and visualization of the overall conversation hours can aid the awareness of each discussant about the core ideas being talked about. Thus, subject data works.

It is noteworthy that the model in Fig. 23.9 has similarity to Fig. 1.2 (Chap. 1) on which our arguments were based, because the target we have been dealing with was the emerging behaviors of people with the awareness of chances in a dynamic social environment.

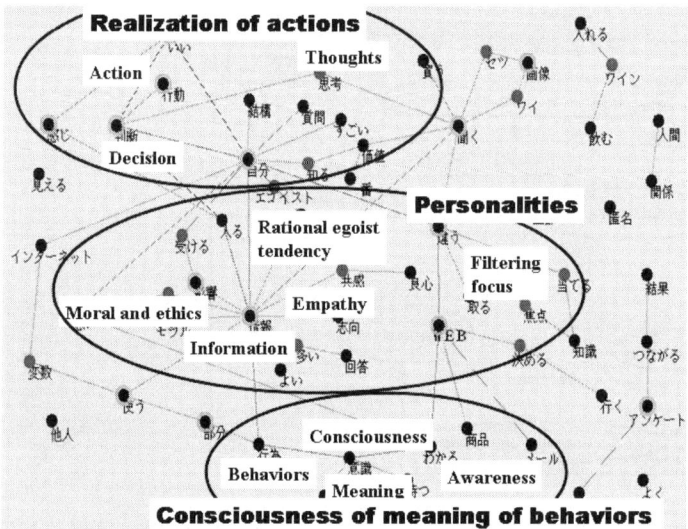

Fig. 23.8. Application of *KeyGraph* to the text of the discussion looking at Fig. 23.2.

23.4 The Final Revision of the Process Model

[The new questionnaire survey]

For validating the latest hypotheses as in Fig. 23.9 as a process model of chance discovery , we organized a new social survey on peoples' decision process in the real world and on the Internet. Corresponding to each arrow in Fig. 23.9, we made 400 questions in total, for 1007 subjects in the USA and 1114 in Japan. Certain parts of substantial questions are owned by Nara [23.15]. The samples here were chosen randomly from the panel data of Nihon Research Center, which had been made as a random set of samples from all over Japan and the USA (excluding the states of Hawaii and Alaska). We made random postal mails for Japan and random digital dialing (RDD) for the USA for these samples respectively, because the postal system is not equally reliable all over the country in the USA.

[The results for US people]

As a result, the result of *KeyGraph* to the answer data for US people was obtained as in Fig. 23.10. This result illustrates a clear cyclic process of chance discovery, validated by the co-occurrence computation among answers. That is, each link and the series of links depict a certain set of people in a specific state, and the connection of these links show the sequence of continuous shifts of the mental contexts of humans. Along this context shift, Fig. 23.10 means that US people shift in order as:

1. Concern with peripheral information.
2. Investigation of central information.
3. Perception of risk factors, first of humans (violation of Internet ethics) and then of nature/society (cancer, traffic accidents, etc).

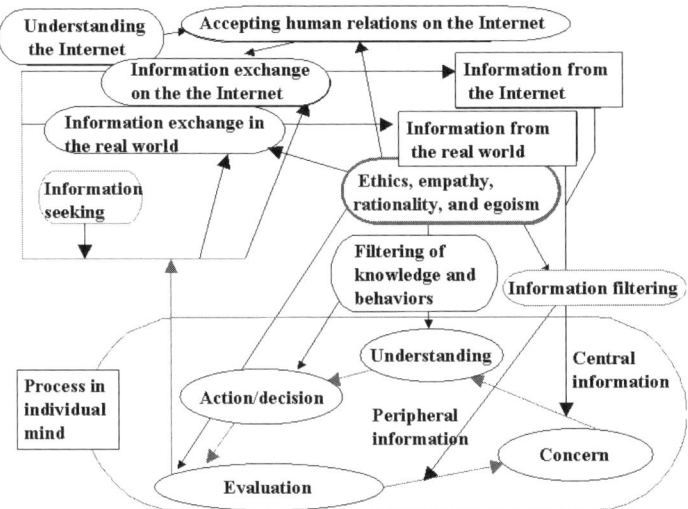

Fig. 23.9. The new model obtained.

4. Perception of and communications for chances, e.g. of buying rare clothes, with strong desire for information.
5. Weakened desire for information.
6. Evaluation of the WWW, and back to new concerns as (1).

Accordingly, we obtained Fig. 23.11 as the validated model of chance discovery, validated in that all links are abstracts of edges in Fig. 23.10.

[The results for Japanese people]

On the other hand, the result of *KeyGraph* to the answers from Japanese people was as in Fig. 23.12. This result is composed of the negative attitudes toward the Internet and new information in general. In comparison with the results of subjects in the USA, the following elements are missed here:

– The cyclic process.
– Concern for new information (information-seeking tendencies).
– Morals and risk perceptions as essential parts of the cycles of mental states,
– Understanding and evaluation of their own decisions.

Even though the authors are Japanese, this result could not be interpreted as implying our orientation for chance discovery. Reflecting this result, we obtained Fig. 23.13 as the validated model based on Fig. 23.11 for Japanese subjects.

23.5 Discussion

As a result, the tendencies of US people had a strong orientation to chance discovery, and matched with the cyclic model we had made as the process model

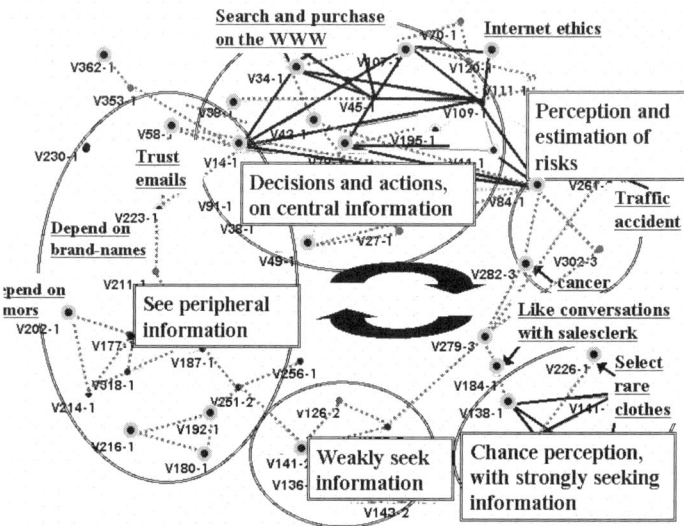

Fig. 23.10. The result of *KeyGraph* for US people (each node represents an answer)

of chance discovery. This result seems co-supporting with another comparison of US and Japanese from literature relevant to chance discovery. That is, fundamental strategies for dealing with significant low-probability changes/events (risks and opportunities) have been developed in the USA as in [23.12, 23.4, 23.13], whereas Japanese literature on chances is mostly dedicated to specific domains, e.g. infant education, e-learning, sales-force tasks, etc.

As in Chap. 1, the model based on the subsumption architecture was proposed for speeding up the double-helical process of chance discovery. However, the new architecture has some risk of missing steps, i.e. of 'concern', 'understanding', 'action', and 'evaluation'. The subsumption will be trustworthy for US people who already have the double helix in mind. On the other hand, people in Japan need some instructions before realizing the helical process by seeing only the multi-interface for aiding the subsumption architecture of human perception of chances. By tracing the double helix sequentially, we are more confident of achieving the discovery of meaningful chances. The lessons from our current results in the final model may enable us to show Japanese people what they are lacking for overcoming the long and fatal depression of the economy.

For the interest of social scientists, the method applied here for chance discovery from the interaction between discoverer, questionnaire data, and the computer is a significant progress if we compare this with the grounded-theory approach [23.3]. The essential common point between our method and the grounded-theory approach is that data is gathered for refining theory and the data is referred to for forming new theory. In the grounded-theory approach, the coding of theory progresses by creating, clustering and, linking factors being involved in the target world for converging to a concrete theory. Then, new data is gathered if required for refining the theory.

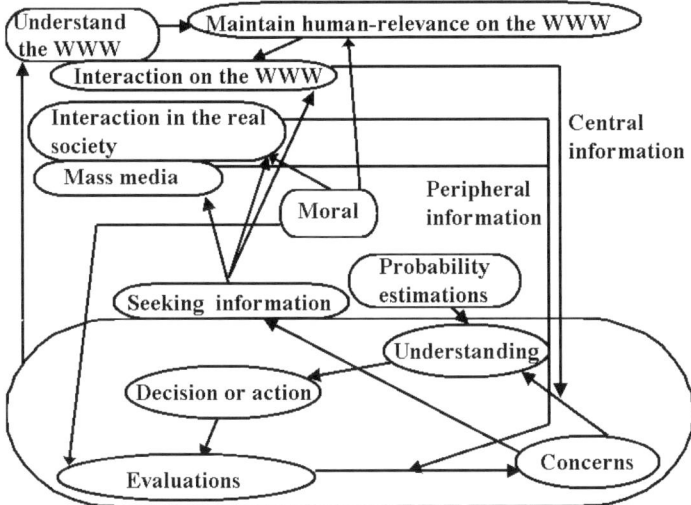

Fig. 23.11. The final model of the decision process for the people in the USA

The theory may be refined eternally. If a new piece of information is given which can not be explained in the current context, the context is changed by introducing new factors to consider.

Thus the grounded-theory approach is oriented for eternal theory revisions. This idea is reflected in practical applications of the grounded-theory approach. For example, if a patient in a hospital claims that he desires strongly to know all information about his own condition, the doctor might change the context from 'how to close facts' to 'how to open facts to the patient'. Then the doctor should gather experiences in the hospital of patients informed totally about conditions.

On the other hand, the double-helix process of chance discovery introduced the step to go from evaluation (of a chance's utility) to new concerns, as the trigger for selecting essential factors from factors the user may newly introduce for explaining rare events to which he had not paid attention to. This mechanism is applied to understand hidden structure underlying data in hand in this chaper, but is also applicable to chances to come later. This corresponds to the selection of new factors in the code-creating step in the grounded-theory approach, and also to the awareness of significant new events that shift the user from a context to another. In other words, by having a user prepare new factors in their own mind, we can expect a new event to trigger the user's awareness of those factors and to become easy to be understood in the face of its rareness, smoothly without rejecting the event or forcing changes in user's attitudes [23.2].

We can say that chance discovery came to introduce this trigger of discovery cycles, for three reasons. First, chance discovery aims at explaining new events as its central purpose, i.e. chances which might be exceptional. Therefore, a new approach for creating factors relevant to a chance was more strongly required. Second, the

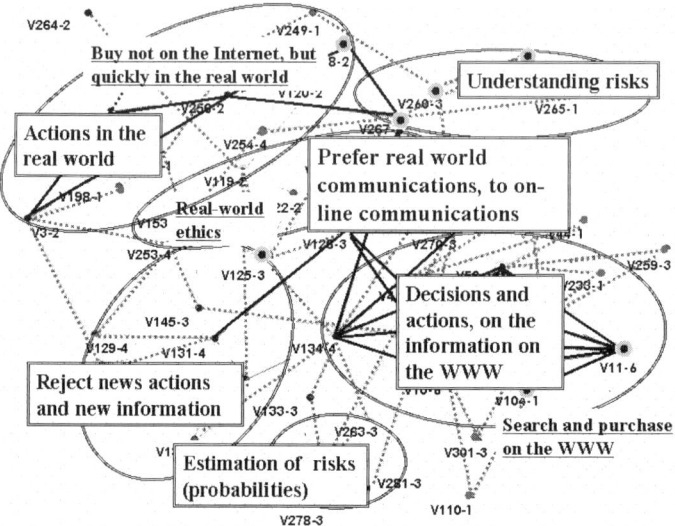

Fig. 23.12. The result of *KeyGraph* for Japanese people

rapid growth of the computational power of computers enabled us to deal with larger data than we could in the days of Glaser and Strauss, and we came to see finely abstracted patterns rather than the data itself. Due to this sophisticated revolution, we also have the risk of noticing too many factors underlying data and need to select a few noteworthy of them. Third, discovery methods, such as in the grounded-theory approach, were previously attended to by researcher(s), whereas the double helix can be processed involving both researchers and people living in the target domain. In this chapter, we introduced business people in step (9-c). In this group, each attendant imported information about experiences of other people, leading to the evaluation of each own life by comparing with others and to new concerns of values in external worlds.

23.6 Conclusion

The double-helix model, for the process in which a human discovers what we call chances, is exemplified by a case of social survey. The insight to deep-level factors of peoples' behaviors in the real and the virtual worlds is achieved, and the proposals based on the new awareness of rare but significant parts of data were acceptable. Finally, the double helix led us to a valid model of the human process of decision making on the information from the real world, mass media, and from the Internet. As in the discussion, we saw that the process model of the double helix itself can be regarded as a chance coming across the progresses of social science, computer engineering, and the real life and business of ordinary people.

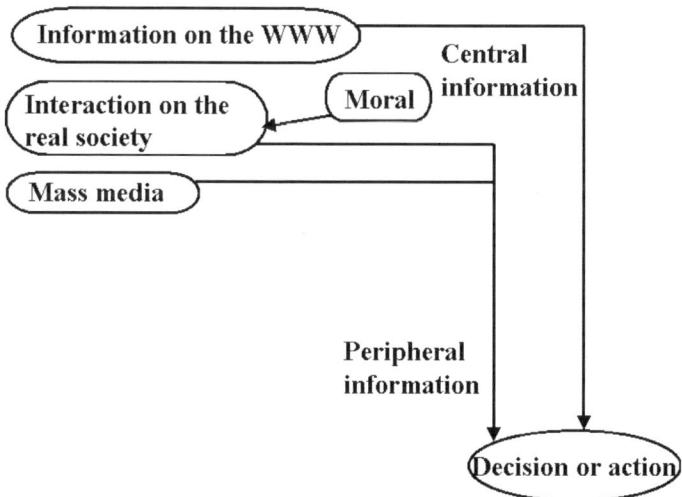

Fig. 23.13. The final model of decisions for Japanese people

References

23.1   Fayyad U, Shapiro GP, Smyth P (1996) From Data Mining to Knowledge Discovery in Databases, *AI magazine*, 17(3):37–54

23.2   Festinger L (1957) A Theory of Cognitive Dissonance, Row, Peterson and Company, Evanston, IL

23.3   Glaser B, Strauss A (1967) Discovery of Grounded Theory, Aldine, Chicago, IL

23.4   Herzenberg SA, Alic JA, Wial H (2000) New Rules for a New Economy : Employment and Opportunity in Postindustrial America, Cornell University Press

23.5   Nara Y (2001) *Abstracted* in Campus Internet Information Ethics in Japan: An Empirical Study, in *Proc. the 13th Annual Meeting of Western Decision Sciences Institute,* Vancouver, p.487

23.6   Ohsawa Y (2002) Chance Discoveries for Making Decisions in Complex Real World, *New Generation Computing,* 20(2):143–163

23.7   Ohsawa Y, Benson NE, Yachida M (1998) KeyGraph: Automatic Indexing by Co-occurrence Graph based on Building Construction Metaphor, Proc. Advanced Digital Library Conference (IEEE ADL'98), pp.12-18

23.8   Ohsawa Y, Fukuda H (2002) Chance Discovery by Stimulated Group of People - An Application to Understanding Rare Consumption of Food, Journal of Contingencies and Crisis Management 10(3):129–138

23.9   Ohsawa Y, Nara Y (2002) Discovery of Virtual Behaviors as Signs of Real Behaviors, *in Proc. KES2001*, pp.1188-1192

23.10 Ohsawa Y, Nara Y (2003) Decision Process Modeling accross Internet and Real World by Double Helical Model of Chance Discovery, *New Generation Computing,* 21(2):109-121

23.11 Ohsawa Y, Yachida M (1999) Discover Risky Active Faults by Indexing an Earthquake Sequence, in Proc. International Conference on Discovery Science, pp.208–219

23.12 Ottoo RE (2000) Valuation of Corporate Growth Opportunities: A Real Options Approach, Garland Publishing, New York, NY

23.13 McGrath R, MacMillan I (2000) The Entrepreneurial Mindset: Strategies for Continuously creating opportunities in an age of uncertainty, Harvard Business School Press

23.14 Simon H (1945) *Administrative behavior: a study of decision-making processes in administrative organizations* (1st edition 1945, 4th edition 1997, Simon & Schuster Inc., New York, NY

23.15 Nara Y (ed) (2002) Survery Report of the International Comparative Study on the Structure of Internet Ethics Behaviors: Focusing on the Differences between Virtual World Ethics and Real World Ethics, the Japan Society for the Promotion of Science (JSPS), and the Research for the Future Program, Electronic Social Systems, Foundations of Information Ethics (FINE) Project

24. Chance Discovery for Consumers

Makoto Mizuno

R&D Division, Hakuhodo Inc., Tokyo 108-8088, Japan
email: mmizuno@hakuhodo.co.jp

Summary.

This chapter focuses on quite a new direction in marketing, the power shift from suppliers to consumers, or the empowerment of consumers. The techniques for chance discovery can be applied to support not only suppliers' but also consumers' chance discovery. These two directions seem completely different but they are based more or less on the same underlying methodology, namely, analyzing consumer preferences and seeking potentially desirable states. We will discuss the background behind our research: recent trends in marketing, the increasing demand for recommender systems, and the applicability of chance discovery there. Then we will propose our own approach to supporting consumers' chance discovery, and show the results of a user test of a prototype.

24.1 Introduction

Chance discovery may sound like an unfamiliar concept to marketers, but once its spirit is understood, it should be strongly supported by marketers because this is what they have been pursuing for a long time. For instance, a textbook on new product development proposes a five-step decision process, i.e. opportunity identification, design, testing, introduction, and life-cycle management [24.13]. Opportunity identification is positioned first and, in that sense, if the decision regarding it is made incorrectly, the whole process of new product development will result in failure. But there seem to be fewer techniques for detecting opportunities than for dealing with other stages of new product development.

The definition of chance discovery by its inventor is 'awareness on and explanation of the significance of a chance, especially if the chance is rare and its significance has been unnoticed' [24.6]. If this is rewritten in the context of marketing, it means detecting opportunities to provide products, services, or communication for which sufficiently large potential demand exists but of which competitors and even consumers themselves are not yet aware. Its importance was recognized a long time ago and so too were the difficulties associated with it. It is no wonder, then, that many marketers are interested in the techniques for chance discovery as proposed in this book.

This chapter, however, focuses on a quite different direction in marketing, the power shift from suppliers to consumers, or the empowerment of consumers, as discussed in Sect. 24.2. The techniques for chance discovery can be applied to support not only suppliers' but also consumers' chance discovery. These two directions seem completely different but they are based more or less on the same underlying methodology, namely, analyzing consumer preferences and seeking potentially desirable states.

In Sect. 24.2, we will discuss the background behind our research: recent trends in marketing, the increasing demand for recommender systems , and the applicability of chance discovery there. Subsequently, in Sect. 24.3, we will propose our own approach to supporting consumers' chance discovery and, in Sect. 24.4, we will show the results of a user test of a prototype. Finally, we will conclude this chapter with a discussion of remaining problems and further development.

24.2 Background

24.2.1 Empowerment of Consumers

As noted above, one trend in marketing that has been discussed often in recent literature is a power shift from suppliers (manufacturers or distributors) to consumers or, in short, empowerment of consumers. This trend has been termed 'reverse marketing' [24.12] or 'customerization' [24.14].

Customerization sounds close to customization, but it is a little different. The latter means sufficiently flexible responses to individualized needs. For instance, it enables a consumer to order a product while specifying details to fit his or her own preferences. On the other hand, customerization is more radical. Rather than responding to the demands of individual consumers, suppliers allow consumers to participate everywhere or somewhere in the total marketing process, from product development to delivery. As actual examples, there are online communities of users transferring new product ideas or design proposals from consumers to manufacturers.

The counterpart in pricing may be reverse auctions, where consumers can choose the most reasonable offer made through auctions involving suppliers, although the business model here has not yet been proved to be sustainable. As consumers are empowered, promotion will change in the direction of emphasizing rational negotiation rather than emotional persuasion [24.1]. Advertising and sales forces may be replaced at least in part by intelligent agents that seek and recommend the best option for each client. An initial version has already been realized in recommender systems built into shopping Web sites such as amazon.com.

These changes seem to be minimizing the future role of market research. Traditional market research aims to know how many and what kinds of consumers want to buy what kinds of products. But as reverse marketing or customerization becomes prevalent, such information can be obtained directly from consumers; moreover, it can be more precise and certain. This seems to suggest that the only thing suppliers need to do is wait for orders from customers.

24.2.2 Education of Consumers

In spite of empowerment of consumers, the techniques used in market research would never die. They could be used to support consumers rather than suppliers, enabling consumers to understand their own preferences, say, 'reverse market re-search'. The assumption behind this is that consumers are sometimes unaware of (the outcomes resulting from) the attributes of new or unfamiliar existing products and hence their preferences regarding them are neither very clear nor consistent. Thus, some marketing scholars claim that there is a shift in the direction of market-ing strategy from 'market-driven' to 'market-driving' [24.2, 24.3]. The implication here is: do not learn from customers but rather educate them. This seems to con-tradict the idea of the empowerment of consumers, but that is not the case. A king always needs smart subjects who provide good advice, since he has authority but not enough knowledge to exercise it.

Consumer research literature has proved that consumers simply construct their preferences on the spot through trial and error to adapt to the environment around them [24.9]. To construct well-organized preferences, consumers need to ask them-selves such questions as:

1. What kinds of attributes should I consider in choosing a product?
2. What outcomes will be delivered to my life by such attributes?
3. How desirable is each outcome for me?
4. How should I integrate the above information to make a choice?

Of course, responding to these questions every time is too bothersome for con-sumers, so more sophisticated techniques are needed to elicit their potential prefer-ences. Note that these questions can be transformed into traditional market research questions by replacing 'you' with 'your customer' and 'should' with 'would' in the above sentences.

As consumers learn about products, their preferences become consistent. Expe-rienced consumers can easily specify what kinds of products they want without any support. Even among these consumers, however, some will not be satisfied with the status quo. They may always seek potentially more desirable options that they have never encountered. Moreover, they may hope to develop preferences for products that give them more pleasurable, admired, or fulfilled lives (for instance, one may want to develop a sophisticated preference or taste in clothing so as to be respected by his or her peers). Applications based on the idea of chance discovery are expected to respond to such widespread existing desires.

24.2.3 Likelihood vs. Chance Discovery

With the penetration of online shopping, a variety of recommender systems have been developed and put into practical use (see [24.4] for a review). These systems aim at helping empowered but not sufficiently informed consumers to choose bet-ter options. Most of the systems adopt one or more of the following fundamental approaches:

1. *Common sense*: if knowledge of the domain of interest is well established, one can recommend the option that is 'best' in the conventional sense for a certain occasion. The counterparts in industry are expert systems.

2. *Individual preference analysis*: using a user's purchase record or alternative data, one can recommend the option that is statistically predicted to be most preferred. Many decision support systems or decision aids have applied this approach to user-stated preferences.
3. *Clustering*: based on similarities between a given user and other users, one can recommend the option that has been most preferred by his or her peers. This approach is also known as 'collaborative filtering', with its best-known example being Firefly [24.5].

While each of the above approaches has distinctive features, all share the common concept of 'likelihood' (i.e. the recommended option is most likely to be preferred in general, personally, or by peers). A likelihood-oriented recommender system may be effective for novice consumers, since it can complement elementary knowledge of the domain. However, it is less effective for experienced consumers since they may already know the options that are recommended to them, whether they have already purchased them or not. Such customers would seek something new.

Chance discovery can be an alternative concept for recommender systems. We propose applying this concept to recommender systems, particularly ones that target experienced consumers. Our main purpose is to offer consumers unexpected, surprising, but exciting options. In other words, the systems should respond to 'variety seeking', which is the tendency for consumers to seek something new [24.11] and seems to be more prominent among experienced consumers. We discuss how to do this in the next section.

24.3 Proposed Approach

In this section, we will present a prototype of a chance discovery support system for consumers making a choice of wine. First, we will briefly present the logic of *Key-Graph* and the procedure for applying it to expert knowledge. Then we will explain how to incorporate the knowledge extracted by *KeyGraph* into application software called *WineNavi* so that users can easily operate the system and enjoy chance discovery with regard to wine.

24.3.1 Basic Logic of *KeyGraph*

The most famous technique for chance discovery is *KeyGraph*, which has been successfully applied to tasks such as the extraction of keywords from scientific papers [24.7] and the detection of risky faults based on earthquake data [24.8]. *KeyGraph* deals with any kind of data involving records ('sentences' in a text), each of which is composed of elements ('words'). *KeyGraph* processes data as follows (see Chap. 18 for details):

1. The elements that occur frequently in the whole body of data are selected and represented by black nodes, as in Fig. 24.1.

2. These elements are linked based on the frequency of concurrence across records or another derived measure (e.g. Jaccard's coefficient), with these links represented here by the black lines in the graph. Linked elements form clusters , which are interpreted as 'foundations', indicating the underlying components of the data. Elements that are not linked with any other elements are treated as clusters containing just one element.

3. Elements that frequently concur with clusters are selected. Among these elements, ones not contained in foundations (i.e. less frequently occurring ones) are represented by the gray nodes in the graph. They are interpreted as 'roofs', indicating 'chances', i.e. critical information supported by all foundations.

4. The elements extracted in the above step are linked again with the elements in foundations based on a concurrence or another derived measure. These links are represented by the gray broken lines and interpreted as 'columns', since they bridge foundations and roofs. Columns indicate which elements, although less frequently occurring, could influence the fundamental structure of frequently occurring elements.

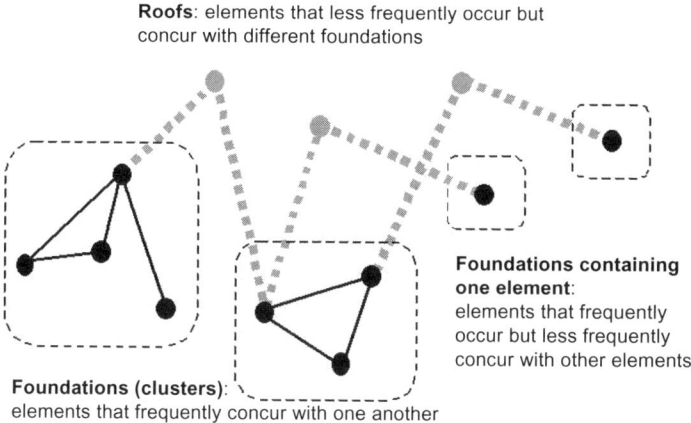

Roofs: elements that less frequently occur but concur with different foundations

Foundations containing one element: elements that frequently occur but less frequently concur with other elements

Foundations (clusters): elements that frequently concur with one another

Fig. 24.1. An illustrative graph drawn by *KeyGraph*

24.3.2 Application of *KeyGraph* to Expert Knowledge

To develop a prototype of a chance discovery support system, we asked experts to evaluate the congeniality of combinations of 208 wine categories and 288 foods. Wine categories are defined by production area (e.g. Bordeaux), type of grape (e.g. red), type of taste/flavor (e.g. full-bodied), and price level (e.g. 1000–1500 yen/750 ml). Each record of the data consists of a wine category and the foods matched to the wine. For instance,

Record #1: W_1, F_1, F_2

Record #2: W_2, F_1, F_3
Record #3: W_3, F_2, F_4, F_5, F_6
....

where W represents a wine category and F represents a food. Unfortunately, *Key-Graph* is not applicable to these data because of a lack of concurrence between wine categories and foods and between different wine categories. To overcome this problem, we multiplied each record so that its number would be in proportion to the number of foods in the record, where a certain number of foods were randomly selected from the original record. For instance, the first record above could be multiplied as:

Record #11: W_1: F_1, F_2
Record #12: W_1: F_2, F_2
Record #13: W_1: F_1, F_2

Other records could also be multiplied in the same way. Applying *KeyGraph* to the augmented data, we were able to extract foundations or roofs regarding wine–food combinations. The graph obtained given a certain set of parameters is shown in Fig. 24.2.

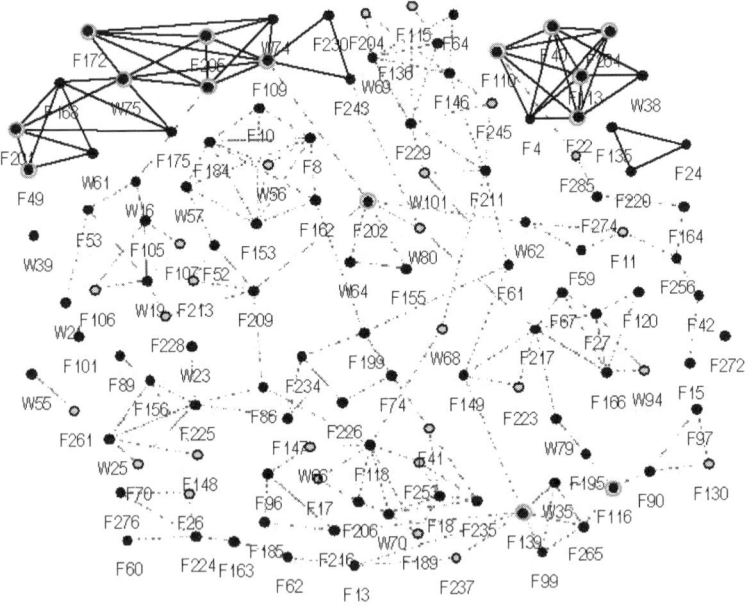

Fig. 24.2. Graph of wines and foods drawn by *KeyGraph*

In this figure, nodes marked with a W are assigned to wines and nodes marked with an F are assigned to foods. The black nodes are the wines/foods frequently endorsed by experts. When these are frequently endorsed concurrently with one another, they are linked by the black lines to form foundations. The gray nodes represent roofs, which are rarely endorsed but are concurrent with foundations; the broken lines represent columns, which are links between a foundation and a roof or between two foundations.

As shown in Fig. 24.2, only three foundations containing more than one wine/ food were formed (black nodes connected by black lines), while most of the frequently endorsed wines/foods were not very concurrent with one another. Many of those wines/foods, however, were linked to overall foundations (and hence connected by gray lines). On the other hand, some less frequently endorsed wines/foods were also linked to foundations (gray nodes connected by gray lines). Hence, this map provides us with little information on 'common sense' but a great deal on 'chances' through gray nodes and lines.

24.3.3 Clustering Wines and Foods

To make it more approachable for consumers, it is desirable to decompose the graph full of complicated connections into a tractable number of clusters. To this end, neglecting the difference in types of nodes and lines (black or gray), we applied the small world clustering method to the graph [24.10] and obtained 21 clusters. An example of these clusters is as follows:

Burgundy – White wine – Dry – Superexpensive
Loire basin – White wine – Dry – Expensive
Sole a la meuniere, Rock lobster terrine, Coquilles Saint-Jacques au gratin, Grilled yellowtail, Shabu-shabu ...

In the above list of foods, the first three are not surprising since they are typical French seafood dishes for Japanese diners and seem to match relatively expensive white wines in France. However, the next two foods are Japanese. These appear unexpected and possibly strange for conservative consumers but interesting for chance-seeking consumers. In other words, these two foods could be chances for consumers.

Each cluster was modified to be more convenient for consumers. The categorization of wine, which was introduced to limit our analyses to a reasonable range, is inconvenient for consumers to use as information for shopping. We therefore decomposed each category into brands and selected some representative ones based on expert advice. We also named each cluster to make it more attractive; for instance, the above cluster was named 'Overlooking the graceful Loire River'. Such names are subjective, but we believe that they could be very effective for stimulating users' imaginations.

24.3.4 Visualizing Clusters

We put all the necessary information about clusters into a visual interface called *WineNavi*. This system offers two types of interfaces for navigation: going through

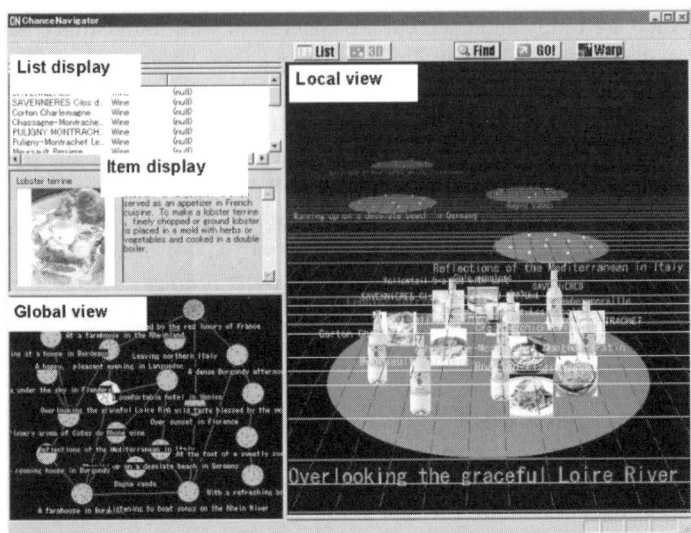

Fig. 24.3. The global view subwindow

several lists or traveling in a three-dimensional space. As visualization, the latter type may be more impressive, with the window consisting of four subwindows (see Fig. 24.3): global view (lower left), local view (right), list display (upper left), and item display (middle left).

Global view: This subwindow shows the overall relationship among clusters (see Fig. 24.4). Two clusters are linked as if there is at least a path between them. A user can drag each cluster on the screen to change its location freely without cutting the links. By clicking on a cluster, he or she can jump into that spot in the local view subwindow.

Local view: a user of *WineNavi* can see here what wines and foods are contained in a specific cluster (see Fig. 24.5). The wines and foods are placed on a turntable that can be rotated freely by hand to find the perspective that offers the most convenient view. When the user clicks on a wine or food, the specified item is presented in a close-up in the same subwindow and presented in item display. If the user drags it outside the cluster, the entire map around the cluster is rotated so that the user can see what the neighbouring clusters are. One can choose the next destination directly by clicking on it or get there automatically by clicking a 'GO!' button. In the latter case, the destination is selected at random from nodes linked to the present node.

Item display: a photo and a brief history or anecdote about the wine and a recipe of foods are presented.

List display: the items in a specified cluster are listed. It is possible to specify a wine or food here.

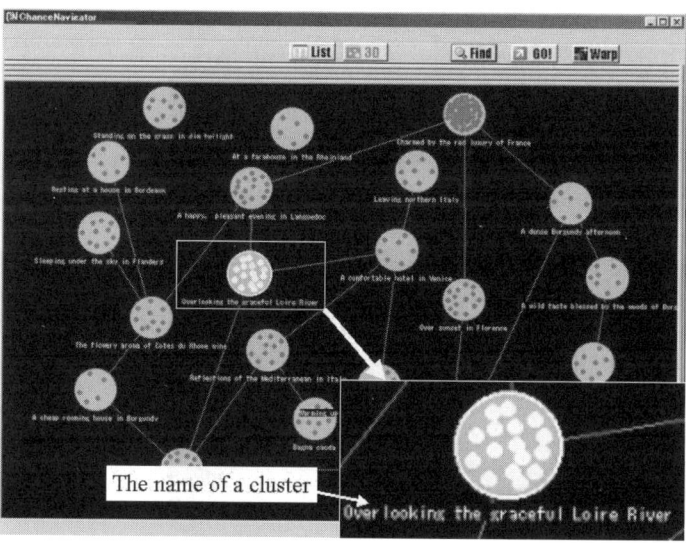

Fig. 24.4. The *WineNavi* window

Users can browse the relative configurations of clusters of wines and foods through global view and learn about items belonging to specific clusters and their neighbors through local view. They can travel from cluster to cluster, wine to wine, or wine to food as they like. In this navigation, users of *WineNavi* are able to experience chance discovery, firstly by staying in a particular cluster and secondly by moving between clusters. In our approach, clusters can include not only foundations but also roofs, in the terminology of *KeyGraph*. As shown in Sect. 24.3.3, clusters can involve more or less unexpected, stimulating combinations of wines and foods. Moreover, there is the possibility of another type of chance discovery, namely, between clusters that are linked in a weak sense. For instance, the cluster 'overlooking the graceful Loire river' is linked to the following three clusters:

1. '*A farmhouse in Burgundy.*': expensive dry white wines produced in Burgundy; snails a la bourguignonne, roasted poulet, fried chicken, etc.
2. '*A comfortable hotel in Venice.*': inexpensive Italian dry white wines; sashimi; grilled sardines; tempura
3. '*A happy, pleasant evening in Languedoc.*': inexpensive medium-bodied red wines produced in Languedoc and super expensive medium-bodied wines of Bordeaux; pot au feu; confit de canard; sukiyaki

Moving from the present cluster to the first two clusters is relatively understandable, since both of the latter are composed of dry white wines. On the other hand, moving to the third cluster seems more adventurous, since the type of wine and

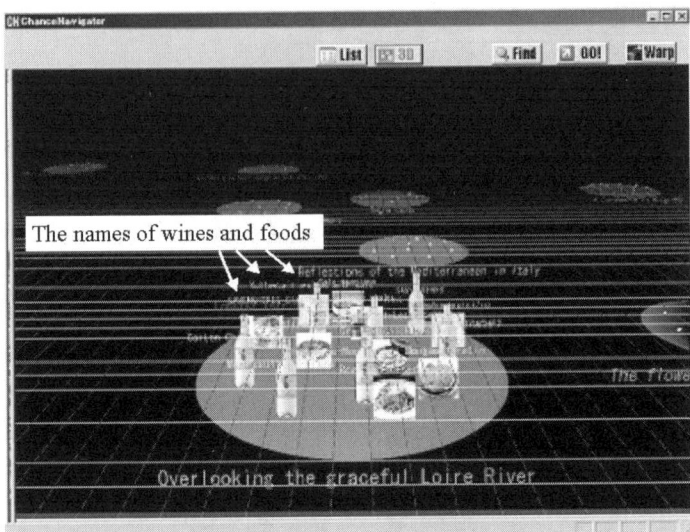

Fig. 24.5. The local view subwindow

the area are different. Nevertheless, these three clusters are all linked to the present cluster by at least one wine or food.

Another type of navigation in *WineNavi* is guided by a list interface (Fig. 24.6), which is composed of the three subwindows of cluster list (upper left in Fig. 24.6), member list (lower left), and item display (right). Clicking on a row in cluster list highlights a specific cluster and the wines and foods in that cluster are listed in member list, where clicking on a wine or food will cause detailed information about it to appear in item display. This interface does not provide as rich a presentation as the three-dimensional interface does, but it can work smoothly even on the Web, since it does not impose a burden on a system's visual capabilities. In any case, users can choose whichever they like by clicking on the 'list' or '3D' buttons at the top of the window.

24.4 User Evaluation

To validate *WineNavi* appropriately, we conducted a user test to compare the evaluations of *WineNavi* with those of a benchmark, which could be composed, for instance, of the most conventional approach to recommending wine–food combinations. Strictly speaking, we needed to test the hypothesis that our approach would outperform the benchmark. We will start this section by discussing the selection of a benchmark for the comparative experiment. Then we will discuss how the experiment was designed to test the hypothesis and what results were obtained. Finally, we will examine whether or not the hypothesis was validated.

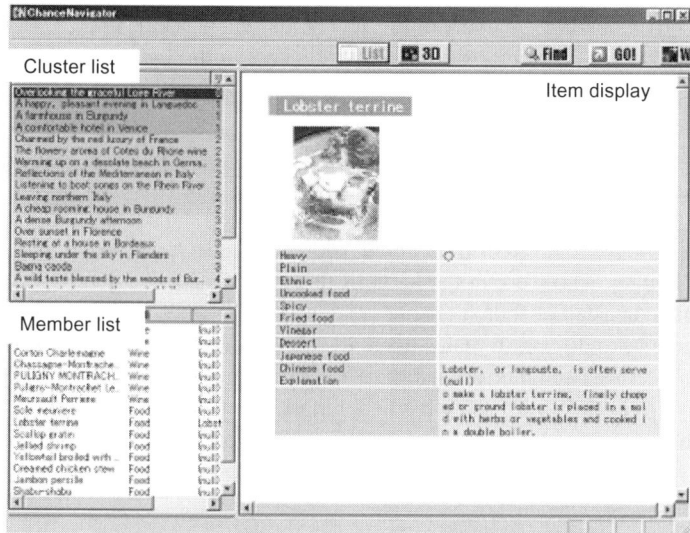

Fig. 24.6. *WineNavi*'s list interface

24.4.1 Constructing a Benchmark

In order to design an exact experiment, we needed to focus on the most essential part of *WineNavi* and replace it with an approach that could be used in conventional counterparts. We only changed the approach to extracting clusters and linking them with one another from the *KeyGraph*-based method to a likelihood-oriented one; as mentioned in Sect. 24.2.3, likelihood is the basic principle of many existing recommender systems. The remaining parts of *WineNavi*, i.e. most of the interface, were taken over to the benchmark system.

For extracting clusters based on likelihood, there is a tremendous number of clustering methods in the toolbox of data analysis. Before discussing methodology, we need to go back to our underlying data, i.e. expert judgments on wine–food combinations. These data can be seen as a matrix of 208 wine categories by 288 foods, with each cell taking the value of 0 or 1, i.e. whether or not this wine goes with that food. The resulting matrix is fairly sparse, such that there seem to be too many rows and columns. We began by excluding some wines and foods from the data in advance so as to match the composition of the original *WineNavi* data. We then performed correspondence analysis to reduce the dimensionality of the data. As a result, we obtained 11 axes whose cumulative contribution marginally exceeds 80%. This means that hundreds of variables were reduced into just 11 axes without losing much of the variation in the data.

We applied the well-known hierarchical clustering method to the coordinates of the foods in the 11-dimensional space and obtained 21 clusters, or the same number as in the original *WineNavi*. Subsequently, we allocated wine categories to clusters based on congeniality with the foods contained in them. Some adjustment was made to prevent clusters from having too many or too few items. There seems to be a very slight correlation between the compositions of the clusters in the two approaches; it certainly is not very strong. Each cluster was named as well and some clusters actually share the same name with the original *WineNavi*.

24.4.2 Experiment Design

For a user test of *WineNavi*, we recruited about 100 subjects from 30 to 49 years of age, male and female, who were in the habit of drinking wine more than once a month. The sample consisted of four segments: males aged 30 to 39, males aged 40 to 49, females aged 30 to 39, and females aged 40 to 49 (each comprising 25% of the total). Each subject was assigned at random to either the group using the original *WineNavi* (test group) or the group using the benchmark (control group).

We needed to encourage the subjects to imagine situations for choosing wine that were as realistic as possible. For this purpose, we provided the subjects with the following three scenarios:

1. At a dinner party at home to which you have invited close friends or relatives.
2. At a restaurant that you and your wine-loving friends like.
3. At dinner alone on your own or with your family on holiday.

Given one of these scenarios, each subject was asked to use *WineNavi*, choose their preferred wine for the situation, and answer questions to evaluate the system. These questions measured the familiarity of the scenarios, the degree of chance discovery experienced, its impact on choices, and overall satisfaction on a four-point scale. This task was repeated under all three scenarios for each subject. To remove the order effect, the subjects in each group were assigned at random to one of six possible orders of encounter of the scenarios. After finishing all of the tasks, the subjects were asked about the meaningfulness of the clusters, the performance of the global view, local view, and list windows, and finally, about their overall evaluation and intention to use the system if it were available.

24.4.3 Results of the Experiment

The results of the user evaluation by scenario are shown in Table 24.1. Among the three scenarios, scenario 3 (dinner alone/with family) was more familiar to the subjects than scenario 1 (dinner party at home with invited friends/relatives) or scenario 2 (restaurant with knowledgeable friends). But the experience of unexpected discovery happened with scenario 2. Moreover, the subjects reported that such discovery was more useful and influential in their choices for this scenario than in the other ones. Thus, in a situation that occurs less frequently and seems to be a bit more formal, there is demand for chance discovery support.

Our interest is in comparison of user evaluations between the test and control groups. Student's t test was applied to the differences in means between the two groups. However, there were no significant differences in any of the evaluation items by scenario.

Table 24.1. User evaluations by scenario: means and standard deviations of four-point-scale items

Items	Test group ($N = 48$)		Control group ($N = 48$)	
	Mean	Std.dev.	Mean	Std.dev.
Scenario 1. At a dinner party at home with close friends or relatives				
I find myself in this situation frequently.	2.56	0.823	2.46	0.849
I frequently experienced unexpected discoveries.	2.67	0.724	2.67	0.808
Such discoveries were useful for me.	1.19	1.734	1.04	1.675
Such discoveries influenced my choise.	1.81	1.580	1.85	1.701
I am satisfied with the wine that I finally chose.	3.23	0.627	3.21	0.743
Scenario 2. At a restaurant that you and your wine-loving friends like				
I find myself in this situation frequently.	2.60	0.765	2.77	0.831
I frequently experienced unexpected discoveries.	2.92	0.647	2.88	0.761
Such discoveries were useful for me.	2.58	1.442	2.48	1.598
Such discoveries influenced my choise.	2.33	1.358	2.17	1.492
I am satisfied with the wine that I finally chose.	3.35	0.526	3.29	0.582
Scenario 3. At dinner on your own or with your family on holiday				
I find myself in this situation frequently.	3.17	0.781	3.06	0.885
I frequently experienced unexpected discoveries.	2.65	0.635	2.81	0.790
Such discoveries were useful for me.	1.90	1.614	2.13	1.721
Such discoveries influenced my choise.	1.69	1.490	1.94	1.643
I am satisfied with the wine that I finally chose.	3.35	0.565	3.29	0.582

User evaluations by aspect were measured again on a four-point scale after all tasks had been completed. The results are shown in Table 24.2. The subjects regarded the clusters presented in both systems as providing useful information and unexpected discoveries. They also felt that it was difficult to understand the output of global view and local view, particularly for the *KeyGraph*-based system. In contrast, the list window received a relatively high evaluation. Overall, both systems seem to be able to support consumers' chance discovery, although the advantages of the three-dimensional interface have not yet been proved.

Note: ** indicates 5% significance and * indicates 10% significance

24.5 Discussion

Recent marketing literature suggests two seemingly conflicting trends: empowerment of consumers and market-driving strategy (or education of consumers). Onc effective way of responding to these changes is to develop recommender systems, many of which are based on the principle of likelihood. However, experienced consumers are not likely to be satisfied with the advice of such systems because they like to seek variety and sometimes intend to develop their own preferences (tastes) to make them more sophisticated. This momentum seems to share the same spirit as the concept of chance discovery.

Table 24.2. User evaluations by aspect: means and standard deviations of four-point-scale items

Items	Test group (N = 48)		Control group (N = 48)	
	Mean	Std.dev.	Mean	Std.dev.
Clusters of wines and foods				
-invoke many appropriate combinations.	2.75	0.636	2.83	0.559
-invoke many unexpected combinations.	2.44	0.769	2.58	0.679
-provide useful information.	3.13	0.640	3.21	0.683
-provide unexpected discovery.	2.88	0.866	3.10	0.722
Global View				
-is easy to operate.	2.25	0.934	2.33	1.018
-is easy to understand..	1.79	0.771	2.10	0.928
-provide useful information.	2.29	0.874	2.27	0.844
-provide unexpected discovery.	2.25	0.838	2.38	0.914
Local View				
-is easy to operate.	2.21	0.849	2.29	0.874
-is easy to understand..	1.92	0.942	2.31	0.903
-provide useful information.	2.46	0.849	2.73	0.765
-provide unexpected discovery.	2.40	0.765	2.75	0.887
List-like window				
-is easy to operate.	2.65	0.887	2.67	0.996
-is easy to understand..	2.67	0.907	2.75	0.957
-provide useful information.	3.25	0.729	3.10	0.778
-provide unexpected discovery.	2.88	0.841	2.92	0.794
Overall				
-is easy to operate.	2.23	0.778	2.46	0.824
-is easy to understand..	2.33	0.834	2.50	0.875
-provide useful information.	3.23	0.592	3.10	0.722
-provide unexpected discovery.	3.04	0.617	2.98	0.812
Intention to use	2.96	0.683	2.83	0.907

Based on the principle of chance discovery, we applied *KeyGraph* to expert judgments on wine–food combinations and extracted clusters from the complicated network with the help of small world clustering. The output from these processes was input into *WineNavi*, which was designed to visualize the relationships between wines and foods. Choosing an interface involving three-dimensional space, users can experience something like travel from one cluster to another. Otherwise, they can operate a list window more rapidly and obtain a sufficient amount of information.

As a benchmark, a replica of *WineNavi* was prepared that differed from the original *WineNavi* only in how clusters were extracted and linked. For this purpose, correspondence analysis and the hierarchical clustering method were applied to the expert knowledge. We conducted a user test and compared user evaluations of the original *WineNavi* and the benchmark version of it.

The results of the experiment are not perfectly supportive of the advantages of chance discovery approaches. The differences between the two versions in almost all of the evaluation items are not statistically significant. That is, it has not been proved that our proposed approach outperforms the conventional approach. The possible reasons are:

1. The subjects recruited for this experiment were not as experienced as expected in the hypothesis. (More experienced or other types of consumers should be recruited next time user tests are undertaken, if possible.)
2. The graphs with clusters and their links are not so different between the two groups. (The input data or the methodology of clustering itself should be reconsidered.)
3. The presentation of clusters in *WineNavi* rather than the logic of clustering might have an impact on user evaluation. (Even so, it should be noted that a list interface was preferred.)

It is, of course, necessary to identify which factor actually caused the lack of a difference. In any case, high absolute scores for overall evaluation and intention to use seem to suggest that *WineNavi* contributed to chance discovery for consumers and gave them certain kinds of satisfaction.

There is considerable room to extend this system. For instance, at the moment, the graph of clusters is common for all users. Taking into account the heterogeneity of consumers, personalization of the graph would be one of the most challenging extensions of *WineNavi*. Relatively realistic goals may be to incorporate an automatic navigation function into the system and to personalize the navigation process. That is, our research is still on the road toward building a chance discovery support system in practice.

Acknowledgements. This chapter is based on joint research with Hiroko Shoji (Kawamura Women's University), Yukio Ohsawa (University of Tsukuba), Yutaka Matsuo (NIAIST), Naohiro Matsumura and Naoaki Okazaki (University of Tokyo). I am grateful to them for their collaboration, to Yuzo Miyake (Kozo Keikaku Engineering) for his excellent contribution in programming, and to Akira Kaneichi, Toshikazu Tani, Osamu Tsutahara , Tetsuji Iidaka (Hakuhodo), Shota Hattori, Kayoko Kimura, and Masaki Tamada (Kozo Keikaku Engineering) for their support in data collection, interface design, and so on.

References

24.1 Blattberg RC, Glazer R, Little JDC (1994) The Marketing Information Revolution. Cambridge: Harvard Business School Press
24.2 Carpenter GS, Glazer R, Nakamoto K (1997) Toward a New Theory of Competitive Advantage. in Carpenter, G.S., Glazer, R. and Nakamoto, K. eds.: Readings

on Market-Driving Strategies: Towards a New Theory of Competitive Advantage, Addison-Wesley, Reading, MA, pp.527-540

24.3 Carpenter GS, Glazer R, Nakamoto K (2000) Market-Driving Strategies: Toward a New Theory of Competitive Advantage. in Iacobucci D (ed) Kellogg on Marketing, Wiley, New York, NY, pp.103-129

24.4 Kautz H (ed) (1998) Recommender Systems. Technical Report WS-98-08, AAAI Press, Menlo Park, CA

24.5 Maes P, Guttman RH, Moukas AG (1999) Agents That Buy and Sell, Communications of the ACM, 42(3): 79-91

24.6 Ohsawa Y (2002) Chance Discoveries for Making Decisions in Complex Real World, *New Generation Computing*, 20(2):143–163

24.7 Ohsawa Y, Benson NE, Yachida M (1998) KeyGraph: Automatic Indexing by Co-occurrence Graph based on Building Construction Metaphor, Proc. Advanced Digital Library Conference (IEEE ADL'98), pp.12-18

24.8 Ohsawa Y, Yachida M (1999) Discover Risky Active Faults by Indexing an Earthquake Sequence, in Proc. International Conference on Discovery Science, pp.208–219

24.9 Payne JW, Bettman JR, Johnson EJ (1993) The Adaptive Decision Maker, Cambridge University Press, Cambridge, UK

24.10 Matsuo Y (2002) Clustering using Small World Structure. Proc. 6th International Conference on Knowledge-based Intelligent Information Engineering Systems and Applied Technologies (KES2002), pp.1252-1256

24.11 McAlister L, Pessemier E (1982) Variety Seeking Behavior: An Interdisciplinary Review. Journal of Consumer Research, 9:311-322

24.12 Sawhney M, Kotler M (2000) Marketing in the Age of Information Democracy, in D. Iacobucci (ed), Kellogg on Marketing, Wiley, New York, NY, pp.386-408

24.13 Urban G, Hauser JR (1993) Design and Marketing of New Products (2nd edition), Prentice-Hall, NJ

24.14 Wind J, Rangaswamy A (2001) Customerization: The Next Revolution in Mass Customization, Journal of Interactive Marketing, 15:13-32

25. Application to Understanding Consumers' Latent Desires

Hisashi Fukuda

Applied Meteorology Department, CRC Solutions Corp., Kotoh, Tokyo 136-8581, Japan
email: h-fukuda@crc.co.jp

Summary.

This chapter introduces a method of chance discovery, i.e. awareness of and explanation of a significant situation for decision making, to a group interview of housewives who daily make decisions to serve meals to their families. The experimental stimulus was a simple visual representation of family food consumption relations. The visual representation of rarely consumed foods worked as chances for the housewives' awareness of unnoticed fundamental issues to be considered in meal service, and for their proposals for new services. Meaningful proposals then grew into acceptable service decisions, in the discussion which followed presentation of the visual stimulus.

25.1 Latent Desires as Fountains of Chances

To remain competitive in the retailing business, it is important for companies to develop services oriented to the satisfaction of consumers. In order to achieve this responsiveness to market needs, it is essential to understand consumers' behaviors, and their underlying latent desires . Many retailers are now aware of the methods of database marketing. These methods rely on information systems, including sales data repositories, and make availabile fine and precise data of sales performance. Vendors at every level of the market have utilized these systems for client management and for sales promotions. However, there is still untapped potential in the use of marketing databases.

In recent marketing data mining activities, identifying potential customers for expensive goods or for multiple purchases has been a major concern. However, another goal of database marketing could be to raise awareness with potential consumers of products of which they are not currently aware. For example, in food promotion to consumers, attention could be focused on the design of a set of foods forming a meaningful breakfast, lunch, or dinner for customers' situations. Here, companies desire chances for promoting a food or a food set to latent customers, especially if the significance of some food in real eating situations is hidden from consumer's conscious awareness. For consumers, being encouraged to eat in a meaningful manner (for health or communications with eating) may support their health care or enjoyable meals.

An essential issue in the process of acquiring knowledge from data, rather than just describing superficial relations among items in the data, is the description of the latent interest of consumers inherent in the relations. But it is difficult to obtain the potential desires of consumers using only the objective appraisal methods of data mining technology. These technologies learn those patterns which have the highest frequencies, and such patterns are assumed to continue also into the future. Chance discovery in retail businesses, on the other hand, aims to discover consumer needs, and to satisfy these with goods or services. If good or services which satisfy consumer needs do not yet exist, it will be hard to discover the needs for these. Therefore, even if a lot of data exists, a chance can not be discovered with such technology.

Such deep meaning of data cannot be discovered only by the manipuations of data by a computer. Human effort is also requuired here. If we can aid the human task, the winner in the retailing marketplace will be that company able to discover real and useful knowledge about consumer desires.

In market research, qualitative investigation methods are used in order to discover latent desires. In these investigations, since the question asked by a moderator and the interpretation of any results are subjective, the quality largely depends on the investigation process. Consumers may not always be conscious of their own needs, and so may not be able to answer accurately even direct questions about the benefits provided by certain products or services. Discovery of the latent needs for foodstuffs, being products purchased by consumers who may not necessarily have conscious awareness of the needs being satisfied, may be difficult for market researchers.

In this chapter, I propose a method to aid human discovery of consumers' latent desires based on the visualized information arising from desires. Here, the main participant of the system is a human who evaluates/validates/interprets the knowledge to be acquired. The aim in this chapter is to present the entire system and methods to evaluate its effects. It is quite noteworthy that the intuitive level of human decision making is a significant matter in considering the value and the meaning of information.

The method adopted in this study has been to:

1. First, obtain the data on food consumption by families, produced by Shoku-MAPR , a marketing-branch project of NTT-DATA.
2. Visualize the relations among dishes involved in meals using *KeyGraph*.
3. Organize group interviews to housewives, stimulating them with the visualized information in step (2) and guiding their discussion to discover the meanings of dishes in the family.

Through the experiment in step (3), I validated the method as an effective enhancement of the awareness and the discovery of consumers' latent desires for foods.

25.2 A Method of Chance Discovery: Communications with Data-Based Stimulation

Communications looking at *Key*Graph

In this chapter we present a method for aiding chance discovery embodying all three keys in steps (1) to (3). In this method a visual data-mining algorithm *KeyGraph* [25.13, 25.14] is applied to food-consumption data of a number of Japanese families (Key 3), and the output figure of *KeyGraph* was used as information stimulating discussion among a group of housewives talking in a room. The visualized output was presented to a group of housewives and they chatted with each other looking at the output (Key 1). First, they imagined the scenes of family meals guided by the moderator introducing discussions about family meals (Key 2), and, second, they discovered essential values of rarely consumed foods in breakfasts, lunches, and dinners. However, a more significant point is that they became aware of deep desires underlying the consumption behaviors of their family members. New proposals for serving meals arose in the discussion and grew into acceptable meal designs, by their possible ability to satisfy the latent desires which were identified.

KeyGraph as a visualization method for event-relations

KeyGraph was first proposed for indexing a document [25.13], and extended to finding risky active faults of earthquakes using the analogy between a series of words in a document and a time series of active faults where earthquakes occurred [25.14]. *KeyGraph* outperformed previous data-mining methods in these applications.

KeyGraph step 1: clusters of co-occurring frequent items (foods of a single meal) are obtained as basic clusters, called foundations. That is, items appearing many times in the data are extracted, and each pair of such items occurring often in the same sequence unit (each breakfast/lunch/dinner in a food-consumption sequence) is linked to each other. Each connected graph of those links and items forms one foundation, implying the existence of a common cause for the occurrence of included items. In the case of consumption data, a foundation can be viewed as arising from an underlying desire of consumers.

KeyGraph step 2: items not as frequent as those in clusters but co-occurring with multiple clusters are obtained as roofs. We regard roofs as candidates of chances, i.e. rare significant items (new satisfaction of consumers in consumption data) in the structure of relations among items. For example, an event from multiple strong causes (implied by foundations) [25.14], or a novel food satisfying different desires of consumers, is expected to be obtained as a roof. (Refer to part IV for details of data mining methods, including *KeyGraph*, for chance discovery).

As illustrated in Fig. 25.1, *KeyGraph* obtains only some candidates of chances, from which humans are expected to select real chances. Here we show a method for realizing this interaction between human and *KeyGraph*, employing an example applied to the market of food. The input data D to *KeyGraph* is to be given in the form of a document where each sequence unit in *KeyGraph* step 1 above is expressed as a sentence, i.e. an item sequence between period ('.')s. For example, a food-consumption history can be given as document D expressed in the plain-text part below:

Breakfast of family X on December 22: yogurt, fruits, salad, scrambled egg.
Breakfast of family X on December 23: rice, natto, grilled fish, soybean soup.
Breakfast of family X on 24 December 24: yogurt, fruits, salad.
. . .

Here, each line is a sentence made of foods eaten in each meal (e.g. in one breakfast of a certain family). In *KeyGraph* step 1, the cluster of nodes {yogurt, fruits, salad, ...} will be obtained as in Fig. 25.1 from co-occurring pairs in the text data D above, meaning a sequence of eating meals, and {rice, natto}, etc. A basic cluster of a single node such as 'sweet' can also be made. In *KeyGraph* step 2, items co-occurring with items in multiple clusters, like 'vitamin supplement', appear as novel topics.

The group interview with *KeyGraph*

(0) There is one moderator and seven interviewees (housewives) seated around a table in a room as shown in Fig. 25.2. Only the moderator knows the aim and the schedule of the interview.
(1) The moderator ('she' below) freely talked with participants about their usual meals.
(2) She explained in ten minutes what *KeyGraph* is, i.e. what the nodes and lines mean.
(3) She asked interviewees to write down in the cartoon balloons as shown in Fig. 25.2 what they imagined or thought of when they looked at the graphical output of *KeyGraph*.
(4) All interviewees talked freely to each other, adding lines, nodes, and words to the output figure of *KeyGraph*.

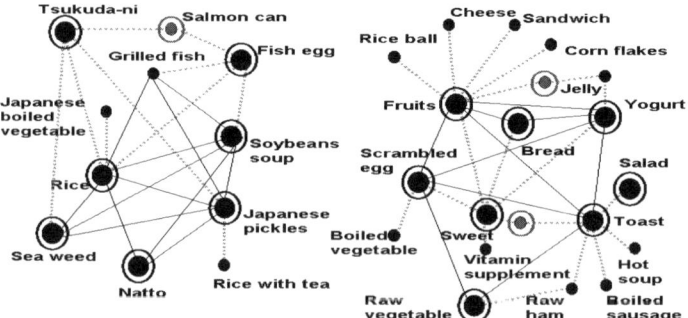

Fig. 25.1. An example output of *KeyGraph* to an eating sequence of meals

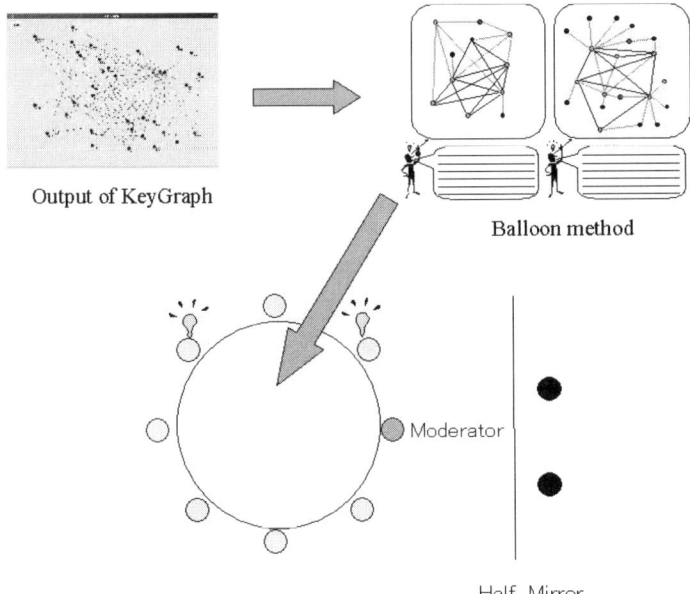

Output of KeyGraph

Balloon method

Moderator

Half Mirror

Fig. 25.2. The design of group interviews. The cartoon-like presentation of *KeyGraph* results helped interviewees imagine the meal scenes and the balloons helped them describe what they imagined

25.3 Application to Discovering Rare but Significant Consumption of Food on Weekdays or Sunday

KeyGraph was applied to data of the sequence of food consumption of 88 families during a year in the south-east district of Japan. This data was supplied to us by Shoku-MAP (trademark registered in Japan) data from NTT Data, Inc. A result from a sequence of one-year breakfasts is shown in Fig. 25.1. We find two black (solid)-lined clusters (one on the right and the other on the left-hand side of the figure), corresponding to the usual Japanese- and Western-style breakfasts respectively. The dotted links can be regarded as the motivations to eat rare significant foods (the double-circled roof nodes, 'vitamin supplement', 'salmon can', and 'jelly') in the usual breakfasts, i.e. with foods in the clustered foundations. Thus, 'vitamin supplement', 'jelly', and 'salmon can' are the candidates of essential rare desires.

We showed six such output figures to two groups each having seven interviewees of housewives, who daily serve meals to their families, aged in their forties and fifties (the same ages as the subjects in the Shoku-MAP data). For each of the six cases corresponding to {usual or Sunday} × {breakfast, lunch, dinner}, a figure as in Fig. 25.1 was used with attached cartoon balloons as in Fig. 25.3. For each figure, the interview proceeded as previously described.

Table 25.1. Food Consumption Data. This data was supplied by Shoku-MAP (trademark registered in Japan) data from NTT Data,Inc.

	Number of table	Number of menu	Number of material
All	99 686	487 963	1 065 155
Breakfast	27 077	154 558	291 240
Lunch	18 665	77 149	152 104
Dinner	24 785	172 018	487 417
Snack	22 324	55 368	69 345
Box lunch	4 646	25 313	52 369
etc.	2 189	3 557	12 680

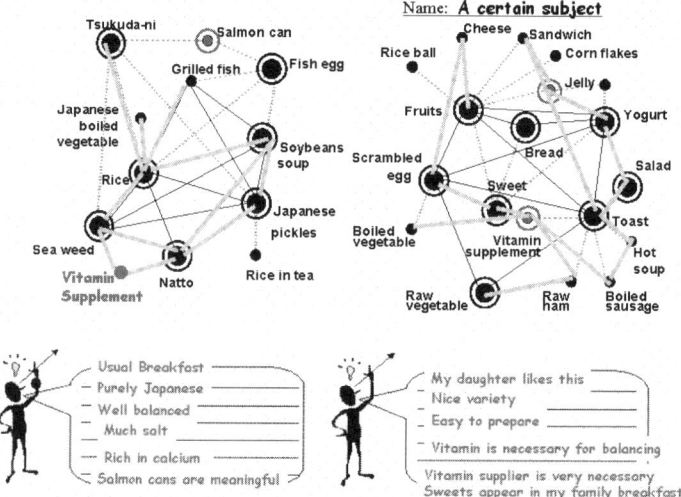

Fig. 25.3. An interviewee's result of the group interviews. The thicker lines, the node 'Vitamin Supplement', and the words in balloons were added by the interviewee

In step (4) of the group interview, proposals for meal designs were obtained in the communication, e.g.

– 'I will use a vitamin supplement for breakfast, for supporting the health of family members without wasting time for cooking.'
– 'I recollect that my daughter likes wine after my taste. I should consider the mutual influences between the tastes of family members.'
– 'The roof nodes of *KeyGraph* are health-care foods in breakfasts and entertaining foods in dinners. I can design each meal for my family meal based on such core concepts, i.e. health and entertainment.'
– ' Since our son has an allergic constitution and is short, he does not miss soybean protein. On the other hand, since I tend to become anemic, I do not miss a prune and fermented soybeans. '

- ' I am careful about obtaining nutrition, such as obtaining calcium with milk. Since a calcium agent is not delicious, and so may not be eaten by children, I think that I want them to obtain it. '
- 'It is the same as in my home. The vitamin compound is placed as a nutritional item, and our daughters go out in the morning, after drinking it. '
- 'Since nutrition and potassium can be given to a husband, a banana should be eaten instead of a Japanese-style dainty. '
- 'Although it is better to take vegetable salad and vitamin, neither of them are eaten in the morning. '
- 'The vitamin compound is also required at my age. '
- 'It is amusing to eat jelly and a sweet roll and to eat vitamins. Parents' handmade dish and vegetables should be eaten and vitamin supply should be carried out.' (Paradoxical opinion)

In each interview, subjective opinions by a few interviewees about the roof-node foods were filtered and spread to make opinions prevalent in the group. Some such opinions were accepted in the latter part of the two-hour interview: 'vitamin' appeared in the latter part of the balloons filled by each interviewee (Fig. 25.3). All participants interpreted the links around 'vitamin supplement' as a breakfast that is desired to supply energy quickly, in the busy situations of family members. For example, Japanese young girls tend to eat a 'sweet' linked to a 'vitamin supplement' in Fig. 25.3, because it is smoothly and quickly edible. However, this is insufficient from the point of nutrition. Some are aware of this insufficiency, and eat vitamin supplements.

In the process of conversation, the significance of 'vitamin' itself prevailed in all groups. However, according to the record of conversations, the frequency of 'vitamin' was not as high as words such as 'potato', 'salad', and so on, i.e. frequently consumed foods. Rather, the word worked as a trigger for participants' thoughts about their busy situations in the morning. Let me show frequent words in each ten minutes of step 3 from the beginning of the conversations:

- The first ten minutes: 'good', 'rice', 'family'.
- The second ten minutes: 'potato', 'sea weed', 'always prepare'.
- The third ten minutes: 'day', 'morning', 'busy'.

'Vitamin supplements' appeared in the third ten minutes, switching the conversation to one about underlying situations of breakfasts. Thus, we can regard 'busy' as a new concept common to the superficially different opinions of participants on 'vitamin'. Thus, using *KeyGraph* as the centre of discussion allowed the identification of the underlying desires.

Counting both groups, let us show the numerical evaluation of this method with respect to the criteria in Sect. 1.3.5 of Chapter 1. The unnoticeability U of words (e.g. 'busy' or proposals such as quick service of breakfast) having appeared first after/with the first appearance of roof-node words (e.g. 'vitamin supplement' in Fig. 25.3) was 7.0 for the first ten minutes, i.e. only one of seven in each group quickly noticed the significance of 'vitamin' and foods eaten for the busy situations. Here we set τ (see Sect. 1.3.5 of Chapter 1) to 10 minutes. The average growability G for these words was 6.3. Finally, the average proposability P was 0.93. On the other hand, the average values of U, G, and P for the black node, i.e. frequent words, were 1.0, 1.0, and 0.34 respectively. As a whole, the roof nodes were found to be more valuable chances, i.e. significant information for housewives' creative decisions, than frequent consumption.

25.4 Application to Discovering Rare but Significant Consumption of Food under Various Weather Conditions

We also conducted a similar experiment with *KeyGraph* for food consumption data under various weather conditions, because the influence of the weather on consumer behavior, especially temperature, is generally said to be large. First, the knowledge acquired from data processing is shown, and the result of an ordinary group interview is shown next. Finally, the knowledge acquired by the method proposed in this chapter is presented, and the validity of the proposed method is shown for this method of acquiring rare but effective knowledge.

Knowledge through computer data analysis

As weather data, the one-day average temperature data for one year was used. This corresponds to the object period and object area of the Shoku-MAP data. The temperature data included the absolute value of temperature each day, the size of each one-day temperature change, the average temperature for the past ten days, and the normal year's temperature. Relationships between these values and the frequency of appearance of different food menus was investigated. This is based on a 'memory-based model' of consumers, which assumes that a consumer will hold the experience as knowledge, for certain experiences he or she has. Each temperature data item was assumed to correspond to data of very short-term memory, short-term memory, and long-term memory.

Although I do not present details, it turns out that the frequency of appearance of food menus is related to the average temperature for the past ten days. We show the hodographs of the average temperature for the past ten days and the 10-day frequency of appearance of dinner menus in Fig. 25.4. In a case upwardly slanting to the right when the temperature is high, this is a menu with a high frequency of appearance, and this is defined as a summer menu type. On the other hand, in a case when the graph is sloping downwards to the right, this is defined as a winter menu type. Clearly, beer is a summer type and it turns out that noodles are a winter type. It is thought that these arise from consumers feeling of seasonality, namely long-term memory. For retail businesses, this knowledge, as shown in Table 25.2, is a commonplace rather than a novelty, and is well-known experientially.

Furthermore, although the food menu showed the influence of the average temperature for the past ten days, I assumed that it was also related to the deviation from the average temperature for the past ten days, and so analyzed this. The data in each table was categorized, having been extracted summer (June – August) and

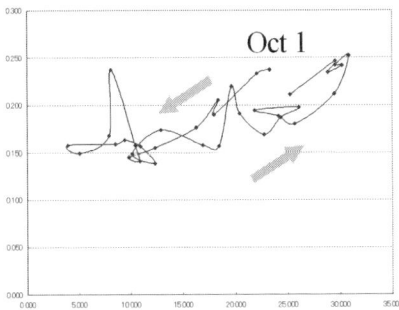 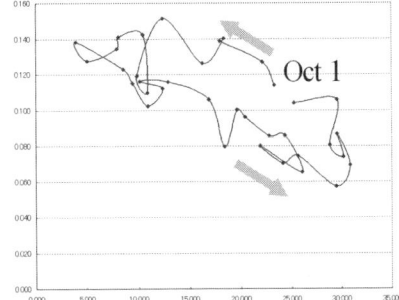

Fig. 25.4. Hodograph of the average temperature for the past ten days and the 10-day frequency of appearance of a dinner menu. The left-hand figure shows beer and the right-hand figure shows noodles

winter (December – February), and having used ±3 degrees as the threshold for the difference from the average temperature for the past ten days. The number of dinner menus is shown in Table 25.3. Although details are not shown, when the frequency of appearance of the most frequent 20 menu items was analyzed, the relation between the deviation from the average temperature for the past ten days and the frequency of appearance of a menu was not understood clearly. This shows that, using only data analysis, novel knowledge could not be extracted from these findings about a consumer's reaction to temperature information, namely short-term memory in the assumed model of consumer behaviour.

Knowledge by the common group interview

I had the housewives mention menus in specific conditions in an ordinary group interview. In summer, three menus were mentioned as ones which softened the impact of heat: the menu 'it is thought that is cold'; the 'clean' menu; and the menu which 'has much acidity'. In contrast, 'hot' menus included menus for stimulating appetite or physical strength, or those 'whose nutritive value is high'. In winter, mention was made only of the menu 'it can be thought that it is warm', which seemed strongly to be conscious of warming the body. In discussing these, the following were mentioned as specific conditions of determining the menu:

1. The participants in the meal (e.g. a husband is late tonight and does not eat).
2. The tastes and health-concerns of the participants.
3. The convenience of the person preparing the meal (e.g. if he or she is busy, it may be necessary to make an easy meal).
4. What foods are in the house.
5. The weather conditions.

Therefore, the housewives consider preferentially first 'who eats or for whom the meal is made' as decisive factors for a menu. Naturally, each one of tastes, health conditions, etc, is taken into consideration. It is thought that weather conditions are the decisive factors after accounting for the above-mentioned factors. Although

Table 25.2. The frequency of food items listed in ranked order, from first to twentieth place, for both summer and winter menu types

	Summer type	Winter type
1	Chinese Tea	Tea
2	Fresh Vegetable	Rice cake
3	Ice Cream	Miso Soup
4	Tofu dish	Fruits
5	Beer	Noodle
6	Wheat noodles	Boiled rice in tea, rice gruel
7	Fruit drink	Japanese noodles
8	Milk	Cocoa
9	Boiled dish	Pot dish
10	Vegetable salad	Vinegared dish
11	Yogurt	Japanese pot-au-feu
12	Rice ball	Western-style soup
13	Chilled summer noodles	Simmered dishes
14	Lactic acid drink	Cod roe
15	Jelly	Japanese sweets
16	Carbonated drink	Stew fo meat
17	Serial one flakes	Japanese-style soup
18	Vegetable juice	Chinese steamed filled dumplings
19	Japanese-style pottery of vegetables	Cookies, biscuit, crakers
20	Japanese-style cold sweets	Cakes

Table 25.3. The number of dinner menus classified according to the deviation from the average temperature for the past ten days

	ALL	<-3 Degrees	>3 Degrees
Summer	5849	977	405
Winter	6341	1077	581

priority may occasionally be given to weather conditions, 'who eats or for whom the meal is made' and weather conditions are compared and it is guessed that weather conditions are not the most important factor in many cases.

Knowledge by the method of chance discovery

We segmented the data (D) by the changes of temperature, e.g. D1 obtained for days as hot in summer (cold in winter) as the past ten days, or D2 for days 3 or more degrees hotter (colder) than the past ten days. As a result, for all the six figures presented, each corresponding to a weather condition, both of two groups (different members from the first interview evaluated above) correctly guessed the season in which each item of data was recorded.

– 'If the dried seaweed is placed over boiled greens or is put into fermented soybeans, it will become delicious.'
– 'The dried seaweed is placed over long-potato or a Japanese-radish salad.'
– 'The dried seaweed is always set at the table. A husband eats it as a snack and a child eats it instead of a snack.'
– 'Only foods in the house are eaten without cooking when cold.'

– 'When busy, when husband does not want to help with shopping, or when not carrying out shopping, he always eats dried seaweed.'
– 'Since I do not want to go shopping when cold or busy, I cook much potato salad and much curry.'
– 'Since I do not want to go outside when cold, potato salad is cooked using the foods that are always in a house.'

Fig. 25.7 shows the visualized dinner menu of the days when weather conditions are severe in winter (for days 3 or more degrees colder than the past ten days). Potential nodes include potato salad and dried seaweed. The housewives looked at this visual menu, and understood immediately that it was the menu for winter. Furthermore, new knowledge was created, not only similar to this general knowledge but also like in the manner of the earlier use of *KeyGraph* mentioned above. Knowledge was stimulated by the potential node and the needs of 'wanting to decorate a table with one menu which can moreover be reserved with potluck since she does not want go outside on a cold winter's day'. In contrast, on the dinner menu for the days when winter weather conditions are quiet (for days 3 or more degrees hotter than past ten days), there were a cabbage roll, sushi, and boiled fish as a potential node in Fig. 25.8. This menu was stimulated by the visual information and the needs of 'wanting to cook over many hours and to make a table rich also in winter since a warm day has the easy body'. That is, when weather conditions changed rapidly, the taste of eating habits changed, and the novel knowledge of this relationship could be acquired using these methods.

A consumer's reaction to the weather conditions is physiological, and since her desires do not directly depend on these reactions, knowledge of their impact can be difficult to obtain. Thus these new items of knowledge cannot be acquired by methods which extract high-frequency patterns or group interviews; instead, they are latent needs brought to the surface by means of isualized information.

These examples show how the discussion stimulated by *KeyGraph* caused interviewees to become aware of hidden situations underlying the data, and to suggest new proposals for decisions. With respect to the process model of chance discovery of Fig. 25.2, the moderator worked to have interviewees be concerned with meal design at step (1), and exchanging new ideas with each other in the discussions stimulated by *KeyGraph*, so as to build their awareness and explanation of the significance of chance candidates depicted by roof-nodes. Finally, they made proposals leading to meal-service decisions. As noted in in [25.7], on a design information repository, design proposals from a group of people with different ideas also worked here for considering the significance of rarely consumed foods, based on the causal structures surrounding each rarely consumed food.

However, the selection of interviewees remains to be resolved. If people without direct experience corresponding to the context of data are chosen, then meaningful data will not likely surface, as the articulation of meaningful information requires the combination of visualized information and relevant experience. Significant results were obtained in this research project because housewives who have experience of menu planning were chosen to participate in the group interviews. The exis-

394   Hisashi Fukuda

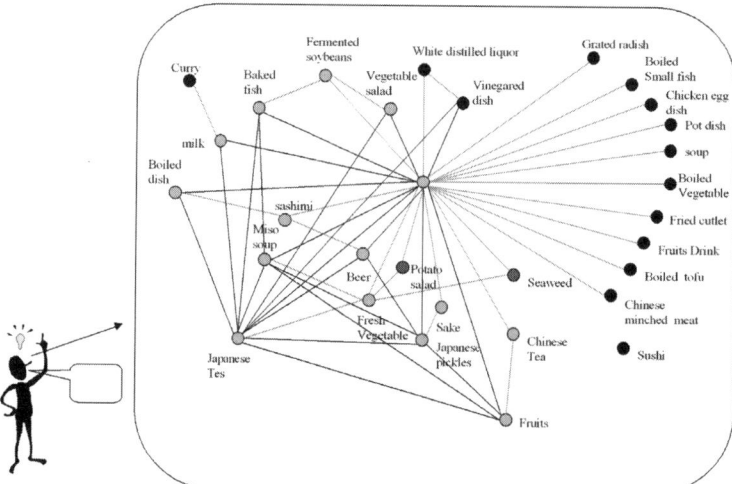

Fig. 25.5. An example output of *KeyGraph* for days 3 or more degrees colder than the past ten days in winter

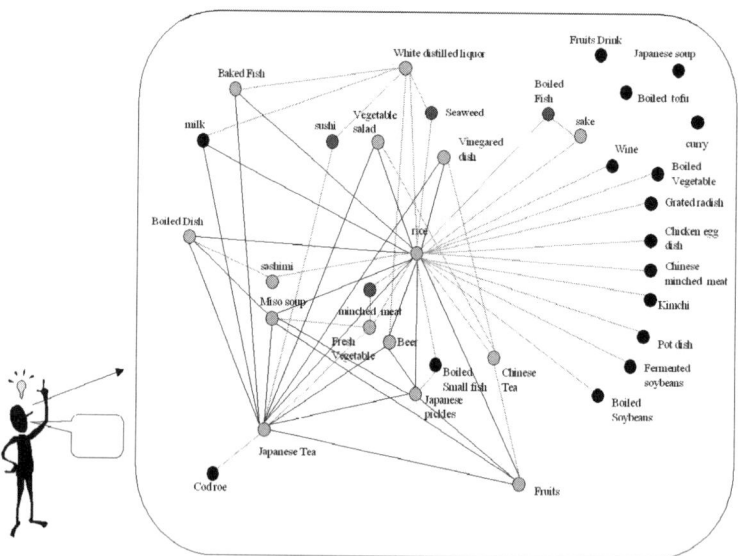

Fig. 25.6. An example output of *KeyGraph* for days 3 or more degrees hotter than the past ten days in winter

tence of an opinion leader in each group was also important. Typically, this person saw the *whole* visual picture, not just partially, making remarks such as 'the menu which existed on the edges until now has come into the middle', or 'this seems to be a gay menu'; these remarks stimulated the other housewives, and remarks such as these led to the surfacing of latent needs.

25.5 Conclusion

This paper introduced chance discovery — awareness of and explanation of a significant situation for decision making — and presented key issues exemplified by a method stimulating the chance discovery process of a group of housewives who daily make decisions to serve meals to their families. The stimulus was made by a simple visual representation of relations among food consumption by families. Rarely consumed foods in the visual output worked as chances for the housewives' awareness of unnoticed fundamental issues to be considered in meal service, and for their proposals of new services. Meaningful proposals then grew into acceptable service decisions, after the discussions arising from the visual stimulus.

This type of application is not restricted to marketing, i.e. market research based on opinions rising from a community of people. The methodology of chance discovery covers the explorations of significant events affecting human life — contingencies and crises, as well as opportunities.

References

25.1 Agrawal R (1993), Mining Association Rules between Sets of Items in Large Databases, *Proceedings of ACM SIGMOD Conf. Management of Data*, pp. 207 – 216.
25.2 Allport GW, Postman L (1947) *The Psychology of Rumor*, H.Holt.*Oxford Advanced Learner's Dictionary* (2000), Oxford University Press
25.3 Embrechts P et al (1997) *Modeling Extremal Events for Finance and Insurance*, Springer Verlag, Heidelberg, Germany
25.4 Fayyad U, Shapiro GP, Smyth P (1996) From Data Mining to Knowledge Discovery in Databases, *AI magazine*, 17(3):37—54
25.5 Festinger L (1957) *A theory of cognitive dissonance*, Row, Peterson, Evanston, IL
25.6 Gero JS (1994) Computational Models of Creative Design Formation, in Artificial Intelligence and Creativity, *Studies in Cognitive Systems* Kluwer Academic Publishers, Boston, MA, 17:269–281
25.7 Kato Y, Yairi T, Hori K (2001) Integrating Data Mining Techniques and Design Information Management for Failure Prevention, *New Frontiers in Artificial Intelligence* LNAI 2253, Springer Verlag, Heidelberg, Germany, pp.475-480
25.8 Kosslyn SM (1980) Image and Mind, Harvard University Press
25.9 Mannila H, Toivonen A, Verkamo I (1995) Discovering Frequent Episodes in Event Sequences, in Proc. First Conf. on Knowledge Discovery and Data Mining (KDD95), pp.210–215
25.10 McBurney P, Parsons S (2001) Chance Discovery Using Dialectical Argumentation, *New Frontiers in Artificial Intelligence* LNAI 2253, Springer Verlag, Heidelberg, Germany, pp.414-424

25.11 Nara Y, Ohsawa Y (2000) Tools for Shifting Human Context into Disasters, in *Proceedings of IEEE conference of KES2000*, pp.655–658

25.12 Nonaka I, Takeuchi H (1995) *The Knowledge Creating Company*, Oxford University Press

25.13 Ohsawa Y, Benson NE, Yachida M (1998) KeyGraph: Automatic Indexing by Co-occurrence Graph based on Building Construction Metaphor, Proc. Advanced Digital Library Conference (IEEE ADL'98), pp.12-18

25.14 Ohsawa Y (2002) *KeyGraph* as Risk Explorer in Earthquake-Sequence, *Journal of Contingencies and Crisis Management*, 10(3):119–128

25.15 Petty RE, and Cacioppo JT (1981), Attitudes and persuasions: Classic and contemporary approaches, WC Brown, Dubuqute, IA

25.16 Rogers EM (1962) Diffusion of Innovations, Free Press, New York, NY

25.17 Schlesselman JJ (1974) Sample size requirements in cohort and case-control studies of disease, American Journal of Epidemiol, 99: 381—384.

25.18 Shibata H, Hori K (2001) A Creativity Support System for Long-term Use in Everyday Life - IdeaManager: Its Concepts and Utilization, *New Frontiers in Artificial Intelligence* LNAI 2253, Springer Verlag, Heidelberg, Germany, pp.455-460

25.19 Shoji H, Hori K (2001) Chance Discovery by Creative Communications Observed in Real Shopping Behaviors, *New Frontiers in Artificial Intelligence* LNAI 2253, Springer Verlag, Heidelberg, Germany, pp.462-410

25.20 Suzuki E, Tsumoto S (2000) Evaluating Hypothesis-Driven Exception-Rule Discovery with Medical Data Sets, in Terano T (ed) Knowledge Discovery and Data Mining, LNAI 1805, Springer Verlag, Heidelberg, Germany, pp.208—211

25.21 Weiss GM, Hirsh H (1998) Learning to Predict Rare Events in Event Sequences, in Proc. of KDD-98, pp. 359-363

Author Index

Subject Index

Druck: Strauss Offsetdruck, Mörlenbach
Verarbeitung: Schäffer, Grünstadt